Studies in Applied Philosophy, Epistemology and Rational Ethics

Volume 23

About this Series

Studies in Applied Philosophy, Epistemology and Rational Ethics (SAPERE) publishes new developments and advances in all the fields of philosophy, epistemology, and ethics, bringing them together with a cluster of scientific disciplines and technological outcomes: from computer science to life sciences, from economics, law, and education to engineering, logic, and mathematics, from medicine to physics, human sciences, and politics. It aims at covering all the challenging philosophical and ethical themes of contemporary society, making them appropriately applicable to contemporary theoretical, methodological, and practical problems, impasses, controversies, and conflicts. The series includes monographs, lecture notes, selected contributions from specialized conferences and workshops as well as selected Ph.D. theses.

More information about this series at http://www.springer.com/series/10087

Margarita Vázquez Campos
Antonio Manuel Liz Gutiérrez
Editors

Temporal Points of View

Subjective and Objective Aspects

 Springer

Editors
Margarita Vázquez Campos
Faculty of Philosophy
University of La Laguna
Tenerife, Canary Islands
Spain

Antonio Manuel Liz Gutiérrez
Faculty of Philosophy
University of La Laguna
Tenerife, Canary Islands
Spain

ISSN 2192-6255 ISSN 2192-6263 (electronic)
Studies in Applied Philosophy, Epistemology and Rational Ethics
ISBN 978-3-319-19814-9 ISBN 978-3-319-19815-6 (eBook)
DOI 10.1007/978-3-319-19815-6

Library of Congress Control Number: 2015945318

Springer Cham Heidelberg New York Dordrecht London

Springer International Publishing AG Switzerland is part of Springer Science+Business Media
(www.springer.com)

Preface

We say "time flies". In fact, we experience time as flowing. But, does time exist? To place events and facts in a fluent time requires to adopt a certain temporal point of view. Fluent time depends on a perspective. Are the past, the present and the future no more than merely subjective epiphenomena? From a scientific perspective they "seem" to be so. Also, it is so from some highly speculative philosophical stances. However, can they "really" be so?

Fluent time performs a crucial structuring role in our epistemic and agentive relations with reality. It is a time with a clear direction: it involves a past, a present, and a future. It also has an essential instability: things that were future become present, and things that are present will become past. We experience that time in our life. And we explicitly embrace a temporal perspective in our plans and strategies for knowing and acting. The existence of such a fluent time is a non-questionable fact of our experience. However, such existence has been disputed both by philosophical and scientific reasons.

From the philosophical side, McTaggart's arguments against the reality of a fluent time, what he called temporal A-series (events or facts ordered according to their "being past", "being present", or "being future"), occupy a central position in the relevant literature. And it is very difficult to articulate a completely satisfactory answer to his arguments. Very often in combination with some version of McTaggart's arguments, that fluent time is also excluded from the scientific picture of reality. The "past", the "present", and the "future" have no clear place in the spatio-temporal continuum of fundamental physics. In the best case, the "present" is relativised to a reference-frame and time is reduced to a mere asymmetric relation between facts or events, what McTaggart's described as B-series (in which the order is defined by the relations "before than", "after than", and "simultaneous to"), or it is even reduced to some kind of C-series (for instance, "to be included in") having no essential connection with the temporal concepts of A-series. The final result is a time with no privileged direction (the "arrow of time"), a time lacking any sort of fluency (the "passage" of time), and a time in which the present (more in general, the "now") has no distinguished position.

Those philosophical and scientific attacks have originated a deep tension between the subjective and the objective aspects of time. The fluent time we experience and the fluent time we consider in our temporal perspectives seems to be only an "epiphenomenon", a merely subjective aspect of our points of view about reality and about ourselves. Facing that situation, this book tries to understand the relationships between the objective and the subjective aspects of time. And it tries to do it from the more basic notion of points of view, or perspectives. Through all the chapters of the book, the notion of points of view is taken to be a pivotal tool to deal with the connections between an external and objective time, for instance, the time conceptualised by science, and the internal, even subjective, time involved in our personal experience.

The emphasis in the notion of points of view is very important in our approach. Chapter 1 (*The Notion of Point of View*) and Chap. 2 (*Subjective and Objective Aspects of Points of View*) offer a detailed analysis of it. There are different ways of approaching the notion. One of them assumes as a model the structure of "propositional attitudes". Another one focuses on the idea that points of view are "ways of having access" to reality, and to ourselves, from certain emplacements. A third one, proposed by Russell some years ago, with a clear Leibnizian inspiration, combines both approaches trying to make sense of the idea of a "space of perspectives". A fourth approach, much more recent, uses the notions of "conceptual spaces" and "state function" to analyse the most prominent features of points of view. In any case, points of view reveal to be complex entities with a robust modal nature and a strongly relational mode of existence. Because of that, points of view turn out not easily reducible either to information, or to psychology, or to physics.

These first two chapters establish a very important distinction that, more or less explicitly, is present in all the rest of the book: the distinction between what is external to points of view and what is internal to points of view without being merely internal to the subjects having those points of view. What is internal to the points of view without being internal to the subjects displays a very important sort of objectivity and intersubjectivity different from the objectivity and intersubjectivity offered by science. This entails that points of view can also offer an internal time which is not a merely subjective time. To adopt temporal points of view about reality and about ourselves would consist in having in perspective such a time.

Chapter 3 (*Temporal Aspects of Points of View*) and Chap. 4 (*Fluent Time, Minds, and Points of View*) are specially devoted to the connections between points of view and time. The first one discusses in depth McTaggart's arguments against the reality of a fluent time and also the claim that fluent time is merely a subjective epiphenomenon. The second one argues that the existence of a fluent time, i.e. a time with a past, a present, and a future, is linked to the existence of experiential points of view with non-conceptual contents. Fluent time would be internal to some experiential points of view without being internal to the subjects having those experiences. In this chapter, it is also argued that statements, or thoughts, about the past and about the future can have truth-makers in the present. In one way or another, both chapters continue analysing the notion of temporal points of view.

Chapter 5 (*Branching Time Structures and Points of View*) and Chap. 6 (*Change, Event, and Temporal Points of View*) have a logical character. They suggest different ways in which time can be formalised. The first one focuses on ways to formalise a branching, indeterminist, sometimes called Ockhamist, time. A Priorian way of seeing time is adopted. And the need to use multimodal and bi-dimensional logical strategies is defended. Moreover, the need of "hybridising" our systems extending the basic logic with mechanisms capable of referring to concrete points in time is suggested. The second one uses a "conceptual spaces" framework to define the notion of temporal points of view. The framework of conceptual spaces is powerful and with direct technical applications. In this chapter, a continuous and processual time is formalised and a logic for such a time is sketched. The temporal element is introduced by adding a time variable to state functions that map entities into conceptual spaces. That way, states may have some permanency or stability around time instances. Following Aristotle's intuitions, changes and events will not be necessarily instant phenomena. They can be processual and interval dependent.

Chapter 7 (*Grounding Qualitative Dimensions*) also makes use of the framework of "conceptual spaces". This time, the focus is the conception of points of view as ways of having access to the world. Points of view always involve a choice among different qualitative dimensions. Moreover, the distinction determinable/determinate is applied inside each dimension to generate different perspectives. It is claimed that we can assume all of that without embracing relativist conclusions.

The last three chapters of the book address particular topics. Chapter 8 (*Kinds, Laws and Perspectives*) is about the notion of "natural kinds", a very important issue in metaphysics and philosophy of science. The three main approaches to natural kinds are analysed: the essentialist, the constructivist, and the causal. It is argued that the third is the most adequate one. The distinction between causal laws and useful generalisations is highly perspective. But this would not be an unsurmountable obstacle for the objectivity of the identifications of natural kinds. Chapter 9 (*Synchronic and Diachronic Luck*) is about "luck attributions", a particular topic that has come to have great relevance in epistemology and theory of action. It is argued that luck attributions always have a perspectival character. Moreover, they depend on the either synchronic or diachronic temporal perspective adopted. It is argued that no such temporal point of view enjoys any kind of logical priority, or metaphysical privilege, with independence from all perspectives. Finally, Chap. 10 (*Presentism, Non-Presentism and the Possibility of Time Travel*) discusses time travel paradoxes. It is argued that a realistic notion of time is capable of accepting the possibility of time travel in such a way that "the grandfather paradox" does not arise. It is distinguished between changing the past by subtraction and changing the past by addition. The first sort of change is impossible: what has happened cannot be changed. But the second sort of change is possible. It is possible to open new causal lines anytime that someone travels in time, to the future as well as to the past, which allows the occurrence of different and new facts.

The need to integrate representations of an internal, but not merely subjective, fluent time with representations of other more objective and more external time is relevant in many fields. One of them is the field of modelling and simulation of complex behaviours involving time. Another one is the analysis of reasonings involving different temporal perspectives. With respect to them, the discussions and analyses contained in the book can offer suggestive insights.

Points of view can make sense of the objective and subjective aspects of time because points of view have, themselves, a strong bipolar character. Points of view are ways of having access to objective reality. But, crucially, they also involve our subjectivity. We have to assume that points of view are parts of the objective reality they display before us. However, points of view also have features that do not find a comfortable place in the objective world described by science. And it is plausible to claim that not all of these features can be discarded as being merely subjective.

The book contains very detailed discussions of current issues about points of view and time. The results are tentative, but the problems under discussion are really important. Because of that, the book is also appropriate for use in graduate and upper-level undergraduate courses. Moreover, it would be of interest for people approaching these issues for the first time.

Most of the chapters of the book have been elaborated in the context of several research projects and grants supported by the Spanish Government. The most important of these research projects are the following:

FFI2008-01205 (*Points of View. A Philosophical Investigation*)
FFI2011-24549 (*Points of View and Temporal Structures*)
FFI2014-57409-R (*Points of View, Dispositions, and Time. Perspectives in a World of Dispositions*)

The topics taken up here by no means exhaust the field. In fact, the field is an open field. And much further work has to be done.

We want to conclude this brief preface by expressing our sincere gratitude to Prof. Lorenzo Magnani and Dr. Leontina Di Cecco, for all their support and patient confidence in the various stages of the book.

Margarita Vázquez Campos
Antonio Manuel Liz Gutiérrez

Contents

Contents

Chapter 1
The Notion of Point of View

Margarita Vázquez Campos and Antonio Manuel Liz Gutiérrez

Abstract This chapter introduces the notion of points of view, or perspectives. First, some classical antecedents are commented. Secondly, we make explicit two general approaches in order to analyze the structure of points of view. One of them uses propositional attitudes as a model. The other one focuses on the notions of location and access. After doing that, we will analyse the peculiar nature and mode of existence of points of view. It is argued that points of view are essentially relational entities with quite a strong modal character. This will entail that they are not reducible either to subjectivity, or to information, or to current physics. A number of relevant topics are also discussed: the distinction between individual and collective points of view, the notion of personal points of view, the different senses in which points of view can be said to have conceptual and non-conceptual contents, and the notion of reflective points of view.

1 Points of View

We cannot have access to reality, and to ourselves, except by adopting a point of view. In fact, the notion of point of view is crucial in many contexts and discourses. But, what is a point of view? Even though there are some classical antecedents for a

This work has been granted by Spanish Government, "Ministerio de Economía y Competividad", Research Projects FFI2008-01205 (*Points of View. A Philosophical Investigation*), FFI2011-24549 (*Points of View and Temporal Structures*), and FFI2014-57409-R (*Points of View, Dispositions, and Time. Perspectives in a World of Dispositions*).

M. Vázquez Campos (✉) · A.M. Liz Gutiérrez
University of La Laguna, Tenerife, Canary Islands, Spain
e-mail: mvazquez@ull.es

A.M. Liz Gutiérrez
e-mail: manuliz@ull.es

M. Vázquez Campos and A.M. Liz Gutiérrez (eds.), *Temporal Points of View*,
Studies in Applied Philosophy, Epistemology and Rational Ethics 23,
DOI 10.1007/978-3-319-19815-6_1

philosophical treatment of the notion of point of view, and in spite of its great importance, there is a deep lacuna about what can the structure of points of view be, and about what can their nature and mode of existence be. In this work, we present two general approaches in order to analyse the structure of points of view. One of them takes propositional attitudes as a model. The other focusses on the notions of location and access. To some extent, the structure of points of view can be analysed without engaging with the difficult issue of the peculiar nature and mode of existence of points of view. But we also face these questions. We argue that points of view are essentially relational entities with quite a strong modal character. This will entail that points of view are not reducible either to subjectivity, or to information, or to current physics.

There are many things that are said to be relative to a point of view or perspective (we will take these notions as equivalent). We say, for instance, that beauty is in the eye of the beholder; or that there are many different moral and cultural perspectives; or that we can adopt distinct conceptual frameworks with respect to a given issue. What is known, thought, said, intended, etc., is very often relativized to a point of view. Even the whole of reality is sometimes so relativized.

The ordinary notion of point of view is highly ambiguous. There is a first ambiguity about what is the crucial element of a point of view. Sometimes the emphasis is placed on the bearer of the point of view. Other times, the emphasis is on the location or emplacement from which the point of view is adopted. There is a second ambiguity about what are the effects, or powers, that points of view can provide. Sometimes, points of view are understood as having a very strong constitutive power over things like content, truth, knowledge, etc. Other times, they are understood as having a much weaker power regarding exactly the same things.

Beyond the ordinary usage, the notion of points of view is relevant in many fields, from fine arts to science, and all these fields are full of very suggestive insights. However, and in spite of being so important a notion, the lack of deep philosophical theories or even tentative analyses about points of view is remarkable. Points of view have rarely become a subject of philosophical attention.

1.1 Points of View in Ordinary Language

We speak of an "engaged point of view", of a "scientific point of view", of a "logical point of view", etc. Also, we contrast "the point of view of self-interest" with "the point of view of morality", or with "the point of view of society". We can take "a large perspective", or "a narrow one". We can take "the point of view of the man in the street", or "a religious point of view". We can even take a "transcendental point of view".

In that variety of uses, the notion of point of view may have two distinct meanings. In one of them, points of view are part of a mental life. They are connected to the mental life of some subjects with a personal character. In that sense, the expression "point of view" is interchangeable with words like "view", "opinion", "belief", "attitude", "feeling", "sentiment", "thought", etc. Points of view in that sense could not exist without a subject with quite a rich mental life.

There is another quite important meaning in the ordinary notion of point of view. In that second sense, points of view could exist without any actual subject exemplifying them. Here, points of view explicitly have a strong relational and modal, specially subjunctive, character. Points of view offer possibilities of having access to the world. They offer possibilities of seeing things (hearing them, touching them, etc.), possibilities of thinking about them (considering them, imagining them, etc.), and possibilities of valuing them (assessing them, pondering them, etc.). Moreover, points of view offer possibilities of seeing, thinking about and valuing our own process of seeing, thinking and valuing any other thing.

It is in that second sense that we use expressions like "The point of view of politics", "The point of view of science", "A utilitarian point of view", etc. In that sense, we also say things like "The point of view you can get from the top of that mountain is really wonderful". In all these cases, we presuppose that there are possible perspectives that are objectively there. We talk about the points of view offered by telescopes and microscopes, or about the point of view of a book, of a picture or, in general, of any work of art. Here, the expression "point of view" is interchangeable with words like "viewpoint", "standpoint", "position", "outlook", etc.

Many times, the ordinary notion of point of view is ambiguous in relation to the above two meanings. Moreover, as we will see, the philosophical approaches to the structure of points of view also can be distinguished according to whether they place the emphasis on the first meaning or on the second one.

There is another very important ambiguity in the ordinary notion of point of view, this time having to do with the sort of effects, or powers, that points of view can provide. Sometimes, points of view are understood as having a strong constitutive power over things like content, truth or knowledge. This has been an important source of scepticism. Also, it is the main idea underlying the connections between points of view and relativism. In that sense, to say, "p, from my point of view" would entail a kind of redundancy. Every content p has to be the content of some point of view. Moreover, for relativism there could not be any content (any truth, any knowledge) that is not determined, or construed, or shaped, by a point of view. Here, the constitutive power of points of view is extreme. Other times, however, points of view are interpreted in the opposite way: as weakening our attachment with a certain content, truth or knowledge. In this second sense, to say, "p, from my point of view" would not be redundant. It would express that my propositional attitude towards p is less engaged with the truth than if I were to believe that p in a strict sense, and of course much less engaged with the truth than if I were to know that p. In that second sense, when I say, "p, from my point of view", it is even questionable that p is the real content of some of my beliefs.

The philosophical approaches to the notion of points of view also mirror this second ambiguity. Sometimes, the philosophical uses of the notion of points of view are associated with very radical forms of scepticism or relativism. Other times, they are associated with much weaker forms of perspectivist compromises.

We have indicated some connections between the notion of points of view and two families of other notions. When points of view are understood as part of a mental life, the expression "point of view" is connected with words like "view", "opinion", "belief", "attitude", "feeling", "sentiment", "thought", etc. When points of view are understood as possibilities of access to something from a certain location or emplacement, the expression "point of view" is connected with words like "viewpoint", "standpoint", "position", "outlook", etc. The notion of points of view is ambiguous in that respect. Now, we have to add a third family of notions. When the point of view has a certain structural complexity and its scope is large enough, we can also speak of things like "conceptual schemes", "frameworks", "logical spaces", "paradigms", "forms of life", etc.

That third family of related notions is connected with the second ambiguity indicated before. When points of view are understood as constituting any possible sort of content, or truth, or knowledge, they become very close to something that can be described as a conceptual scheme, a framework, a logical space, a paradigm, or a form of life.

There are, however, important differences among all the last notions. Conceptual schemes, frameworks, and logical spaces fall short of determining particular sets of contents. Moreover, they tend to take into account only contents of a conceptual kind. In contrast, paradigms and forms of life entail the idea that there is something determining particular sets of contents, both of a conceptual and non-conceptual kind. That idea is central in the relativistic ways of understanding the notion of points of view.

To speak of paradigms and forms of life introduces other problems. A paradigm involves objects, instruments, and strategies in ways that are not entailed by the notion of point of view. Paradigms also are normative in senses that are not exactly the same in which we can say that points of view are normative. The idea that paradigms include "models", or "standards", that have to be imitated is not directly entailed by the notion of points of view. Similar things can be said of forms of life.

1.2 Beyond the Ordinary Usage

The notion of points of view is used in literature, theatre, painting, photography, movies, architecture, design, etc. And it is also very important in social sciences and humanities, and in many technical disciplines. All these uses are full of very promising insights.

In literature and theatre, for example, it is common to distinguish among a great variety of points of view: the one of the author of the text, the one of the narrator, the one of the audience of the narration, the one of the protagonist, the ones of the other characters of the narration, etc. Also, with respect to a narration, it is usual to distinguish between an "external time" and an "internal time". Both kinds of times would be defined by the text, in the sense that none of them is the real "physical time" in which the text is placed. The internal time is determined by what the text says and the external time by how the text says what it says.

There may be serious discrepancies among all the above mentioned points of view. Also, the teller, the protagonist, the other characters, etc., can adopt a first-person point of view or a third-person point of view. Moreover, in some cases, they can even embrace a second-person point of view. And these personal points of view can be expressed either in singular or in plural.

In theatre and movies we would have to add many other points of view: the ones of the actors with respect to the play, with respect to the other actors, with respect to the director, with respect to the public, etc. There are also very different interpretative schools, and very different ways to interpret a certain play. All of that leads to distinct narrative structures linked to points of view, and to distinct narrative rhythms.[1]

Performances, both in theatre and in fine arts, show the peculiar constitutive character that some peculiar points of view are capable of having. A performance is essentially something "intending to be a performance". Performances also make crudely explicit the crucial tension, existing in every work of art, between the points of view of the artist and the points of view of the public.

The fields of fine arts, photography and movies are full of references to the notion of points of view. We can find them in the discovery of perspective, something also related to maps and cartography, or in movements like impressionism, conceptual art, and cubism. Impressionism can be taken as claiming that the authentic work of art is always "inside our eyes". Going a step further, the so called "conceptual art" maintains that the authentic work of art only exists "inside our mind". Cubism can be understood as an attempt to get a perspective over all perspectives. There are also works of art trying to create the illusion of impossible perspectives. The aesthetic claim of artists like Esher consists precisely in that it is possible "to show" the impossible.

The notion of "frame" in a work of fine art is also very relevant. The frame is both a limit and a means to identify the work of art. But some works of art try to transcend the limits of their very frames. In a similar sense, some plays in theatre try to transcend the limits between the scene and the audience, or the limits between the scene and what is behind the scene. And in a similar sense, many musical creations try to transcend the limits between the sounds produced by the expert musicians and the environmental sounds, or between the programmed sounds and some occasional or aleatory sounds.

Some particular works of art have been the subject of philosophical analyses explicitly involving the notion of points of view. Foucault offers a paradigmatic example in his commentaries on *Las Meninas*, of Diego Velazquez. Another example is Heidegger's commentaries on van Gogh's *Shoes*. There are many others.[2]

Photography and movies introduce new elements. Are pictures simply a kind of painting? Furthermore, are movies simply pictures in motion, a collection of pictures displayed in a certain temporal order? No doubt, there are crucial differences. But there are also cases which would be very difficult to classify.

---

[1]See Lloid [65] and Linhares-Días [63].
[2]See Foucault [32] and Heidegger [54].

Literature, theatre, fine arts, pictures, and movies are very good indicators of the complexity of our points of view, and of how complex the relationships among them can be. In relation to the contents of points of view, the distinction between "to say something" and "to show something" always is very important. There are contents that are shown but not said. And many of these contents shown but not said are non-conceptual contents.

The complexity of our points of view, and the density of their relationships, is manifest in other related areas. Architecture and design are full of references to points of view. Any design, or project, or plan, etc., can be understood as the solution of a conflict between many different points of view, the points of view of the producers not always being the more important ones.

Points of view also are very important in social sciences, humanities and technical disciplines. Very often, it is said that social sciences and humanities have to follow methodologies completely different from those that are employed in natural sciences. The problem of "methodological dualism" (more generally, "methodological pluralism") can be seen as a conflict between alternative methodological points of view. Methodological points of view are a variety of practical points of view. The aim of a methodological point of view is to offer ways of achieving, preserving and communicating knowledge and skills about a certain field.

There is an old contrast between the methodological points of view of natural sciences and the methodological points of view of social sciences and humanities. The latter ones are usually understood as involving things like comprehension, interpretation, rationality, empathy, engagement, etc. In other words, in social sciences and humanities, but supposedly not in natural sciences, there is the need to integrate some subjective and personal points of view.

Sometimes, the point of view of science also is contrasted with the point of view of engineering. The first one would not have the sort of practical constraints that are relevant in the second one. The recent evolution of both science and engineering has made this issue much more complex.

There is a technical field in which points of view have a distinguished role. It is the field of the construction, and use, of computer simulation models using some kind of "expert knowledge". Simulation models are particularly relevant for prediction and control when there is no scientific knowledge available. Simulation models also are capable of offering a peculiar sort of understanding different from the one provided by theories. The construction of models always has to rely on some expert knowledge, and in many cases that expert knowledge comes from an acquaintance with the systems that are trying to be modelled. There are many problems that can be faced only with the help of simulation models. Also, there are many different techniques and methodologies to construe simulation models.[3]

---

[3]The so called "System Dynamics" is among the most important ones. The connections between System Dynamics, as a paradigmatic methodology for the construction of simulation models, and the notion of points of view are very interesting and full of relevant insights. See Vázquez et al. [123], Vázquez and Liz [121, 122].

In the construction and use of computer simulation models, there are always three different kinds of points of view involved: 1) the points of view of the experts in the systems that are going to be modelled, 2) the points of view of the people who are constructing the model, who have the operational skills to apply successfully a certain technique and methodology of simulation, and 3) the points of view possessed by the users of the models. Sometimes, the three kinds of points of view are adopted by the same subjects. Normally, that is not so. In any case, the three kinds of points of view would entail different attitudes and different contents, both non-conceptual and conceptual.

1.3 The Philosophical Relevance of Points of View

Scepticism, relativism, and perspectivism are three very important philosophical positions crucially involving the notion of points of view. Many discussions in philosophy of language and philosophy of mind also appeal to points of view. It is easy to show the philosophical relevance of the notion of points of view. What is not so easy to understand is the lack of elaborated philosophical theories about what the structure and nature of points of view is. The notion of points of view has rarely been the target of a direct philosophical analysis.

Nevertheless, many of the problems, ideas, and distinctions we will discuss have antecedents in the philosophical tradition. And it would be useful to remember briefly some of them. Let us go back to Parmenides and Heraclitus. According to Parmenides, beyond the unstable and misleading point of view of "appearances", we can take another point of view guided by "reason". From that rational point of view, Parmenides argued, it is possible to know the reality in itself. In contrast, Heraclitus claimed that we never can go beyond appearances. There is no other point of view apart from the point of view of appearances because there is no stable reality beyond appearances. There is no reality apart from appearances. All reality is in flux, like a river. And we are part of that flux.

Heraclitus's perspective leads to a certain sort of "relativism". However, it is very important to distinguish that sort of relativism from other varieties of relativism, in particular from the relativism inspired in Protagoras's thesis that the human being is the measure of all things. Both positions are radical forms of relativism. However, their use of the notion of point of view is very different.

Common to Heraclitus and Parmenides is the assumption that we are able to know how our points of view are connected with reality. This entails another kind of point of view that we can call "transcendental". It is a certain kind of reflective point of view from which we can obtain a perspective over all the other points of view, even over every possible point of view.

The dualism of Parmenides between appearances and reality was assumed by Plato, by Aristotle and by most of the philosophers in Western philosophy. The relativistic claims of Heraclitus and the relativist claims of Protagoras had important echoes in Nietzsche. Also, we can find them in the empirical scepticism of Hume.

In one way or another, all relativist approaches are under the influence of both Heraclitus and Protagoras.

Protagoras's relativism goes against all doctrines, including the one of Heraclitus. In his dialogue *Protagoras*, Plato criticises the position of Protagoras as "self-refuting". However, according to Protagoras, in Plato's dialogue, even if the claim that non-relative truth does not exist is self-refuting, we have to assume it. Moreover, in fact we live assuming it.[4]

The need to achieve a transcendental perspective about how our points of view are connected with reality also became a classical subject matter. Either in a naturalistic or in a non-naturalistic guise, we can find the search for such transcendental perspective in philosophers as distant from each other as Plato, Kant, and Wittgenstein.

Plato distinguished sharply between the empirical point of view of the "appearances" and the rational point of view of the "ideas". Ideas constitute reality. They are some kind of "pattern" or "standard". Every reality has an ideal nature to the extent that it develops, or pursues, or points at, a certain pattern or standard. But it is not easy to change from the point of view of the appearances to the point of view of the ideas. It requires a serious reflective effort.

Aristotle rejected that "ideal forms" can exist independently of "matter", i.e., prior to the matter. Matter and form cannot be separated. Any reality is constituted by some matter having certain forms, and some of those forms possess a teleological dimension. This supports the distinction between "potentialities" and "actualisation of those potentialities". Also, some of those forms are "essential" for something to be just the sort of thing it is, and other forms are only "accidental". The first ones constitute "individual substances" and kinds of substances, the traditionally so called "genus" and "species". The second ones are involved in all sorts of non-substantial changes.

According to Aristotle, we are rational animals constituted by a structured set of "capacities" that can be actualised in one way or another. The actualisation of the higher ones depends on the actualisation of the lower ones. All knowledge of the world begins with the actualisation of some congenital or innate discriminative capacities. Our intellect is able to grasp what is common to the things, and to express it in language. But, what is particular to each thing cannot be grasped by the intellect, nor non-demonstratively expressed in language. It only can be known non-conceptually through the senses, and it only can be expressed demonstratively.

The ideas of Plato and Aristotle were crucial in the evolution of Hellenist and Medieval thought. The dualism between the point of view of appearances and the point of view of reality was decisive, either in the Platonic way or in the Aristotelian one. Also determinant was the possibility of adopting a teleological perspective over the whole of reality, or over parts of it. The same relevance had the two different ways, the Platonic one and the Aristotelian one, of adopting a transcendental point of

---

[4]About the "self-refuting" character of Protagoras's relativism, see Burnyeat [10]. See also Siegel [110].

view. In addition, the relativism of Heraclitus and the relativism of Protagoras maintained a hidden but continuous influence. In connection with scepticism, both forms of relativism would come back with force in the Renaissance and they would contribute to configure Modern and Postmodern thought.

The Renaissance entailed new religious attitudes, new political aspirations and new ways of conceiving reality and our knowledge of it through science. In the18th century, those changes gave way to the Enlightenment. Two features are specially relevant for our subject: 1) the search for a solid "foundation" for knowledge, in particular for the scientific ways to achieve knowledge, and 2) the "universal" value of reason, both in a theoretical and practical sense, over any particular point of view.

For rationalists like Descartes, Leibniz or Spinoza, both the foundation of knowledge and the universal value of reason are derived from something everyone can find in "reflection". For empiricists like Locke, Berkeley or Hume, they have to be found in "experience".

Descartes offers the details of a reflective, internal search. His approach is full of insights concerning what has been called "the point of view of the subject",[5] and also concerning what has been called "the absolute conception of the world".[6] The Cartesian search for certainty is radically subjective in its methodology, but it has as its final aim something maximally objective.

In Leibniz and Spinoza we can find two different sorts of "perspectival metaphysics". For Leibniz, reality is constituted by some infinitesimal parts: the so called "monads". Monads have no interconnections among them. They do not, say Leibniz, have "windows". Indeed, relations do not exist at all. However, each monad mirrors all the entire universe from a certain "internal point of view". For Spinoza, reality has many aspects (he calls them, "modes") of infinite kinds (he calls them, "attributes"). We, human beings, can only have access either to those aspects that are of a "spatial" or "material" kind, and to those aspects that are of a "mental" or "spiritual" kind. Each particular thing, including ourselves, is a part of reality having both spatial or material aspects and mental or spiritual aspects. Reality is the whole of all those parts, aspects (modes), and kinds of aspects (attributes). And it can be understood either as Good or as Nature.

The debates between rationalists and empiricists are also relevant regarding the two kinds of explicit contents we can identify in points of view. A point of view can explicitly have either "non-conceptual contents" or it can have "conceptual ones". Non-conceptual contents are the contents of experience: what we see, hear, smell, touch, feel, etc. Conceptual contents are the contents of thought: what we believe, desire, remember, etc. The rationalists, and particularly Descartes, considered that the contents of experience also are some sorts of thoughts. However, they usually made room for other sorts of non-conceptual contents linked to agency and the experience of acting. Examples are Descartes treatment of the "will", Spinoza's

---

[5]About that, see the interesting book of Farkas [30].
[6]With respect to this, see Williams [125].

notion of "impetus", and Leibniz's notion of "appetites". In any case, in contrast with the rationalists, the empiricists considered that, in one way or another, the explicit non-conceptual contents of our experiences are at the origin, and they constitute the foundations, of any conceptual content.

Both rationalism and Empiricism originated sceptical lines of thought. The Cartesian "point of view of the subject" derives in an application of the method of the doubt to whatever can be supposed to exist in an "external world". Hume is also an important example of quite a radical scepticism. In an Heraclitean way, Hume claims that we cannot go beyond the contingent presence of some sense impressions "inside our minds", being our own mind no more that a cluster of sense impressions. This puts severe limits on our knowledge. There is no justification for our beliefs in an external world of substances, or for our beliefs about our very existence as a peculiar sort of substantial "egos", or for our inductive beliefs, or for our confidence in miracles, or for our reliance in God.

Very often, as we have said, scepticism is closely connected with the recognition of the crucial role of points of view in our relations with reality. This is shared with relativism and with perspectivism. However, there are important differences among these three positions. Whereas relativism and perspectivism contain claims of knowledge, scepticism does not. Relativism claims to know the complete relativity of some intended knowledge, or the complete relativity of all intended knowledge, or the complete relativity of any possible knowledge. Perspectivism claims that those things only are "partially relative". Scepticism consists in a, more or less active, refusal to make "any claim" about a certain field of intended knowledge, or about all intended knowledge, or about any possible knowledge.

The so called "Kantian Turn" can be understood as the (transcendental) assumption that there is no access to the world that does not consist in the adoption of a certain point of view. We cannot have access to the world such as it may be "in itself". There is a world in itself, Kant assumes, but we cannot absolutely know anything about how it is.

The Critique of Pure Reason can be taken as an analysis of the general structure of our epistemic points of view, from sensibility to understanding, and from understanding to reason. The other two major Kantian works, *The Critique of Practical Reason* and *The Critique of Judgement* can also be seen as offering analyses, respectively, of the general structure of our practical and moral points of view, and of the general structure of our teleological and aesthetic points of view.

Conceptualism is the thesis that all our mental contents, even the contents of our sensible experiences, are in the last analysis conceptual. Let us say that a content is conceptual if it is a thinkable content, or a content that can be the content of some belief, or a content that can be true or false in the way that propositions are, or a content that can be logically connected with other contents, or a content that can be non-demonstratively expressed in public languages. It is a matter of discussion whether Kant argued for a conceptualist way of understanding all the contents of our points of view, or whether he preserved some room for non-conceptuality. In any case, Kant's approach has had a strong and direct influence on recent conceptualism.

The Kantian approach is explicitly transcendental. We cannot know reality in itself, but we can know how we are epistemically related with it. We can know it by "critical" reflection. That way, we can discover that any knowable reality has to be shaped by the structures of our subjectivity. From that transcendental point of view, we have to adopt an idealist position. From an empirical point of view, however, we can only be realists. Our thoughts have to respond to the contents of our sensible experiences. This combination of "transcendental idealism" and "empirical realism" is one of the most important aspects of the Kantian heritage.

Kant was very influential in both Shopenhauer and Hegel, who were in conflict with each other. According to Shopenhauer, the world is constituted by "will" and "representation". Shopenhauer's conception of the world as representation follows quite literally Kantian approach. However, the world as will introduces some new ingredients. We can find in Shopenhauer important insights about the volitive, agentive and expressive aspects of our points of view. Those non-conceptual aspects also will be emphasised by Nietzsche.

Hegel gives up the Kantian notion of "thing in itself". The nature of every entity depends on all the relationships it maintains with us and with other entities. Hence, the only truly substantial entity is the reality as a whole.

Hegel makes abundant use of the notion of "collective" points of view evolving in time. This is shared by other philosophers in the Romantic era. Many of the discussions about the existence and nature of collective subjects have their origin here, including all the discussions in social sciences about the existence and nature of national identities, social classes and other collective entities, and about their peculiar points of view.

In fact, Hegel was an active protagonist of the emergence of the cultural movement known as Romanticism. In many senses, it was the "antithesis" of the Enlightenment. It rejected the need to search for a solid foundation for knowledge and the universal value of reason. This movement is also relevant for our topic because of its dramatic emphasis on "singularity". There are unique points of view linked to each particular subject. We did find that idea in Leibniz. Also, there are idiosyncratic points of view linked to very special kinds of collective subjects. Moreover, those idiosyncratic points of view have a crucial role in structuring a certain "world vision".

Nietzsche is another very important author at the end of this period. As we have said, he offers a mixture of the two sorts of radical relativism that we have found in Heraclitus and Protagoras. On the one hand, Nietzsche presents a dynamic world in which there is nothing stable. On the other hand, Nietzsche assumes an image of reality in which everything is constitutively relative to some point of view. Every reality is understood as the expression of the "will" of a certain subject, or as the expression of a struggle among different kinds of subjects. It is a problem whether both conceptions are ultimately consistent with each other. In any case, Nietzsche rejects the Platonic dualism between the changing world of "appearances" and the timeless world of "ideas". There is no hidden reality behind the world of appearances, and there is no way of having access to reality apart from our points of view about appearances.

Nietzsche's position confront us again with the problem of the "self-refutation" of radical relativism. It is not easy to deal with that problem. Relativism is not self-refuted if it is maintained only in a local sense. Moreover, even if such a self-refutation can be easily obtained in the case of a general relativism, like the one proposed by Protagoras, a general relativism in the line of Heraclitus cannot be so easily accused of self-refutation.[7]

Historicism, pragmatism and existentialism constitute three further important ways to give substance to the claim that our points of view are dependent on something else. That dependence can be understood only in a perspectival sense, or it can be understood in a strong relativist sense. In the last case, we would be faced with three relevant varieties of the kind of relativism suggested by Protagoras.

According to historicism, the historical period, geographical place, social group, cultural traditions, etc., have the power to configure our points of view about reality and about ourselves. Antecedents of historicism are G.B. Vico and M. de Montaigne. It was a position suggested by Hegel and developed by Dilthey, Spencer and F. Boas. Marxism has very often maintained quite strong historicist approaches.[8]

For pragmatism, all our connections with reality have an "agentive" and "practical" dimension with a constitutive power. This leads to anti-Cartesianism and to a rejection of the dichotomies mental/physical and fact/value. It also leads to pluralism. There can be as many different truths as there are different practical aims, and ways to satisfy them.[9]

Pluralism, even a certain conceptual relativism, is adopted by many exponents of pragmatism, both classical and recent ones.[10] However, it is very important to note that according to a pragmatist position, the thesis of conceptual relativism could not be claimed as an "a priori" truth, but only as a matter of fact closely connected with practical issues.

The wide cultural and philosophical movement known as existentialism is also of interest. In one way or another, existentialism puts the ontological emphasis on the "contingencies" of particular and concrete existent beings over any essence, or nature, or law; and the normative emphasis in the ultimate value of "authenticity". Important existentialist authors are S. Kierkegaard, J.P. Sartre, M. Heidegger, A. Camus, and K. Jaspers, among others.[11]

---

[7]About the self-refutation of Protagoras' relativism, see again Burnyeat [10]. See also Solomon [112], Clark [16] and Haack [40]. About a perspectivist interpretation of Nietzsche, see Hales and Welshon [47]. About relativism in general, see Hales [46, 45]. About the notion of "relative truth", see García-Carpintero and Kölbel [35].

[8]A classical critique of historicism is Popper [85, 86].

[9]About pragmatism in general, see Goodman [37]; and Haack [41]. About the pluralist consequences of pragmatism, see James [59]. Recent, and very important, pragmatist positions are Brandom [5, 6], Rescher [95] and Rorty [97]. The connections between action and perception have been emphasised by Hurley [57] and Nöe [77].

[10]See, for instance, Putnam [89, 90, 91, 94], Rorty [97–99] and Brandom [7].

[11]About existentialism, see Earnshaw [29], Cooper [17], Flynn [31] and Guignon [38].

Each subject entails a "maximally singular point of view" entering into conflict with the points of view of science, morality, politics, etc. And the only way to solve the tension is through the exercise of our "free will". This can have quite a direct relativist interpretation. In any case, existentialism is an important source of inspiration regarding personal points of view and practical points of view with a strong "singular" character. A very relevant author in that sense is the Spanish existentialist philosopher M. de Unamuno.[12]

We have to make special mention of another Spanish philosopher of the last century, José Ortega y Gasset. He was influenced by neo-Kantism, historicism, and pragmatism. Also, he worked on the metaphysics of Leibniz. For the first time, we can find in him a philosophical position which is explicitly called "perspectivism".[13]

According to Ortega, the only way to understand any reality is from the perspectives offered by concrete circumstances. Moreover, the world itself is constituted by a "variety of perspectives". It is neither material nor spiritual. Ortega rejects both materialism and idealism. He frequently illustrates his theses with examples involving different views of a landscape. When a spectator sees something, the organisation of the objects and their relative importance is determined by his perspective. In the case of sights which are mutually contradictory, it makes no sense to declare that one of them is false and the other one is right, nor does it makes sense either to declare both false. We have to admit both of them as true. To reject one of them as false, or both of them, would entail another perspective which would enter into conflict with at least one of the previous perspectives, perhaps with both of them. But this would bring us back to the very same problem of the beginning. The only possibility of avoiding such a regression would be to reject the existence of a privileged "absolute sight" made from no point of view at all. However, Ortega claims that that does not make sense. Reality is ultimately multiple and perspectival.

Ortega's perspectivism has important implications of a normative and evaluative kind. Many parts or aspects of reality can be identified and appreciated only from certain singular perspectives. And this constitutes a very strong *prima facie* reason to value positively personal and cultural diversity.

According to Ortega, all perspectives are in principle (but only in principle) "equally valid" or true. Moreover, the rejection of a perspective is something that always has to be made from another perspective. Perspectives are our only way of having access to the world. Perspectives are determined by the place each one occupies in the world. Reality can only be understood and described from particular emplacements. Reality is not an invention, but neither is it something independent

---

[12]See de Unamuno [23].

[13]From 1914 to 1923, he elaborated that position in a number of essays. See especially *Meditaciones del Quijote (Meditations of the Quixote)*, of 1914; *El Espectador (The Spectator)*, of 1916, specially the essay "*Verdad y perspectiva*" ("*Truth and Perspective*"), included in the first volume; and *El Tema de Nuestro Tiempo (The Subject of our Time)*, of 1923. All these essays are (re-)edited in Spanish in Ortega y Gasset [78].

from our position in it. There are as many "realities" as points of view. Truth is simply that description of the world "faithful to a perspective". The only false perspective is the one that claims to be the only valid one, i.e., the one that pretends not to be based on some point of view.

Knowledge is anchored in a point of view, and points of view depend on a certain emplacement in the world. For Ortega, the idea of a reality structured with independently of all points of view makes no sense. There is no such thing as an "absolute", fully objective and completely independent reality. And this is true regarding the physical world as well as regarding other dimensions of reality such as values. Ortega's claims are in many relevant respects very close to pragmatism. They are also very close to some recent approaches like that of Putnam's "internal realism".[14]

Another very important author for our subject is Wittgenstein. He is relevant for various reasons. The first one has to do with his *Tractatus*.[15] This work offers a very clear, and brilliant, example of "transcendentalism" concerning the relationships between points of view and reality. Moreover, this position is articulated in such a way that if it is true, then it cannot be strictly described as such. In Wittgenstein's own terms, that position can only be "shown", it cannot be "said".

Many of the particular discussions of the *Tractatus* are of interest regarding the notion of points of view. The most important are the following ones: 1) the claim that both the "bearer" of a point of view and the "shape", or structure, of a point of view can be only "implicitly" present in the point of view itself, they cannot be "explicit contents" of the point of view; 2) the thesis that the "limits of the world" are coincident with the "limits of our points of view" over the world, and so they cannot be explicitly contained in the point of view either; 3) the claim that there is a sense in which "solipsism" is a necessary and obvious truth; 4) the claim that the "attitudes" involved in a point of view can change, even offering a radically different world, without any change in the explicit contents of the point of view, so that "the world of a happy person can become a different world than the world of an unhappy person"; and 5) the claim that what is "beyond the limits" of all points of view, as for instance what we are just saying in these lines, is something that strictly cannot be "said" but only "shown". For Wittgenstein, points of view are ultimately "ineffable". In his own words, they belong to "the mystical".

The second major work of Wittgenstein, *Philosophical Investigations*,[16] also is relevant. It contains two important lines of thought concerning the notion of point of view. This time, they go in opposite directions. One of them is the relativist, even in many cases sceptical, interpretation that can be given to theses like that "meaning is use", or to notions like "forms of life" and "language games". The other line of thought comes from Wittgenstein's criticism of the notion of a "private language" in the context of "the problem of rule-following". The point of view of the subject,

---

[14]See Putnam [87, 88, 91, 92].

[15]Wittgenstein [126].

[16]Wittgenstein [127].

such as it was conceptualised by Descartes, is a subjectivist point of view that, even if it looks for certainty, or perhaps because of that, very easily leads to relativism and to scepticism. By the time of his *Philosophical Investigations*, Wittgenstein was very critical of the coherence of that point of view. Wittgenstein's arguments based on "rule following" focus on the need to achieve a non-sceptical and non-relativist stance from which it may make sense to say that "a rule is being followed".

The relativist and sceptical tendencies linked to the first line of thought are in sharp contrast with the anti-relativist and anti-sceptical implications of the second one. Moreover, it can be argued that there is an important tension between both lines of thought. Wittgenstein's main argument against private languages can be extended to a social context. The need to distinguish between "following a rule" and "merely thinking that a rule is followed" can be posed both in an individual and in a collective sense. Both situations would be similar in that respect. The anti-relativistic implications of Wittgenstein's criticism of private languages can be applied to theses like "meaning is use", and to notions like "forms of life" and "language games". But things also can be the other way around. The relativist interpretation that can be given of "meaning as use", "forms of life" and "language games" can be also applied to the problem of private languages and rule following.[17]

In the last period of his life, Wittgenstein initiated other approaches, and some of them are relevant again for our subject matter. In *On certainty*,[18] Wittgenstein argues that there are some questions that we cannot ask. They are beyond what can be answered because they refer to claims establishing the frameworks inside which questions and answers make sense. This suggests the need for a background in order to have points of view, a background that cannot be itself another point of view. Searle has argued recently for something closely similar.[19] Any conceptual interpretation and ascription of content requires the previous existence of a background of abilities and capacities that cannot have only a conceptual character.

Wittgenstein's last approach also has similarities with Carnap's distinction between "internal" and "external" questions.[20] Only the first ones make sense. They are questions articulated inside a certain conceptual framework. Strictly speaking, the second ones do not make sense. They cannot make sense, or in any case they cannot have the same kind of sense, because they are questions about what defines what makes and does not make sense. They are questions about the conceptual frameworks. Wittgenstein appears to suggest that there is a very basic framework defined by ordinary notions and ordinary possibilities of action. There is, however, an important difference between Wittgenstein and Carnap. Carnap's "Tolerance

---

[17]The first, relativist and sceptical, line of thought has been followed by Kripke [61]; the second one, anti-relativist and anti-sceptical, has been explored by Cavel [12], Diamon [28] and Putnam [93]. About the first line, see the analyses of the problem of rule following included in Boghossian [4].

[18]Wittgenstein [128].

[19]With his notion of "the background". See Searle [105].

[20]See Carnap [11].

Principle" emphasises the optional character of all the frameworks. We decide which one to adopt. For Wittgenstein, in contrast, the ultimate frameworks are not optional.

Let us finish our historical commentaries by referring briefly to hermeneutics. Classically, hermeneutics was the study of the interpretation of written texts, especially in the areas of religion, literature and law. G. Vico, M. Luther, B. Spinoza and W. Leibniz are usually cited as precedents. F. Schleiermacher is the more representative figure in the period of Romanticism. Between the 19th and 20th centuries, hermeneutical approaches were applied to other meaningful phenomena, mainly in the fields of history, culture, and society. W. Dilthey and M. Weber are important authors in that area. The problem of methodological dualism is closely linked to that movement. The point of view of natural sciences is opposed to the interpretative point of view that is required in social sciences and humanities. Contemporary hermeneutics has taken a further step, dealing with everything involved in any process of interpretation. In particular, philosophical reflection is understood as a hermeneutical exercise. K-O Apel, H-G- Gadamer, J. Habermas, M. Heidegger, and P. Ricoeur are the main authors of reference inside the so called Continental Philosophy.[21] The nuclear ideas of hermeneutics can be summarised in the following way: 1) every interpretative process comes from a "fusion of horizons", the one that constitutes the context of the interpreter and the one that constitutes the context from which the interpreted phenomena come from; 2) there is an unavoidable circularity, sometimes called "hermeneutical circle", between that what makes the interpretation possible and the result of the interpretative process, so that what makes the interpretation possible presupposes certain results, which in turn presuppose some peculiar possibility conditions; 3) there is no meaningful reality without interpretation, or in other words meanings never are simply "given"; and 4) every interpretation is always a "re-interpretation", an interpretation made over the materials coming from other interpretations.

One of the most important problems for hermeneutics is to attach a robust sense to notions like truth, objectivity and rationality. Even the very notion of reality appears to be at serious risk. It is difficult not to see all those notions as answering to something that comes only, in the best case, from our "intersubjective perspectives".

Recent analytical philosophy also contains many proposals very close to the ideas of hermeneutics. The most remarkable ones would be those of G. von Wright, D. Davidson, and D. Dennett. The problem of methodological dualism is crucial in the approaches of von Wright and Davidson. And the problems of relativism are important problems in the approaches of Davidson and Dennett. Dennett's "intentional stance" also puts in the foreground the question of realism concerning what can be identified only from a certain perspective. According to him, this is what happens with our own minds, and with all its products.[22]

---

[21]See Apel [1], Gadamer [33] and [34], Habermas [42], Heidegger [53] and Ricoeur [96].

[22]About these authors, see von Wright [124], Davidson [21, 22] and Dennett [24, 26].

2 Two Approaches to the Notion of Points of View

As we have said, there is a serious lack of elaborated philosophical theories about what the structure and nature of points of view is. However, we can detect two main approaches about their structure. The first one assumes as a model the structure of propositional attitudes. The other one is based on the notions of location and access.[23]

2.1 The Model of Propositional Attitudes

Propositional attitudes offer a model for the internal structure of points of view. Points of view can be understood as having an internal structure similar to the one we can find in "propositional attitudes". That structure is constituted by a subject, a set of contents, and a set of relations connecting in certain conditions the subject with those contents. This model can have either a conceptualist reading or a non-conceptualist reading. We will introduce briefly both interpretations. Then, we will offer a general reconstruction.

2.1.1 A Conceptualist Reading

According to a very common analysis, propositional attitudes are constituted by a subject maintaining a certain psychological attitude toward a certain proposition. Examples of such psychological attitudes are "to believe something", "to desire something", "to perceive something", "to remember something", "to imagine something", "to guess something", etc. The proposal here is that points of view can be understood as structured sets of propositional attitudes.

Propositional attitudes always have as their content a certain proposition: the something that is believed, desired, perceived, remembered, etc. Propositions have a conceptual structure and an inferential articulation. Propositions are constituted by combinations of concepts, perhaps also in combination with other referential devices. And they are interconnected through many sorts of inferential relations in a wide sense: deductive, inductive, abductive, presuppositional, justificatory, etc. In the conceptualist reading of the model of propositional attitudes, points of view can have only conceptual contents of that kind.

The thesis that points of view are structured sets of propositional attitudes is particularly suggestive when we focus on maximally large propositional fields, constituted by conceptual structures and inferential relations able to determine all possible contents.

---

[23]See a preliminary account of them in Liz [64].

Some propositional fields are larger than other ones. A propositional field is maximally large if there would not be any need of new concepts in any situation. Points of view constituted by maximally large propositional fields become something very close to Carnap's "conceptual frames", Davidson's "conceptual schemes", and Putnam's "conceptual relativism". Certain ways of understanding concept possession would also allow inclusion of Wittgenstein's "forms of live", Kuhn's "paradigms", and Quine's "manuals of translation". In all these approaches, in one way or another, the model of propositional attitudes is adopted.[24]

2.1.2 A Non-conceptualist Reading: Christopher Peacocke

The model of propositional attitudes also can take another very different direction. It can be combined with a vindication of the existence of "non-conceptual contents", something not reducible to conceptual content. The model is not abandoned because the analysis of the structure of points of view regarding such non-conceptual contents is done in close connection with the structure of propositional attitudes.

There are many ways of characterising non-conceptual content. Sometimes, it is said that non-conceptual content is a content specified in terms of concepts that the subject to whom the non-conceptual content is attributed does not need to possess. Other options involve notions like "acquaintance", "continuous homogeneity", "finesse of grain", "analogical content", the "what-it-is like" of experience, "demonstrativeness", "contradictory character", "know-how" and "agency", etc.[25] In any case, defenders of non-conceptual content claim that our experience cannot be completely characterised by a description of its conceptual contents.

Christopher Peacocke has offered an account of non-conceptual content which could be easily applied to the analysis of points of view of a non-conceptual kind.[26] He distinguishes between representational and sensational, or phenomenological, properties of experience, arguing that the last ones, the "what-it-is-like" to have that experience, are indispensable to characterise the experience. Suppose, for example, that you are seeing two trees. One of them is one hundred yards away from the other one. Your experience represents them as being the same size. However, one takes more space in your visual field than the other one. According to Peacocke, this would be a clear non-representational, sensational or phenomenological, aspect of your experience.

---

[24]See, for instance, Davidson [20], and Kuhn [62].

[25]For the notion of "acquaintance", see Russell [100]; for "continuous homogeneity", see Sellars [107]; for "finesse of grain", see Peacocke [79]; for "analogical content", see Peacocke [80], for the "what-it-is like" of experience, see Nagel [73], Jackson [58], and also Peacocke [82]; for "demonstrativeness", see Perry [83]; for "contradictory character", see Crane [19]; and for "know-how" and "agency", see Nöe [76, 77].

[26]See especially Peacocke [79, Chap.1].

Peacocke has modified some aspects of his previous position.[27] Sensational properties of experience are assumed to be representational, but not in a conceptual way. Also, there are some proposals about the "analogical" and "fine-grained" representational character of the non-conceptual content of experience. However, the most important development has been the notion of "scenario content" introduced more recently.[28]

Let us explain that notion. Peacocke assumes that perceptual experience represents the world as being "in a certain way". Perceptual experience has a representational content. However, it is a content of a non-conceptual kind. Moreover, it is the most fundamental kind of representational content. All other sorts of representational properties presuppose the existence of this first type of content.[29] In order to make that kind of representational content clear, Peacocke proposes the notion of "scenario content". Scenario contents are individuated "by specifying which ways of filling out the space around the perceiver are consistent with the representational content's being correct".[30]

Scenario contents are "spacial types" defined by certain ways of "filling out the space around the perceiver". Scenario contents represent certain "types of space" filled with things having properties. The correctness of a scenario content is a matter of instantiation in the real world around the perceiver of the spatial type which corresponds to what the scenario content is "representing". Peacocke proposes a full specification of scenario contents in two steps:

1. The first step consists in fixing an "origin" and "axes". It has to be possible that many different spatial portions of the real world can find a position in relation with that origin and axes. For instance, one such origin could be given by the centre of our chest, with three axes defined by the directions back/front, left/right, and up/down defined with respect to that origin. The adequate origin and axes would be different if we consider other kinds of perceivers. Also, the adequate origin and axes would be relative to each different "mode of sensorial perception". The important point is that the choice of an origin and axes has to facilitate distinctions in the phenomenology of experience.

2. The second step in the specification of scenario contents consists in specifying the ways in which the space around the origin can be "filled out". One such specification, orientated to ordinary macroscopic objects, could be the following: for each point identified by a distance and direction from the origin, it is specified whether there is a surface there, and if so what orientation, texture, hue, saturation, brightness, temperature and solidity it has at that point. Of course, the specifications can be more or less precise or imprecise.

---

[27]See Peacocke [80, 81].

[28]Peacocke [82]

[29]Peacocke [82: 105].

[30]Peacocke [82: 105].

In any case, the concepts used in making the specifications do not need to be concepts possessed by the perceiver. This entails that, in principle, there would be no problem in using a very sophisticated conceptual or theoretical apparatus to fix the "scenario contents" of conceptually very primitive perceivers.

Any spacial type specified in the preceding way is called by Peacocke a "scenario", and the volume of the real world around the perceiver at the time of a certain experience is called a "scene". Scenario contents are properly defined as classes of scenarios, and scenes are defined as what is really happening around the perceiver at a certain time.

The "non-conceptual content" of an experience is a scenario content. And that content is correct if and only if there is a scene taking place in one of the relevant scenarios. That is, the content is correct if and only if there is a real scene at the time of the experience fulfilling the way of filling out the space around the perceiver which constitutes the scenario content.

Peacocke's analysis is very suggestive in order to understand the structure of sensorial or experiential points of view with non-conceptual contents. We can represent experience as defining a number of classes of scenarios. The real world offers the scenes. And the experience is correct when some real scene is displayed in some of those scenarios.

There are further developments. Peacocke has come to distinguish "positioned content" from scenario content.[31] A positioned content is a scenario content together with 1) an assignment to its origins and axes of real places and directions in the world, and 2) an assigned time. The positioned scenario would count literally as the content of a real experience, not as a representation or specification of its possible content.

Also, he introduces the notion of a "proto-propositional content".[32] It would be a second layer of non-conceptual content. It is not determined by positioned content, but it does not require the possession of concepts either. Proto-propositions contain individuals exemplifying properties or relations. One example is to see two unfamiliar shapes as being "the same" before elaborating any judgement. The experiences before and after seeing the identity of shape would represent the world as being in the same way at the level of positioned content. However, there is a change in non-conceptual content. And that change is not the result of any conceptual intervention. It is a change in "the ways things are seen". It is a change in the non-conceptual contents of the point of view produced by a change in "the attitudes involved". Proto-propositional content is assessable as objectively true or false. However, it does not require mastery of concepts.

Peacocke argues that the identity of conceptual contents depends upon the nature of their links with all those sorts of non-conceptual contents.[33] In general, scenarios offer a promising recourse for anchoring conceptual content in the more basic levels

---

[31]Peacocke [82: 107–108].
[32]Peacocke [82: 119].
[33]See mainly Peacocke [82: 111–135].

of positioned scenario content and proto-propositional content. At those levels, non-conceptual content can be defined independently of concept possession. Non-conceptual content makes it possible to attribute semantical properties to experience without the need to appeal to the conceptual recourses of the perceiving subject. Because of that, non-conceptual content makes it possible to "ground" concepts in a non-circular way.[34]

2.1.3   A General Reconstruction

According to the model of propositional attitudes, <u>a point of view PoV</u> can be seen as having the following canonical structure:

$$PoV = \langle B, R, non\text{-}CC, CC, Cp \rangle,$$

where

1. B is the bearer, or the possessor, or the titular, of the PoV.
2. R is a set of relations connecting B with the explicit contents of the PoV, $non\text{-}CC$ and CC.
3. $non\text{-}CC$ and CC are the two kinds of contents that can be explicitly included in the PoV; respectively, $non\text{-}CC$ is a set of non-conceptual contents and CC is a set of conceptual contents.
4. Cp is a set of possession conditions for having the PoV.

B is the entity to which the point of view is attributed. The point of view is anchored in reality through the relations R that the bearer of the point of view B is capable of maintaining with the explicit contents $non\text{-}CC$ and CC of the point of view. There are various relevant possibilities for B. It can be a personal subject, or it can be a psychological subject without the status of a person, or it can even be a non-personal and non-psychological entity.

Not always the bearer of a point of view is taken to be a "subject". Sometime, we say that a novel or a poem involves a point of view over the world, that a tree displays a point of view over its environment, and over its past and future possibilities, or that a propositionally structured database offers a point of view from which we can obtain many inferential conclusions.

When the bearer of a point of view is taken to be a subject, it is relevant to distinguish between "individual" and "collective" points of view. Whereas individual points of view have an individual subject as their bearer, collective points of view have a collective subject.

---

[34]There are important similarities between Peacocke's analyses, in particular his notion of "scenario content", and Gärdenfors [36] notion of "conceptual spaces". In both cases we find a peculiar sort of "geometrical approach". Moreover, in both cases, some kind of "bridge" between non-conceptuality and conceptuality is intended.

When can a subject be said to be collective? In a trivial sense, every entity, or almost every entity, would be collective in the sense of being "composed" of other entities. In a not so trivial sense, we can establish two alternative ways to understand collective subjects:

A collective subject is an entity such that 1) we can attribute to it some point of view, and 2) it is constituted, at least in part, by a collectivity of subjects having, each one of them, a certain mental life.

A collective subject is an entity such that 1) we can attribute to it some point of view, 2) it is constituted, at least in part, by a collectivity of subjects having, each one of them, a certain mental life, and 3) that collectivity of subjects also has by itself a mental life independent of the attributed point of view.

The second sense is stronger than the first one. However, the first sense also introduces a powerful notion of collective subject. Societies, groups and nations are collective subjects in the first sense, Sometimes, they have been understood as being also collective subjects in the second sense. However, we normally assume that societies, groups and nations do not have by themselves any "independent" mental life, and this looks reasonable.

But, even in the first case, we need a further distinction between "mere aggregates of subjects" and "genuine collectivities". What kind of constitution can originate a collective subject in the weak, first sense? One very direct answer has been given recently by J. Searle. Collective subjects are possible because the individual subjects constituting the collectivity are capable of having "collective intentions", or "us-intentions".[35]

In parallel with the above characterisations of a weak and a strong sense of the notion of collective subject, we can understand "individual subjects" in the following two ways:

An individual subject is an entity such that 1) we can attribute to it some point of view, and 2) it is not constituted by a collectivity of subjects having, each one of them, a certain mental life.

An individual subject is an entity such that 1) we can attribute to it some point of view, 2) it is not constituted by a collectivity of subjects having, each one of them, a certain mental life, and 3) it has by itself a certain mental life independent of the attributed point of view.

Again, the second sense is stronger than the first one. But, now, only the second sense seems to be capable of giving a robust sense to the notion of individual subject. If the only way to be an individual subject were the first one, then subjects could not be different from other non-subjective bearers of points of view except for some differences in the peculiar kinds of points of view attributed.

---

[35]See Searle [105]. See also Tuomela [118, 119]. Wilfrid Sellars is an important precedent of that idea. See Sellars [108, VII].

According to some authors, this would be the case. Individual subjects would exist only in the first sense. If it is possible to attribute to an entity a certain number of points of view, with enough variety, richness and flexibility, then that entity is a psychological subject. This has been maintained, for instance, by Daniel Dennett. Individual subjects exist only from an "intentional stance". They exist only because it is possible to attribute to them a certain number of points of view.[36]

However, we are usually more demanding in regard to the existence of individual subjects. And from positions like that of Dennett is very difficult to do justice to that demand. Individual subjects become something like "useful fictions".

It can be claimed that when the bearer of points of view is a subject, individual or collective, the points of view have very important features:

1. their contents tend to be full of "qualitative", phenomenal aspects;
2. these contents also tend to be full of "indexical" ingredients[37];
3. many times, the points of view are quite "idiosyncratic"; and
4. some points of view entail a peculiar "constitutive relevance" for the subjects that are adopting them.[38]

We can equate the notion of a "psychological subject" with the notion of a "person". Alternatively, we can introduce some differences. In any case, the peculiar "personal" character of a point of view would depend on the sort of subjectivity present in the bearer of the point of view.

"Personal points of view" can be defined directly as follows:

A personal point of view is a point of view with a person as its bearer.

We noted that when a point of view has a subject as its bearer, the point of view displays some important features. Now, in personal points of view those features would have a very much stronger relevance. Furthermore, personal points of view have other important features:

5. typically, they have both *non-CC* and *CC*, and a distinction between explicit and implicit contents makes sense in relation to them[39];
6. frequently, they are "reflective" points of view, i.e., points of view over other points of view of the same subjects;

---

[36]See, for instance, Dennett [24–26].

[37]Moreover, as many authors have noted, human beings seem to be the only animals clearly able of "pointing at".

[38]With respect to all these features, see Nagel [73, 74], McGinn [69] and Farkas [30].

[39]In a moment, we will say more about the distinction between the explicit and the implicit contents of a point of view. In personal points of view, the attitudes of the bearers can be said to have a very important implicit *non-CC*. In other sense, we can also say that many of the consequences of the contents explicitly contained in our points of view are "implicitly" contained in them, and that more or less easily they can become contents "explicitly" contained in the point of view. These things can be properly said only of personal points of view.

7. very often, their contents involve "normative" aspects, i.e., they are contents sensible or responsive to rules,[40] and
8. they also include essentially "temporal" perspectives.[41]

Many times, features 5–8 have been claimed to be necessary for a subject to acquire a "personal status". Hence, even if we understand persons in continuity with other non-personal subjects, and these in turn in continuity with other bearers of points of view which are not subjects, for something to be a person it would have to be capable of taking very peculiar and sophisticated points of view.

We can be the bearers of non-personal points of view. Also, we can even be the bearers of points of view that do not need from us anything psychological. However, many of our points of view are personal ones. They only can be ascribed to us "as persons". They have the four mentioned features 1–4 corresponding to points of view with a subject as their bearer and, in addition, they have the four features 5–8 corresponding to personal points of view.

Now, let us focus on feature 6. Many of the points of view of personal subjects seem to be "open" to reflection. Personal subjects can adopt reflective perspectives over their points of view. How to understand that? How to understand "reflection"? There are three big options:

1. Reflection as a partial portion of explosive sequences of new perspectival facts: According to this approach, reflection always generates new perspectival facts of a strongly subjective kind. There is a sort of recursive explosion that cannot be apprehended in any objective way. The situation is one of an infinite regress. So, reflection only is possible as a partial portion, or a proper part, of those explosive sequences of new perspectival facts which are irreducibly subjective.
2. Direct and immediate reflection: Here, reflection operates in a direct and immediate way. It can be understood as a kind of "pure intuition", or "pure apprehension". When a personal point of view is the object of a reflective move, that point of view becomes, let us say, a "self-presenting transparent state" for that very subject. Moreover, the subject conceptually knows something about that state.
3. Reflection as an objective point of view over our own points of view: According to that approach, the supposed recursive explosion of perspectival facts in reflection would not pose any especial problem. It would lose all its problematic character if we assume that reflective points of view can be completely objective. That way, even though it is always possible to repeat a reflective movement, the new perspectival facts originated would be at the same level as the

---

[40]Personal points of view do not merely produce behaviours that can be described "as if" certain rules are followed. The behaviour itself consists in "following rules". And this is so because the contents of personal points of view (non-CC or CC, explicit or implicit) are sensible or responsive to rules.

[41]It is not easy to imagine how a person could exist without adopting "temporal perspectives".The combination of this feature with the other ones would make sense of our "personal identity" and "self-knowledge". See Chisholm [15] and Shoemaker [109].

new perspectival facts we can achieve when we change a point of view concerning any matter. Simply, reflection is possible in exactly the same sense in which any other objective point of view is possible. Moreover, it is plausible to suppose that, from a certain point, the new perspectival facts originated in reflection are always of the same basic "types".

So, the alternatives are "perspectival explosion", "self-presentation", and "objectivity". Option 1 embraces an explosion of perspectival facts that, in the end, makes a completely reflective movement impossible. Reflection would entail an infinite regress provoked by the recursive creation of new perspectival facts, always of a subjective kind. But, so understood, reflection becomes an impossible task. If reflection were to create infinite series of subjective perspectival facts in that sense, then there could not be, strictly speaking, any fully reflective point of view.[42]

Faced with the problem of an infinite regress in reflection, option 2 gives a certain sort of "ad hoc" answer. It claims that, under reflection, our personal points of view are capable of becoming "self-presenting transparent states". That was the option favoured by Roderick Chisholm[43] and it was also the option of Russell, Descartes, and Plato. Reflection is a kind of direct and immediate self-knowledge able to produce conceptual knowledge, i.e., knowledge of certain facts about ourselves. In reflection, we become self-presented to ourselves as subjects having certain points of view with a personal character. This approach is able to avoid the problem of a subjective explosion of new perspectival facts only at the cost of introducing a notion of "self-presentation" capable of producing states of conceptual knowledge.

Option 2 has to face two further problems. The first one is the general charge of being involved in some version of the "Myth of the Given". But, the answer to this charge can be also very general. There is not only a "Myth of the Given", but also a "Mythology of the Myth of the Given".[44] And not all forms of the so called "Myth of the Given" are equally dismissible. Apart from that, the sort of conceptual knowledge coming from the self-presentations involved in reflection does not need to have any foundational epistemological relevance.[45]

The second problem is much more concrete. A possible model for reflection in the sense of option 2 would be the kind of knowledge we seem to have of our attitudes in personal points of view. We are aware that we are feeling a certain pain, that we are desiring something, that we are questioning some other thing, etc. However, we cannot take this as a model without paying attention to an important limitation: what I know when I am aware of my attitudes in a so direct and immediate way is always a *non-CC*. Hence, the result of reflection so understood could only be some sort of non-conceptual knowledge. Even when we were dealing

---

[42]The options we are examining have an important role in epistemological discussions. See Sosa [113, 115, 116].

[43]See Chisholm [14, 15].

[44]See Sosa [114].

[45]About that, see Burge [9].

with the *CC* of our points of view, our reflective knowledge would have to consist in some non-conceptual knowledge. But reflexion does not have only that role. Also, it is intended to be capable of generating some kind of conceptual knowledge. Reflection is cognitively relevant. And it is so because it is a very important source of conceptual knowledge.

At this point, option 3 looks plausible. A subject adopting a reflective point of view has a certain point of view, and a second point of view about the first point of view. Reflective points of view are points of view about other points of view. However, not all points of view about other points of view would be reflective ones. Reflective points of view can be understood in the following way:

A point of view PoV_1 about another point of view PoV_2 is reflective if:

1. Some of the components of PoV_2 become part of the contents of PoV_1,
2. PoV_1 contains CC capable of referring to those components.
3. PoV_1 is an "objective point of view",
4. the subject of both PoV_1 and PoV_2 is the same, and
5. there is an explicit and objective CC in PoV_1 according to which the bearer of both points of view is the same.

In that approach, reflection would be a sort of "objective" and "conceptual" point of view about another point of view, including the true and objective conceptual content that the bearer of both of them is the same. There is no need to appeal to the cognitive consequences of the "self-presenting" character, or self-awareness, or even self-knowledge, involved in our personal points of view.[46] And the problem of explosion simply disappears. The crucial point is that the subject would not necessarily be creating new perspectival facts of a "subjective kind" in every reflective movement. The problem of explosion is neutralised. It is transformed in the simple fact that one can always change of perspective, having objective access to new aspects of the world and new senses with which to conceptualise it.

As we have noted, it can be also maintained that reflection does not necessarily create "aspects" and "senses" completely new in type. Beyond a certain point, the recursiveness of reflection in subjects like us would find the same sorts of things, the same "types", again and again.

Any of the sets *non-CC* and *CC* can be empty, but not both of them. There cannot be a point of view without containing explicitly some contents, either non-conceptual or conceptual. The *non-CC* explicitly contained in a point of view can be "construed" in many ways. As we have seen, Peacocke offers a very promising proposal. Other proposals could be based on sense-data, physical properties, etc. In any case, the result for subjects like us has to be 1) a landscape of ordinary objects having their ordinary properties and relations, and 2) an array of experiential states.

---

[46]Using an expression coined by deVries and Triplett [27], we can say that the "self-presenting" character, or self-awareness, or even self-knowledge, involved in our personal points of view would not need to have any peculiar "epistemic efficaciousness".

The *CC* explicitly contained in a point of view also can be "construed" in many ways. Propositions can be derived from concepts or, alternatively, concepts can be derived from propositions. In any case, the result for subjects like us would have to be something that can have the conceptual structure and inferential articulation we can find in the contents of propositional attitudes.

To take a perspective is to put something into perspective. Points of view are always about something. And to have a point of view about something is to have a point of view with a certain explicit content. There are no points of view without some explicit contents. But, perhaps, points of view can also include other much more "implicit" contents.

First of all, what is it to have a content? Contents show two essential features. The first one is that they have correction conditions. Contents may be correct or incorrect. The notion of content is normative. Moreover, as we have said, the contents of personal points of view are "sensible or responsive to rules".[47] This applies to "linguistic content" as well as to "mental content". The second feature is that contents are mentioned in the "explanation" of behaviour in non-causal terms. Also, there are two other more problematic features which can be used in order to detect the presence of contents. One of them is that contents have a "representational" character. Contents are about something, and that "aboutness" is representational. Contents represent something. The other feature is that, at least when personal subjects are involved, contents are those things we are "conscious" of. That cluster of features provides criteria for attributing content. And points of view have content in all of those senses.

Now, with those criteria in hand, we need to make a distinction. There are <u>two kinds of explicit contents</u> that a point of view can have:

1. Non-conceptual contents
2. Conceptual contents

The expression "non-conceptual contents" suggests a contrast with conceptual contents, as if conceptual contents were the basic blocks of the notion of content. This would be so in the conceptualist approaches, but we do not need to assume that entailment. It is a problem whether there may be a representational content which is not conceptual. Nevertheless, a content which is not representational, or does not serves to articulate other representational contents, cannot be conceptual. In that sense, conceptual content is always representational. And it is arguable that points of view can have contents which are not representational. It is arguable for instance, that when we are having an experience of pain, we are not representing anything at all. We are simply having (suffering) a certain kind of experience.

---

[47]Every content is normative because it can be considered from the perspective of rules (i.e., from a normative point of view). In other words, every content has correction conditions. However, personal points of view are such that their contents are sensible or responsive to rules. It is not only that the dynamics of personal points of view can be described "as if" some rules were followed. Personal points of view are capable of "following rules".

If we have the perceptual experience of seeing a table before us, we are representing something. However, even if we want to defend a representational theory of vision, there is a sense in which "to see" a table before us is not only "to represent" that here is a table before us. When we are seeing a table before us, we are having a certain content in perspective. It has correction conditions, it serves to explain non-causally our behaviour, it is about a table being there, and we are conscious of what we are seeing. However, there is a sense in which we are not, or we are not only, representing something. There is a qualitative, phenomenal sense in which we see a table in the same way in which we "suffer a pain". In that sense, our perceptual experience also has a non-conceptual content.

Furthermore, it can be argued that there are explicit non-conceptual contents of a non-representational kind having to do with agentive abilities, or having to do with ways to cope with reality through a certain "know-how". There are even versions of sense-data theory in which these entities do not represent anything.[48]

Trying to be neutral about the different ways to construe both non-conceptual and conceptual contents, we can say that the final outcomes of non-conceptual contents have to be actual or possible ordinary objects with their properties, together with an array of experiential states, all of them structured and interrelated in the usual ways in which ordinary objects with their properties, and our experiential states, are structured and interrelated through qualitative, causal, topological, etc., features. Also, we can say that the final outcomes of conceptual contents have to be propositions conceptually structured and inferentially articulated.

The distinction between two kinds of explicit contents, non-conceptual and conceptual, suggests a distinction between <u>three kinds of points of view</u>:

1. Non-conceptual points of view
2. Conceptual points of view
3. Hybrid points of view

The explicit contents of "non-conceptual points of view" are only non-conceptual contents. The explicit contents of "conceptual points of view" are only conceptual contents. The explicit contents of "hybrid points of view" are blends of non-conceptual and conceptual contents. However, it can be claimed that a hybrid point of view is not simply the juxtaposition of a non-conceptual point of view and a conceptual one. The <u>structure of a hybrid point of view</u>, in a proper sense, has to be more complex. For instance, in the following two ways:

- Some of the *non-CC* of the point of view are <u>conceptualised</u> giving place to some *CC*.
- Some of the *CC* of the point of view are <u>expressed</u> through some *non-CC*.

---

[48]About conceptual contents, see McDowell [67, 68] and Brandom [6, 8]. About non-conceptual contents, see Bermudez [3], Crane [18, 19], Gunther [39], Harman [48], Heck [52] and Peacocke [79–82].

This articulation is suggestive. According to it, any entity able to adopt hybrid points of view has to be capable either of "conceptualising" some of its *non-CC*, or of "expressing" some of its *CC*.

In points of view with psychological subjects as their bearers, particularly in personal points of view, we can make another crucial distinction. We can distinguish between

- The contents of the point of view that are <u>contained in the point of view</u>, either as explicit contents or as some direct consequences of the explicit contents of the points of view.[49]
- The contents of the point of view that are only <u>implicitly involved in the point of view</u>.

The <u>contents "of" a point of view</u> may include both sorts of contents. But the second ones would only appear in points of view with psychological subjects as their bearers, and they would be very important in personal points of view. They are generated by the "relations" that the subject maintains with what is contained in the points of view (either explicit contents or some direct consequences of the explicit contents).

Those relations are "psychological attitudes" having correction conditions. Moreover, they have an important role in non-causal explanations of the subject's behaviour. Also, even though they do not have a representational character, many times the subject is conscious of them. This would be enough for claiming that, according to the criteria introduced above for having content, those attitudes "have" a certain content. The attitudes have a content, and the points of view structured through those attitudes involve that content in an implicit way.

Furthermore, whereas the contents that are contained in a point of view can be either conceptual or non-conceptual, the implicit contents linked to attitudes have always to have a "non-conceptual character". Our believing something, desiring something, hoping something, feeling something, sensing something, etc., has always a peculiar kind of implicit non-conceptual content. It is a non-representational, and hence non-conceptual, content. But it is a kind of content. It is a non-conceptual content exerting "pressure" over the other contents of our points of view. That pressure come from our psychological attitudes of believing, desiring, hoping, feeling, sensing, etc. Certainly, these attitudes can become explicit contents of other (perhaps reflexive) points of view. But, in that case, they would be affected by the pressures of other further attitudes involving other sorts of implicit non-conceptual contents.

The distinction "conscious/non-conscious" is orthogonal to the distinction we are making between contents "contained in" a point of view" and implicit contents

---

[49]For instance, many (but surely not all) of the inferential consequences of our explicit *CC* would count as contents of that kind. Even if they are not "explicitly" contained in the point of view, they are implicitly "contained" in it. The emphasis we want to make concerns a distinction between all those contents and other implicit non-conceptual contents "involved" in points of view that are linked to psychological attitudes. Sometimes, we will speak in general of the first kind of contents as contents "explicitly contained in the point of view", including in that kind some of the consequences of other more clearly explicit contents.

"involved in" the point of view. On the one hand, we are not always conscious of the contents contained in a point of view. Even in personal cases, not all the explicit contents of a point of view have to be contents the subject is conscious of. On the other hand, as we have said, we are very often conscious of many contents "of" our points of view that are not explicitly contained "in" them. We are conscious of many of the contents implicitly present in our attitudes towards the contents contained in our points of view.

The contents "of" a point of view that are not contained "in" the point of view, and that are only implicitly involved in it, namely the *non-CC* linked to attitudes, could be very important in relation to a number of issues.

1. Many times, some of the *CC* have an indexical character. More generally, they may be dependent on the intentional context, or dependent on the real environment, in one sense or another. When this happens, the *non-CC* linked to the attitudes could be capable of selecting the particular *CC* that are "active". They could fill in the "gaps" in indexical *CC*, producing "complete" *CC*. Or, at least, they could contribute to do it

2. Also, those implicit *non-CC* could be decisive for explaining the "generation", "dynamics", and occasional "malfunctioning" of personal points of view. Moreover, perhaps they could explain how the peculiar "intentionality" of both explicit *non-CC* and explicit *CC*, in personal points of view, is originated from quite basic psychological attitudes.[50]

3. The implicit *non-CC* linked to attitudes also offer interesting explanations of how, in personal points of view, there may be changes in the point of view without any significant change in the contents "explicitly contained" in the point of view. The notion of "changing one's point of view" in its more radical sense would be an effect of such pressures. More particularly, those pressures could explain why only "some" of all the possible inferential consequences (in a broad sense of "inferential consequence") of a given set of explicit conceptual contents get to become explicit conceptual contents, but not other ones.

4. The *non-CC* linked to attitudes also could be important in order to explain some sort of "self-knowledge" without any need to appeal to reflective moves. We could explain our direct and immediate self-acquaintance paying attention to those kinds of *non-CC*. We have a direct awareness of ourselves that does not seem to consist simply in having in perspective any kind of explicit *non-CC* or *CC*. We can say that it consists in some sort of "acquaintance" with the psychological attitudes we adopt towards any kind of content explicitly contained in our points of view, and with the ways those psychological attitudes are articulated.[51]

---

[50]That interesting idea is explored in Horgan and Tienson [55], Horgan et al. [56], Loar [66], Siewert [111], Strawson [117] and Pitt [84].

[51]This would be closely connected with issues about "knowledge by acquaintance" and the "direct awareness of the self". See, respectively, Russell (1910; and 1912, chap. 5); and Chisholm (1976, chap. 1). We are suggesting a very interesting distinction between "self-knowledge" and "reflection". Whereas reflection is conceptual and explicit, self-knowledge (in the sense of

We are presenting a certain way of approaching the structure of points of view. It is based on the model of propositional attitudes. The analysis of this model has been very useful. It has facilitated the introduction of a number of relevant notions and distinctions. The most important ones are the distinction between non-conceptual and conceptual contents; the distinction between non-conceptual, conceptual, and hybrid points of view; and the distinction between the contents explicitly contained in a point of view and the implicit contents involved in it. These distinctions have not only an epistemological, logical, or methodological value. They intend to have a crucial "ontological" significance.

Now, let us finish focussing on the possession conditions Cp of the point of view. They are a constitutive part of it, and a very important part. Every point of view has to have some Cp. The possession conditions are "internal" to the point of view in the sense that if they were to change, then the point of view also would have to change.

But, we can also introduce the weaker notion of "attribution conditions". They would allow the attribution of points of view without facing the question whether these points of view are really possessed by the relevant bearers. Attribution conditions would not be constitutive of the points of view. They would be "external" to them.

2.2 The Model of Location and Access

A second general approach about the structure of points of view focuses on the notions of location and access. We can speak of the model of location and access. Here, points of view are not internally analysed. They are identified by a certain "role". Points of view, or perspectives, are ways of having access to the world and to ourselves. A point of view is constituted by a location offering a certain perspective.

We will introduce three different accounts of points of view following this model. Then, we will offer again a general reconstruction.

(Footnote 51 continued)

self-acquaintance, or self-awareness) is basically non-conceptual and implicit. Because of that, even if self-knowledge has correction conditions, it does not have the sort of correction conditions that are appropriate for conceptual contents. In other words, there may be something "given" in self-knowledge, we seem to be directly and immediately aware of the non-conceptual contents implicitly involved in our attitudes, and we can even describe that as a sort of non-conceptual self-knowledge, but this cannot have any direct or immediate "epistemic efficaciousness" concerning conceptual knowledge. About this, see deVries and Triplett [27].

2.2.1   A Behaviourist Approach: Jon Moline

Some years ago, Jon Moline offered a very interesting analysis of what it is to take a point of view.[52] According to him, points of view are "ways of viewing things and events from certain locations".

Moline emphasises the theoretic and practical importance of giving an adequate account of what it is to take a point of view, and points out that no such account has been given. He mentions a book of Baier's, *The Moral Point Of View*,[53] as one of the few such proposals. According to Baier, to adopt a moral point of view of a certain kind is to adopt a particular set of principles, laws, maxims, etc. The important idea is the one of "rules". Baier's proposal is based on the Kantian claim that human behaviour is always a matter of following rules.

Moline rejects that approach. Points of view are ways of seeing, hearing, smelling, touching, feeling, thinking, imagining, etc., the world from certain locations, and those ways of facing the world cannot be reduced to any kind of rules. To take a point of view is not reducible to rules. Moline firmly rejects any Kantian's conception of points of view. Simply put, points of view cannot be defined by rules of any kind (principles, laws, maxims, theories, etc.).

Rhetorically, Moline asks "What 'principle' could be said to define the Negro point of view?".[54] To take such a point of view does not entail any sort of principle, but to adopt certain "attitudes" linked to having a certain colour of the skin in some social contexts. Other times, to take a point of view entails "learning" some special facts, or techniques, or languages. This would be so in cases like that of the physicist's point of view. Other times, to take a point of view entails entertaining certain "experiences" of a very special sort. This would be so in cases like those of a schizophrenic point of view, or a socially engaged point of view, or a religious point of view.

What is Moline's positive account of points of view? As we have said, points of view are not reducible to rules. They are not reducible to some emplacements either. He claims that a point of view is not only a place from which one views things and events, but also the peculiar way in which those things and events "can be viewed" from that emplacement. And that peculiar way of viewing the world has to be manifest in "behaviour".

Before explaining the last claim, let us introduce some very interesting remarks made by Moline. Points of view, he says, can be either taken by only one person at a time, or they are "sharable" by many people. This is a very important remark. Some points of view are like the top of an extremely narrow peak. But others, like some other peaks, are so broad that many people could be said to be looking at things from the same point of view or perspective at the same time.

---

[52]Moline [70].
[53]Baier [2].
[54]Moline [70: 191].

Another important remark is that we can know many things about a point of view without adopting it. Furthermore, there would be two different senses in which a subject can "take", or "adopt", a point of view. There is a sense in which to take a point of view implies a certain sort of "overt behaviour", or action, and there is another sense in which the adoption of the point of view only implies a certain sort of "thought". For instance, a detective in a large department store takes a detective's point of view in the first sense, and takes the point of view of a shoplifter only in the second sense. He does not take the second point of view in overt behaviour by stealing from the store, but only in thought (or so we suppose), as part of a strategy of anticipation.

Moline also maintains that by using the expression "point of view" we can be involved in three important sorts of claims. The first one is constituted by what he calls "comprehension claims". They are made in explanatory contexts. One example would be "If you would try to understand her point of view, you would not find her decisions so incomprehensible". The second sort is constituted by claims about the "relevance or irrelevance of certain considerations" from some point of view or other. One example would be "Psychological matters are irrelevant to logic". The third sort is constituted by "size claims" using qualifiers like "narrow", "restricted", "larger", "broader", or "wider". One example would be "He takes a narrow, economic point of view on all political issues".

The use of the expression "point of view", Moline claims, is restricted by considerations of "personality" and "role". Whereas we speak freely of things like "my personal point of view", "a paranoid point of view", "the point of view of science", "an administrative point of view", or "a parental point of view", it would be very odd to speak of things like "a coffee-break's point of view", "a cello's point of view" or "the point of view of the square root of 2". Other cases occupy an intermediate character, for instance "a dog's point of view", "a computer point of view", or "a clam point of view". Sometimes, the appropriateness of using the expression "point of view" depends on personality (my personality, a paranoid personality, the quasi-personal character of dogs, the metaphorical personality of computers and clams). Other times, it depends on the existence of certain roles (the one exemplified by science, or administration, or parents).

These connections between attributions of points of view and considerations of "personality" and "role" are very important. Both persons and roles suggest what it is given to a certain character to say or do, or even what it is appropriate to feel, or accomplish, or assume.

At this point, Moline argues that it will be helpful to replace the original question "What is it to take a point of view?" with the question "What does one taking a point of view do?".[55] The crucial point is that the new question is not a question about any mysterious relation described as "taking a point of view", but a direct question about a certain expected behaviour. We expect one who takes, or espouses,

---

[55]Moline [70: 194–195].

a point of view to display a set of "behavioural tendencies" such as the following ones:

(a) a tendency to have and pursue certain specifiable interests and aims;
(b) a tendency to use only certain criteria to evaluate actions as conducive to achieving those aims;
(c) a tendency to regard other interests and criteria as largely irrelevant;
(d) a tendency to make certain factual assumptions, but not other ones; and
(e) a tendency to agree with the interests, aims, evaluative criteria and relevance judgments of others taking the same point of view.

Moline offers the following analogies: "Taking a point of view, then, is like picking up and looking through a lens constructed for a particular purpose and having a particular focal length and field of vision. Some objects will be in focus (relevant) and others will simply be excluded from view (irrelevant). Taking a point of view is like adopting a metric standard–it enables one to accomplish certain tasks by referring to certain authoritative marks. It is also like making assumptions in an argument–one can then generate conclusions".[56]

The above mentioned behavioural tendencies a-e are the essential component of a point of view. Moline argues that points of view are "sharable" and "intersubjective" to the extent that subjects can share them.

In many relevant senses, Moline's replacement of the question "What is it to take a point of view?" with the question "What does one taking a point of view do?", is analogous to the behaviouristic change proposed, some years earlier, by Türing with his "simulation game".[57] To take a point of view consists in being capable of behaving in certain peculiar ways, and taking the same point of view, or sharing it, also consists in behaving in those peculiar ways.

Moline does not analyses the explicit contents of points of view. In particular, there is no distinction between points of view with an explicit conceptual content and points of view with an explicit content which is not conceptual. As we have said, Moline's approach is very behaviouristic. He emphasises in a sense the notion of point of view which is largely independent of mental states or mental relations. In the end, points of view are constituted by behavioural tendencies a-e. Because of that, Moline's approach would make it impossible to distinguish between genuinely taking a point of view and simulating taking that point of view, as in the case of an "actor", or as in the case of one who is only tentatively adopting it.

The distinction between conceptual and non-conceptual points of view is very important. So is the distinction between adopting a point of view and simulating such adoption. However, there is no room for them in Moline's behaviourist approach.

Nevertheless, Moline's account is insightful in many other respects. The various senses in which points of view could overlap, when the subjects are able to share all

---

[56]Moline [70: 195].
[57]Türing [120].

or part of the tendencies a–e, for instance, suggest ways to overcome situations of supposed "incommensurability", in particular situations of supposed "relativism", which would not depend on any set of clearly shared contents.

2.2.2 A Logical Approach: Antii Hutamäki

Antii Hautamäki has offered a very elaborated logical work about points of view. To our knowledge, it is the only research in logic dealing directly and in depth with the notion of points of view.[58]

In Hautamäki's approach, "viewpoint" means a way to conceptualise the world. The main idea is that discussions about scientific change, conceptual or linguistic schemes, or frameworks, theoretical perspectives, etc., invite the creation of logics in which truth values depend not only on the world considered but also on the ways to conceptualise the world.

Hautamäki presents a propositional logic of viewpoints leaving the inner structure of points of view unspecified. Points of view are taken as "propositional modal operators". The language of this logic consists of the usual modal operators L (for necessity) and M (for possibility) together with two new operators: A (interpreted as "from all viewpoints") and R (interpreted as "from some viewpoint"). Ap would mean "p from all viewpoints", and it also can be read as "absolutely p", or "invariably p". Rp would mean "from some viewpoint p", and it also can be read as "relatively p".

The semantics for Hautamäki's logic uses Kripke models enriched with viewpoints and with a new relation S defined between pairs of worlds and viewpoints. A model is a structure < W, I, R, S, V >, where W is a set of possible worlds {w, w', ...}, I is a set of viewpoints {i, i',...}, R and S are relations defined in W × I, i.e., they are subsets of (W × I) × (W × I), and V is a function of evaluation from the set of well formed formulas F and W × I to truth values {1, 0}, that is from F × W × I to truth values {0, 1}.

The operator Rp is defined as $\neg A \neg p$, and the truth condition for Ap is defined as follows: V(Ap, w, i) = 1 iff V(p, w, i') = 1 for all i such that $\langle\langle w, i\rangle, \langle w, i'\rangle\rangle$ belongs to S. This parallels the definition of Mp as $\neg L \neg p$, and the truth condition for Lp as V (Lp, w, i) = 1 iff V(p, w', i) = 1 for all w such that $\langle\langle w, i\rangle, \langle w', i\rangle\rangle$ belongs to R. What is relevant for modal operators L and M is how truth values can change in relation to changes in the possible worlds considered. What is relevant for the new modal operators, A and R, is how truth values can change in relation to changes of perspective.

Hautamäki construes various possible axiomatisations of the logic of viewpoints and he proves them to be complete.[59] Let's consider the following set of modal systems {K, T, B, S4, S5}, and let x and y be any element of that set. He presents

---

[58]See, mainly, Hautamäki [49, 50].

[59]Hautamäki [49].

logical systems of viewpoints following the structure (x, y), where the system is an x-system with respect to operators L and M in the axioms of x, and a y-system with respect to operators A and R standing, respectively, for L and M in the axioms of y. According to that, there are in principle 25 different systems for the logic of viewpoints.

The relation S between pairs of worlds-viewpoints is crucial. He proposes to understand it as formalising the intuitive idea of "alternativeness". There would be various ways to do it. According to Hautamäki, the weaker adequate way would be to say that $\langle\langle w, i\rangle, \langle w, i'\rangle\rangle$ belongs to S iff V(p, w, i) = V(p, w, i') for some p. With respect to some propositions p and some world w, a viewpoint i would be "alternative" to another viewpoint i'. So defined, the relation S is reflexive and symmetric. Hence, the weakest adequate system in the set of those 25 different logical systems for viewpoints would be (T, B), a logical system as T with respect to operators L and M, and as B with respect to operators A and R.

Another very important point is the following. Let modality be any sequence of the operators ¬, L, M, A and R, including the empty sequence. Let modalities m and m' be equivalents in a system iff for every p, the co-implication of mp and m'p is a theorem in that system. In every logical system for viewpoints considered by Hautamäki, all the eight modalities LA, AL, LR, RL, MA, AM, MR and RM are distinct or non-reducible, in the sense that equivalence fails for any two of them. Now, if we consider combinations of those eight modalities, this entails infinitely many distinct modalities for the logical systems involving viewpoints.

Hautamäki notes close relations between his logics of viewpoints and two-dimensional modal logic.[60] More concretely, he mentions the work of Segerberg.[61] If we take the set of viewpoints I to be also W, then we get the same kind of models as in Segerberg's logics. There are other connections of interest. Among the applications of his logics, he mentions tense logic and offers an interpretation in that sense.[62] He also notes that he got the idea of using that special set I of viewpoints from Needham's tense logic.[63]

In Hautamäki's approach, points of view are taken as propositional operators giving place to new truth values. This assumes the relativisation of our knowledge claims to a point of view or perspective. Taking the analogy with tense logic a step further, we can ask for the possibility of adopting other alternative approaches in parallel with the sort of "predicate approaches" we can find in temporal logic. Points of view could be treated as hidden variables in predicates used to constitute propositions free of any perspectival relativisation. This is another possible approach. However, in that case we would be analysing the internal structure of points of view. Hence, we would be going out of the model of location and access for points of view, and coming very close to the model of propositional attitudes.

---

[60]Hautamäki [49: 188].

[61]Segerberg [106].

[62]Hautamäki [49: 195].

[63]In Hautamäki [49: 188], Needham [75].

Something in that line is being explored by Hautamäki in more recent researches. He uses Gärdenfors's notion of "conceptual space". Conceptual spaces are defined by Hautamäki as sets of determinables (for instance, "colour", "shape", etc.) acquiring determinate values ("red", "triangular", etc.). The determinables set down the dimensions of the space. Furthermore, the acquisition of determinate values could have a certain structure and a peculiar temporal dynamics. According to Hautamäki, a point of view could be identified with a certain subset of a conceptual space.[64]

2.2.3   A Metaphysical Approach: Adrian Moore

Another analysis of the notion of point of view following the model of location and access, this time of a metaphysical sort, can be found in Adrian Moore.[65]

He defines a point of view as follows: "By a point of view I shall mean a location in the broadest possible sense. Hence points of view include points in space, points in time, frames of reference, historical and cultural contexts, different roles in personal relationships, points of involvement of other kinds, and the sensory apparatuses of different species."[66]

Moore distinguishes "points of view" from "outlooks". An outlook is any way of representing the world, any way of seeing it or thinking about it. When points of view are involved, our representations are dependent on a perspective. However, by itself, to represent the world in accord with an outlook does not entail representing it from a point of view. Moore claims that it is possible to represent the world, and our position in it, from no point of view at all. This would be to have an "absolute representation" coming from an absolute outlook. Even though points of view are always perspectival, there could be representations from no point of view. Some outlooks would be of that kind. They would be "non-perspectival outlooks".

The notion of an absolute representation, or conception, of the world comes from Bernard Williams, and it has been criticised by many authors.[67] In line with Williams, Moore claims that the existence of an absolute conception is a direct consequence of the very notion of representing reality as "representing what there is there anyway". If there is a substantial or stable way, even minimal, in which reality is in itself, then a true, complete representation having "that content" would be an absolute representation of reality, and parts of that representation would be true partial, but absolute, representations of such reality.

That representations are representations of what there is there anyway is called by Moore "The Basic Assumption". The argument for the possibility of absolute

---

[64]See Hautamäki [51] and Kaipainen and Hautamäki [60]. Also, see Charro and Colomina [13]. About the notion of "conceptual space", see Gärdenfors [36].

[65]Moore [71, 72].

[66]Moore [72: 6].

[67]Williams [125]. Among the critics, see Nagel [74] and Putnam [89].

representations, which is called "The Basic Argument", follows quite directly from the Basic Assumption. We can offer a brief sketch:[68]

- To account for how it is possible that a number of representations made from incompatible points of view can be all of them true entails showing how those points of view contribute to the fact that those representations have the true content they have.
- But, those representations could not be integrated by "simple addition", or "direct integration". At least one of them has to refer explicitly to the points of view involved.
- Here, Moore introduces the notion of a "range of points of view". It is defined as a maximal class of points of view any two of which are incompatible with each other, and such that there is no point of view outside the class that is incompatible with each of those in it. Temporal points of view over any property variable in time would offer a good example. The different temporal points of view would produce a range of points of view (occurring in a certain past moment, occurring in the present, occurring in a certain future moment) in the defined sense.
- All the points of view inside a range produce true representations because those representations include an element of perspective. And the only way to account for how that can be so is through a representation which is not produced from any point of view in the range. That representation has to "supersede" the range. It can only be produced, so to speak, stepping up a level.
- Now, if representing is representing what is there anyway (The Basic Assumption), and reality is something substantial, then the processes of accounting for incompatible points of view inside certain ranges have to produce a number of absolute representations. These absolute representations could be combined by simple addition, or direct integration. At the end, what we will obtain is an "absolute conception of reality" (an "absolute outlook").

It can be objected that the representations capable of superseding certain ranges do not need themselves to be free of perspective. Moreover, this could happen indefinitely. Perhaps, we could have fewer and fewer elements of perspective without "ending" in any absolute conception. Moore's answer to this objection is that to assume seriously that possibility would nullify the thought that our representations are made true by "what is there anyway", i.e., The Basic Assumption.

According to Moore, the distinction "absolute/perspectival" is very different from the distinction "objective/subjective". To think that a melody is exquisite is an example of a representation subjective and perspectival. To say that there was a total eclipse of the sun, here, yesterday, is an example of a representation objective but perspectival. And that $e = mc^2$ is an example of a representation which is both objective and absolute. The aim of science is to construe representations that are not only objective, but also absolute.

---

[68]See Moore [72: Chap. 4].

The only way to reject conclusively Moore's argument while accepting, at the same time, The Basic Assumption would be by maintaining a purely Heraclitean conception of reality. If reality has no substantial or stable character, then there would be a way in which reality is. However, there could not be any true representation, or outlook, of "what is there anyway" apart from a representation saying that it has no stable way of being. The absolute conception would not be possible except in a "negative" way, as the conception according to which any intended more robust, or substantial, absolute conception is wrong.

The important point is that, in that case, the absolute conception would not be possible not because The Basic Assumption is wrong, but because there could not be any "stable" absolute conception. That peculiar form of relativism does not make use of any notion of perspective. Perhaps, we can say that it is a relativism because all reality is "relative to time". No part of reality would remain "out of change". Here, time would not be simply a contribution "from the subject's perspective". It would be something strongly constitutive of reality.

In his defence of an absolute conception of reality, Moore argues against transcendental idealism.[69] He finds it not only in Kant, but also in the later work of Wittgenstein. The connections Moore traces between Kant and Wittgenstein are very interesting. According to Moore, there is in both cases some sort of transcendental idealism. However, transcendental idealism is incoherent. What it tries to say, or what it tries to think, cannot be literally said or thought. Transcendental idealism tries to "represent conceptually" something that cannot be represented.

Assuming that, Moore considers another proposal derived from Wittgenstein's earlier work, mainly the *Tractatus*. Even if we cannot state with coherence that transcendental idealism is true, perhaps we can "be shown" that it is. Moore's approach assumes as true the following instances of the scheme "x is shown that y": 1) We are shown that all our representations are perspectival, and 2) that way we are shown that transcendental idealism is true. Moore analyses the relation "x is shown that y" in terms of the notion of an "ineffable knowledge", and that ineffable knowledge as a kind of "practical knowledge". Moore argues that certain states of understanding are examples of that ineffable practical knowledge.

Some critics have noted that many of the ideas and arguments of Moore are not as clear as it would be desirable.[70] This would be especially so regarding ineffable knowledge. However, Moore's analyses of points of view are very provoking, and the argument we have summarised is really powerful.

Moreover, Moore's proposal about "being shown" that transcendental idealism is true, even though we cannot coherently "say" that it is, would explain the great appeal of transcendental idealism. Transcendental idealism promises to put us in the position of "being shown" a very peculiar philosophical truth. And it is a very important truth. Certainly, it is a truth that cannot be coherently said or thought. However, perhaps we can "be shown" that things are so.

---

[69]Moore [72: Chaps. 5 and 6].
[70]See Hales [44].

We will finish by saying something more about the notion of an absolute conception of reality. Many of the ideas linked to that notion can be rephrased in terms of "independence from any particular point of view". Under that interpretation, to have an absolute conception of the reality would not be "to represent it from no point of view", but to represent it "with independence from any particular point of view". The key word in the notion of an absolute conception would be "perspectival invariance". Representations showing perspectival invariance would have many of the characteristic features of absolute representations. For instance, they could be integrated by "simple addition", or "direct integration". Perhaps the aim of science is that perspectival invariance, and not absolute representations "from no point of view".

2.2.4   A General Reconstruction

According to the model of location and access, a point of view is a peculiar location making possible the access to some fields of objects having certain properties or to some fields of propositions.

Using intuitively the notion of possible worlds, and symbolising a possible world as Mi, we can reconstruct that way to conceptualise the notion of points of view defining a point of view PoV as a structure $PoV = \langle Me, Mp \rangle$, where

1. Me is the set of *emplacement-worlds* associated with the PoV, and
2. Mp is the set of *worlds-in-perspective* associated with the PoV.

Each point of view would determine a relation between classes of possible worlds. From certain emplacement-worlds Me we can have access to some worlds-in-perspective Mp.

When an appropriate entity B comes to be emplaced in a world Me, the entity B gets to have in perspective a world Mp, and becomes the bearer of a certain PoV. To a certain extent, it is optional whether to put the emphasis either on features of the entity B or on features of the world Me. Both B and Me can be re-defined in many ways to accommodate this optionality.

In any case, the model of location and access does not analyse the internal structure of points of view. So, it turns out to be very difficult to distinguish between non-conceptual and conceptual contents. Furthermore, we do not have anything like the relations R connecting the bearers B of a point of view with the contents of the point of view.

However, structures $\langle Me, Mp \rangle$ can be very useful in order to analyse the relation between pairs of possible worlds constituted by the fact that to be located in the first of them makes the second one accessible. It is arguable 1) that only in the case of reflective PoV would it be a reflective relation, 2) that it would be symmetrical only in those cases in which it can be reflective, and 3) that it is a transitive relation. In general, we can think of structures $\langle Me, Mp \rangle$ as defining a partial order among possible worlds.

2.3   Comparing the Two Models

So far, we have introduced two main ways of conceptualising the notion of points of view. Each one of them is based on a certain model. Some of the most important differences between them are the following:

1. As we have said, in the model based on propositional attitudes, both the bearer of the point of view and the relations that in particular circumstances that bearer maintains with the explicit non-conceptual and conceptual contents of the point of view play a crucial role. In the model based on location and access, those ingredients do not play any role. At best, they are only in the background.
2. Because of the crucial role attributed to the bearer of the point of view, and to how that bearer is related to some explicit contents in particular circumstances, the first model is akin to those approaches that, like existentialism, emphasise the autonomy and independence of personal points of view from any context. In contrast, the model of location and access is more akin to all sorts of social or historical contextualism and relativism.
3. In the model based on propositional attitudes, the primary idea is that of a point of view "being adopted" by a particular bearer in particular circumstances. In the model based on the notions of location and access, the primary idea is the "kind of access" that a point of view offers from a certain sort of emplacement.
4. The possession conditions of the point of view have a different status in the first model than in the second one. In the model based on propositional attitudes, they can be considered separately. In contrast, in the model based on location and access, the possession conditions are "incorporated" in the way the worlds of emplacement and the worlds in perspective are going to be defined.
5. From the model based on propositional attitudes it is very easy to make room for the peculiarities of "non-conceptual contents". From the model based on the notions of location and access it is very easy to make room for "modal notions". And this would be especially useful regarding the "normative" aspects of points of view.
6. Roughly speaking, the first model could be very useful to analyse the "onto-logical" features of points of view, and the second model could be very useful to analyse their "epistemological" and "logical" aspects.

Despite all these differences, the two models are compatible. There is no opposition between them. Moreover, there is no need to decide between them. Each model places emphasis on different features of points of view.

That there is no opposition between the two models is manifest in some approaches involving elements of both models. We will examine briefly 1) Russell's phenomenalist proposal of constructing logically the world out of sense-data structured as "spaces of perspectives", and 2) the recent proposal of Hautamaki of defining points of view through Gärdenfors's notion of "conceptual spaces".

2.4 Russell's "Space of Perspectives"

Russell's phenomenalism has a very special interest for us. He makes an explicit use of the notion of a "space of perspectives". That space of perspectives is a complex system of points of view. The internal structure of each point of view follows the model of propositional attitudes, being always the contents of these attitudes some sense-data in a broad sense of the term. But the interrelations among those points of view intend to reflect the role of points of view as ways of having access to reality. We can say that the space of perspectives has a "virtual nature". It is a construction made from the inside of each point of view, for the most part unconsciously. But the space of perspectives is not merely subjective. As we will see, it has a crucial sort of "objectivity". However, that objectivity is not reducible to the objectivity of the physical world. There is no such reducibility because ordinary things, physical objects, matter, even time are logically construed out of sense-data.[71]

Phenomenalism can be understood in two main ways. Firstly, it can be taken as the view that physical objects (physical properties, matter, etc.) do not exist at all. There are only sensible or perceptual phenomena. Phenomenalism in that sense can be considered a form of idealism, specifically a form of idealist empiricism.

The ontological phenomenalism of George Berkeley and David Hume were idealist in that sense. For Berkeley, there is no matter. There are only minds and "ideas" (sense-data). God's mind is the supreme reality. He has created us. He is also the cause of our "ideas". Moreover, God assures the existence of objects that are not perceived by us. Hume's phenomenalism is even more radical. For Hume, both objects and minds are no more than bundles of "ideas", and the belief in God has no rational support. Kant's transcendental idealism can also be interpreted as a peculiar variety of phenomenalism, in this case of an epistemological variety. Knowledge is limited to "phenomena". The phenomenalism of John Stuart Mill is both ontological and epistemological. The source of all knowledge is experience, and objects are permanent possibilities of experience. In the 19th and 20th Centuries, other varieties of phenomenalism appear, this time of a logical kind. This is what we can find in Russell, with some ontological and epistemological components, and in other authors as Ernst Mach, Alfred Ayer, Rudolf Carnap, or Nelson Goodman.[72]

---

[71]It is noteworthy that, in spite of its importance for analysing the notion of points of view, Russell's construction of a "space of perspectives" has been rarely taken into account in the relevant literature.

[72]About Russell's phenomenalism, see Russell [101] and, in a more elaborated form, Russell [102]. That phenomenalism can be considered a part of Russell's philosophical position known as "Logical Atomism". Logical Atomism includes 1) the metaphysical view that reality is a plurality of logically independent "atomic facts" consisting in a simple particular exhibiting a quality, or in a number of simple particulars standing in a relation, 2) the semantic view that any truth is supported, or made true, by "atomic truths" about atomic facts; and 3) the methodological view that atomic truths, and the corresponding atomic facts, can be discovered by logical analysis.

Russell understands phenomenalism as the view that sense-data are a very important part of the world such as it is in itself. Furthermore, he claims that all objects and properties of the world can be logically construed from sense-data, with independence of whether these sense-data are experienced or not by a subject. Phenomenalism understood in this second way is compatible with physicalism. Physical objects (physical properties, matter, etc.) are not eliminated. They are identified with certain classes of sense-data. Sense-data are taken to be the physical entities out of which any other entity is constituted.

Russell's approach is based on 1) a methodological maxim, 2) a very broad notion of sense-data, or more precisely "sensibilia", and 3) the existence of some relevant patterns of relations among these sensibilia.

According to Russell, the supreme maxim in scientific philosophising would be this: "Whenever possible, logical constructions are to be substituted for inferred entities".[73] He applied that maxim in the philosophy of mathematics, for instance defining numbers in terms of classes. The aim is always to interpret a body of propositions about some postulated entities as a "logical function" of other propositions about less hypothetical entities. At the end, Russell claims, the most basic propositions would be propositions about the sense-data we can be acquaintance with in our experience.

Sense-data have in Russell a very broad sense. They are not the whole of what is "given" in sense at one time. They are what is "singled out" by attention: particular patches of colour, particular noises, etc. Also, they can be simple particulars or they can be very complex facts, such as that this patch of red squared surface is to the left of that patch of triangular surface. Also, they can be particulars with which we have acquaintance through experience, as this patch of red squared surface, or facts involving a propositional form which provides knowledge of truths, as the knowledge of the fact that this surface is red and squared. All of that would count as "sense-data" in Russell's sense. Moreover, there could be sense-data with which we are not acquainted. In this respect, a possible sense-data would also count as a sense-data. Russell uses the term "sensibilia" for referring to sense-data in that so broad sense.

Some relations among sensibilia make possible to define the notion of "space of perspectives". The pivotal thesis is that every sensibile can have a position in two different places: 1) the place "from" which it appears, and 2) the place "at" which it appears. The first place is the subjectivity of an observer. Every sensibile which becomes a sense-data appears from an act of awareness, and such acts of awareness generate some "private spaces of experience". The second place where a sensibile can have a position is where the sensibile is: a certain point, or region, in a "space of

(Footnote 72 continued)

According to Russell's phenomenalism, the atomic truths and atomic facts would be truths and facts involving sense-data (in the broad sense explained below). About continuities and changes in Russell's philosophy, see Hager [43].

[73]See Russell [102, Sect. 6].

perspectives". The different perspectives we can adopt, or other people can adopt, are structured in a certain space by means of "correlations of sensibilia". We discover by moving and by testimony that different perspectives can contain similar or different sensibilia. Some groups or series of correlated sensibilia coming from different perspectives count as "appearances of one thing". That way, a thing can be defined as a class of appearances (i.e., as a class of sensibilia).

Russell uses the example of a penny seen from different perspectives.[74] There is one series of perspectives in which the appearance of the penny is circular. These perspectives are placed on one straight line, and they have an order determined by the apparent size of the penny. There is another series of perspectives in which the penny appears as a straight line. They are placed upon a plane, and they form concentric circles with the penny. There are other relevant groups and series of appearances. And the penny can be defined as the class of all of these appearances.

The places "at" which sensibilia can appear to an observer constitute a three-dimensional space of perspectives (and a four-dimensional space once an "external time" is introduced in that space). Each one of those perspectives is a point in the space of perspectives.[75] However, each perspective is in itself a three-dimensional "private space of experience" (and a four-dimensional private space if we consider the "internal time" of each perspective).[76] The world so construed would be constituted by a 3-D (or 4-D) space of 3-D (or 4-D) private spaces.[77] Each one of these private spaces of experience is a perspective, and each perspective is a point in the space of perspectives.[78]

Physical space, physical objects, matter, and physical time are defined from here. The "physical space", for instance, is defined as a peculiar, very elaborated, space of perspectives, surely not identical with the ordinary, common sense space of perspectives. "Physical objects" are defined as peculiar classes of appearances,

---

[74]See Russell [102]. Russell insists that he prefers the expression "space of perspectives" to the expression "space of points of view" because the first one suggests a much more objective image.

[75]Is there a continuum of points in a space of perspectives? Is it always possible to introduce a new perspective between any two perspectives? These are interesting questions. Perhaps the answer has to be negative. In any case, it would be important to investigate the possible "geometries" of a space of perspectives.

[76]That spaces of experience are "private" is consistent with Russell's insistence that the construction of the space of perspectives could be made from a solipsist basis. To the extent that the space of perspectives can be "intersubjective", the spaces of experience could not be "purely private".

[77]Note that the "external time" we can introduce in the space of perspectives would be different form the "internal time" we find in the private space of each perspective. According to Russell, in fact they are very different. Whereas we know the second one in a direct way, the first one can only be known indirectly. It is a construction. And the "physical time" is also another kind of construction (a more elaborated external time). We will explain below the Russellian construction of the "physical time".

[78]Russell [102, Sect. 7], recognises close affinities with Leibniz's *Monadology*. There are also differences. For Leibniz, each "monad" contains appearances of each thing. Russell does nor require such completeness in the sense-data available from each "perspective".

surely not identical with ordinary, common sense things. The appearances of one thing are different from different perspectives. And some appearances are more close to the place at where the thing is than other ones. The "matter of a thing" is defined as the "limit of its appearances" (i.e., sensibilia) as the distance from the thing diminishes in the physical space. In general, the appearance of a thing in a given perspective is a function of the matter composing the thing and of the matter placed between the thing and the perspective. The nearer we approach to a thing, the less its appearances are affected by the intervening matter.

There are many points where conventionality is needed in all the above analyses and definitions. The identification of the sense-data we experience depends crucially on attention. Testimony is also an important source of information about other sense-data. Moreover, there is no univocal way of grouping sensibilia and series of sensibilia forming ordinary things, physical objects, matter, etc. In the case of time, the conventionality is even stronger.

Russell claims that we know in a "direct" way the existence of temporal relations of earlier than, later than, or simultaneous with, when two sensibilia belong to the same "person's experience", i.e., when they belong to the same private space of experience. This is used to define a "biography" as everything that is directly earlier than, later than, or simultaneous with a given sensibile. We can say that a biography is a "temporalised perspective". Sensibilia can be grouped into different biographies. Each biography is a temporal series of sensibilia. And the "history of the world" can be defined as a number of mutually exclusive biographies.[79]

We can find here another very disturbing factor of conventionality. In virtue of what can two sensibilia belong to the same "private space of experience"? How to make a clear distinction between, on the one hand, different sensibilia belonging to the same private space of experience, to the same perspective, and, on the other hand, a number of spaces of experience, a number of perspectives, each one of them constituted by different sensibilia? Are memory and imagination the sources of that distinction? Are perception, memory and imagination the sources of the intended "direct knowledge" of temporal relations inside each private space of experience, inside each "biography"?

In any case, Russell claims that things are very different when we try to establish temporal relations among sensibilia belonging to different private spaces of experience, i.e., when they belong to different biographies. There are not direct temporal correlations between sensibilia of different biographies. Russell argues that when we project over the space of perspectives the detection of those non-direct temporal relations, we have to assume that they "take time". That time is an "external time". It is the time taken by a "signal" going from a point to another in the space of perspectives. In the last term, when the space of perspectives is transformed into a physical space, the velocity of light is the ultimate constraint.[80]

---

[79]See Russell [102, Sect. 11].

[80]Russell was always very sensitive of the philosophical implications of Einstein's Special Theory of Relativity. In Russell [102]: section X), he says "The general principle is that the appearances, in different perspectives, which are to be grouped together as constituting what a certain thing is at a certain moment, are not to be all regarded as being at that moment. On the contrary, they spread

We were faced with a crucial problem in the way Russell understands the direct knowledge we can have of temporal relations among sensibilia belonging to our private spaces of experience. Without something like perception, memory and imagination as sources of that direct knowledge, the "internal time" connecting sensibilia inside a perspective, i.e., inside a biography, would be in exactly the same position than the "external time" connecting sensibilia of different perspectives, i.e., of different biographies. Now, we are faced with another important motivation for conventionalism. Such as the space of perspectives is construed, the identification of temporal positions in that space of perspectives has to "take time". Very often, this is not considered. Even inside science, says Russell, the temporal grouping of the appearances belonging to a given thing at a given moment is in large part conventional. We assume simultaneity in order to simplify the formulation of physical laws.

Russell's approach to physical space and physical time is in sharp contrast with the Kantian's approach (which is, we could say, a "transcendentalisation" of the Newtonian's conception of space and time), and it is very close to the Leibnizian's relational way of understanding space and time. For Russell, space and time are not neutral containers, absolute frames where things and events come to be placed. They are relational constructions. They are construed out of relations among sensibilia.

Russell continues defining other relevant notions, as the persistence of ordinary things and physical objects, and motion. The "persistence of ordinary things" is based on continuities of the appearances at ordinary distances in the space of perspectives. It has not complete precision. The "persistence of physical objects" depends on small distances, and needs to be supported by physical laws. "Motion", both in the case of ordinary things and in the case of physical objects, presupposes something persisting through the time of motion.

The possibility of dreams, illusions and hallucinations poses a serious problem for Russell's phenomenalist approach. The problem is to distinguish reality from mere "fiction". His answer is the following.[81] What we apprehend in a dream, or illusion, or hallucination, is just of the same kind than what we apprehend when we are not dreaming, or not having an illusion, or not hallucinating: a set of sensibilia. However, in the cases of dreams, illusions and hallucinations, there is no "adequate assemblage" with other sensibilia in the space of perspectives. That sort of answer can be found in many other authors. At the end of the day, perhaps it is the only possible answer.

Russell's approach has always received a lot of objections. However, not all of them are fair. To offer a logical construction of A in terms of B is not to explain how we can know A, and it is not to explain how we can be capable of having a language able to refer to A either. Strictly, according to Russell's methodological

---

(Footnote 80 continued)

outward from the thing with various velocities according to the nature of the appearances". It can be said that Einstein was also very sensitive of the physical implications of Mach's phenomenalism.

[81]This answer is offered in the last part of Russell [102].

maxim, to offer a logical construction of A in terms of B is only to show that propositions having A as constituent can be understood as a "logical function" of propositions having B as constituent. It is true that Russell's constructions can be seen also as offering epistemological explanations and semantic explanations. Very often, Russell himself goes from one field to the other ones. However, the three projects can be separated. And some objections against the epistemological and semantic projects would not have any force against the logical project. Arguments based on the epistemic priority of our knowledge of ordinary things having properties, over our knowledge of sense-data, would not have efficacy against Russell's logical constructions, neither do they have the arguments, inspired on Wittgenstein, against the existence of private languages. Even though sense-data were to have a highly theoretic status, and even though we could not but speak a public language, or think through an internalised public language, Russell's constructions could maintain much of their logical value.

There is another family of objections vindicating the role of "postulatory inferences", in particular inferences to the best explanation. But, as we have seen, Russell himself makes use of these inferences in the context of his phenomenalist proposal. Not always is it possible to substitute logical constructions for inferred entities. Sensibilia, for instance, are inferred entities in Russell's system. Moreover, in a certain sense, the space of perspectives itself is something inferred from private spaces of experiences. Very often, this is not correctly appreciated by commentators and critics.

Russell's broad conception of sense-data neutralises another group of objections. For instance, the objection that when we experience the world, our experiences have always a conceptual content. We experience that things are is certain ways. As we have said, Russell claims that sense-data are not only particulars we have acquaintance with through senses. The category of "sense-data" includes also the conceptual contents of perception. Moreover, sense-data are not merely "given" to the subject. They are singled out by "attention".

The notion of sensibilia is very peculiar. Sensibilia are sense-data that are not actually perceived by any subject (alternatively, sense-data are sensed sensibilia). According to Russell, they exist. Their existence is crucial for the existence of the space of perspectives, and that way for the existence of ordinary things, physical objects, matter, time, etc.. The postulation of sensibilia has also a crucial role in relation to the problem of "other minds". Unless we had a large number of sense data in our own mind, if we could not appeal to sensibilia we would need to postulate the existence of "other minds" in order to construe a space of perspectives.[82] However, once we have sensibilia, this is dispensable. Once sensibilia are available, all the constructions could be made from a solipsist basis.

---

[82]If each perspective were to contain appearances of each thing, the appearances each thing has from that perspective, and each perspective were also to have enough "activity", so that the appearances contained in it could change relevantly, as it happens in Leibniz's *Monadology*, then sensibilia would not be needed. Each perspective (in a solipsist sense) would be capable of generating a full space of perspectives, and a very sophisticated physical space.

There is a very simple but illuminating way of understanding sensibilia. We can interpret them as "possible sense-data". Sensibilia would exist as some sensorial and perceptual possibilities.[83] Their reality would be the reality of some possibilities. The consequences of this interpretation are important. Some relations among sensibilia define the space of perspectives, and some assemblages among them define ordinary things, physical objects, matter, time, etc. So, the reality of all of that would be in part, in a very relevant part, the reality of some possibilities.

Is there only one space of perspectives? This question is really important for an adequate understanding of Russell's approach. We construe a space of perspectives from the similarities and differences we encounter in our experience. Also, we take into consideration what seems to be the "testimony" of other people. We find patterns of sensibilia, relevant groups of them, assemblages among them, limits, etc. In that way, we construe a physical space and a physical time, physical objects, matter, etc. There is no guarantee of convergency. But we can assume that there is such convergency as a matter of fact. We can assume it in just the same way in which we can assume that we are not dreaming, or having an illusion, or hallucinating, to the extent that there is an adequate "assemblage" with other sensibilia in the space of perspectives.

Russell's concept of a space of perspectives has a strong "virtual nature". Epistemologically, it can be seen as a construction made from the inside of each private point of view, generally in a completely unconscious way. As was noted, Russell claimed that his logical constructions could be made even from a solipsistic basis. However, the space of perspectives is not something merely subjective in the sense of being arbitrary. This is a crucial point. The space of perspectives has a very important sort of objectivity. But, it is not an objectivity reducible to the objectivity of the physical world. Physical space, physical time, physical objects, matter, etc., are things logically construed out of what we encounter to in our experience: sensibilia (sense-data in a broad sense). We can say that the peculiar kind of objectivity we find in the space of perspectives is something "internal to points of view" without being something "internal to the subjects" maintaining those points of view.

There is, however, a gap in Russell's approach. And it is a very big gap. We can illustrate it with the help of his notion of "biography". Among sensibilia belonging to the same perspective, there is a directly known temporal relation of earlier than, later than, or simultaneous with. So, a "biography" can be defined as everything that is directly earlier than, later than, or simultaneous with a given sensibile. Furthermore, the "history of the world" can be defined as a number of mutually exclusive biographies. However, a "biography" is not a "life". A biography does not contains a distinction between the past, the present and the future. And that

---

[83]Of course, this is to give a "metaphysical" sense to modal notions (possibility, necessity, etc.). Perhaps Russell would reject this interpretation of sensibilia. But it is the most natural one. And we know that perhaps not all modal discourse can receive a merely "linguistic", or a merely "epistemic", interpretation. Moreover, that metaphysical interpretation was also present in Stuart Mill when he defined objects as "permanent possibilities of sensation".

distinction is crucial for life. In written biographies there is no past, there is no present, and there is no future. But written biographies are not identical with the lives they narrate. In these lives there is a past, a present and a future. If the history of the world is a set of biographies, then the history of the world is like a library full of books. The books are the biographies. And we ourselves are simply one of the books (the book of our life).

What is especial in our personal perspectives[84] is that they are experienced as being past, present, and future; and that they are so experienced in a direct way. Only because of that, there is also a direct temporal ordering of the type "earlier than", "later than", or "simultaneous with". If there were not the first type of temporal order, then there would not be the second type of temporal order either. Using McTaggart's known terminology, we can say that the experiences of a person are essentially structured as a temporal order of kind A, and that this makes possible that experiences can have also the structure of a temporal order of kind B.[85]

2.5 Hautamäki-Gärdenfors Notion of "Conceptual Spaces"

Russell's notion of a space of perspectives shares elements of the two models for points of view. The internal structure of each perspective follows the model of propositional attitudes. But the analyses of the interrelations among the perspectives follow the model of location and access. Now, let us examine another more recent and also very interesting approach to points of view that could also be placed between the two models. It has been recently proposed by Hautamäki, and we have mentioned it in other sections.[86]

The key idea is that a point of view can be understood as a certain subset of a "conceptual space" in the sense recently introduced by Gärdenfors.[87] Conceptual spaces are a "geometrical" way of representing information, both of a non-conceptual and of a conceptual kind. Gärdenfors argues that conceptual spaces offer a bridge between symbolic (logical) ways of representing information and connectionist (associationist) ways.[88]

A conceptual space (for instance, the conceptual spaces of colour, of sound, of taste; or the conceptual spaces of moral concepts, of epistemic concepts, etc.) is a geometrical structure based on a number of "quality dimensions". Each quality dimension corresponds to a different way in which stimuli are identified as being

---

[84]As we said, Russell [102], Sect. 11, speaks quite loosely of "experiences of a person".

[85]In other chapters of the book, we will analyse in depth McTaggar's approach.

[86]Häutamaki himself will explain his proposal in other chapter of the book. Here, we will only give a brief sketch.

[87]Mainly, see Gärdenfors [36].

[88]See Gärdenfors [36, Sect. 1.1.1].The name "conceptual" may be misleading. It only means that the essential aspects of concept formation are best described using this kind of "geometrical" representation.

"similar" or "different". For instance, in the case of the conceptual space of colour, we can appreciate similarities and differences in hue, saturation, and brightness. These identifications (for instance, through some judgements) generate a peculiar ordering relation, or even a metric.

Dimensions can be interpreted phenomenally (psychologically) or scientifically (theoretically). So, we can construe a conceptual space for colour in relation to experiential properties of colour (for instance: hue, saturation, and brightness); or we can construe it in relation to physical properties (for instance, light absorption, reflection, etc.). We found the same idea in Peacocke's notion of scenario content. The important thing is that the attribution of a conceptual space is useful in order to understand (predict, explain, control, etc.) the behaviour of a system, including here the system's possible conceptual behaviour.[89]

The conceptual spaces approach would show the compatibility between the two general models for points of view. In that approach we have some kind of "internal structure" for points of view. And we also have a "role" for points of view to play: the generation of a "geometry of content" capable of mediating between the subjects and their real environments.

We have presented two approaches integrating elements of the two models for points of view. Besides all of that, a crucial claim in the two models is that points of view are very peculiar entities with quite a strong relational and modal character. In the next section, we will explain this.

3   The Peculiar Kind of Reality and Mode of Existence of Points of View

Let us go from questions about structure to questions about reality and existence. What is the peculiar kind of reality and mode of existence of points of view? We are going to explore this ontological problem.

3.1   The Non-reducibility of Points of View

The ontological status of points of view really is singular. There are important reasons to consider that points of view are not reducible either to subjectivity, or to information, or to physics.

---

[89]Peacocke [82]'s work is not mentioned in Gärdenfors [36]. But there are great affinities between the two approaches. By the way, Russell's construction of a "space of perspectives" is not mentioned in Gärdenfors [36]; and it is not mentioned in Peacocke's [82] either.

Against first appearances, points of view are not something merely "subjective". They cannot be reduced to subjective mental states and processes because points of view include many ingredients that are not merely subjective. This is particularly clear in the model based on propositional attitudes. Putting aside the possession conditions of a point of view, which obviously cannot be subjective, even if they are internal to the point of view, points of view have to involve either non-conceptual contents or conceptual contents, and it is not at all obvious how those things could have a merely subjective status.

But points of view cannot be subjective in the model of location and access either. That model emphasises the modal character of points of view. Points of view are directly understood as possibilities of accessing the world. When points of view are taken in that way, they clearly override any merely subjective approach. The reason is pretty obvious: the world is not a subjective creation (or, to put it in current words, the world is not a "construction").

Points of view would not be reducible to "information" either. This would be so in spite of any effort to shorten distances between the notion of information and the notion of point of view. Even if we adopt a very general and wide notion of information, according to which we can speak of things like the non-conceptual information linked to analogical representations, or the informational states connected with a "know-how", and even if we assume that points of view can have a certain informational structure, points of view cannot be reducible to information. They cannot be so reducible because points of view include components which are not reducible to information. It is not easy to argue, for instance, that the bearers of points of view, personal subjects in our own case, are no more than informational structures.

Finally, it is also difficult to see how points of view could be reducible to "physics". Again, the problem comes with the highly heterogeneous nature of points of view. Points of view include many informational features, in general many kinds of functional structures. Also, they may include subjective components, of a very peculiar kind in personal cases. Furthermore, points of view have a very important modal dimension full of normative aspects. Unless all those components are shown to be reducible to physics, points of view would not be reducible to physics either.

In all the cases, the extreme heterogeneity of points of view suggests their non-reducibility. In principle, this could be read as an argument against the claim that points of view can constitute a, let us say, "natural kind". However, it is very difficult not to consider them in that way. The relational and modal character of points of view is the background from which we have access to the rest of reality, and to ourselves as part of it. Moreover, it is arguable that the identification of any natural kinds can be made only by adopting some point of view. This suggests a very simple but powerful argument for considering that points of view constitute a natural kind: if there are natural kinds, that what makes possible the identification of natural kinds has to be, itself, also a natural kind.

3.2   The Relational and Modal Character of Points of View

As we have seen, two features are crucial for the nature of points of view and for their peculiar mode of existence:

1. Points of view have an irreducible situated <u>relational character</u>.
2. Points of view also have a very strong and also irreducible <u>modal dimension</u>.

We can do justice to the first feature by characterising points of view as follows:

<u>Points of view</u> are certain kinds of functional structures physically realised in some peculiar ways.

A very relevant part of those functional structures would have to be constituted by the networks of relations that the bearer of the point of view is capable of maintaining with the non-conceptual and conceptual contents of the point of view. The peculiar ways in which those functional structures are physically realised, or implemented, would define the point of view with more precision.

Only functional structures physically realised in the adequate ways, hence satisfying certain substantive and not only formal possession conditions, would constitute a certain point of view. It is reasonable to think that the above combination of functional structures and physical realisation can do justice to the irreducible situated relational character of points of view.

Now, let us go to the second feature. Points of view have to be projected into an irreducible modal dimension. This means that

<u>Points of view</u> have explicit contents, either non-conceptual or conceptual, concerning not only what is actual but also what is possible.

Points of view are situated relational entities modally qualified. Moreover, that modal qualification goes beyond what can be found in the bearers of the point of view individualistically considered. This is especially important in the case of personal points of view. Moreover, it suggests a crucial distinction between "the subjects" and "their points of view". We must distinguish between

1. What can be <u>internal-to-the-subjects</u>.
2. What can be <u>internal-to-their-points-of-view</u>.

To have certain non-conceptual or conceptual contents in perspective could be something "internal to the point of view" of a subject, without being necessarily something "internal to the subject" individualistically considered.

The new distinction we have introduced is particularly relevant in the field of the philosophy of mind. Many times, the appeal to first-person points of view has been crucial. For instance, it has been crucial in the strategies followed by John Searle against computationalism and artificial intelligence, through his mental experiment known as "The Chinesse Room", and in Frank Jackson's argument against physicalist reductionism based on his mental experiment about "Mary the Neuroscientist".[90]

However, do the contents of our first-person points of view have to be "internal-to-ourselves"? Do they have to be "internal" in the sense that is questioned by mental experiments like those of "Twin Earth". It has been argued by some authors that they have to be so.[91] First-person points of view must be completely "internal-to-the-subjects".[92]

That position is, however, puzzling. As we have said, points of view always have a situated relational nature and a strong modal character. They are not reducible to subjectivity. This cannot be obviated. And it would apply also to first-person points of view. They cannot be simply "internal-to.the-subjects".

The issue can be clarified paying attention to the distinction between "internal to a point of view" and "internal to the subject which is the bearer of that point of view". The contents of first-person points of view can be "internal-to-those-points-of-view" without being "internal-to-the-subjects" who are adopting those first-person points of view. Moreover, the status of being a person adopting a first-person point of view could be, itself, something "internal-to-a-peculiar-kind-of-points-of-view", without being something "internal-to-the-subjects" that are adopting those points of view.[93]

In other sections, we have introduced a number of relevant notions and distinctions. We have defined in various ways the notion of points of view. Also, we have defined the notions of individual points of view, collective points of view, reflective points of view, etc. Among the most important distinctions are a distinction between the non-conceptual and the conceptual contents that a point of view can have; a distinction between non-conceptual, conceptual, and hybrid points of view; and a distinction between the contents, more or less explicit, non-conceptual or conceptual, contained in a point of view and the implicit non-conceptual contents linked to the attitudes. All these distinctions are intended to have not only an epistemological, logical, or methodological value, but a crucial ontological value. The last distinction between something "internal to the subjects" and something "internal to their points of view" is also very important. And it has also a crucial ontological value.

To take seriously the notion of point of view leads to a rejection of the claim that points of view are something simply "internal to the subjects". This is a completely mistaken picture. Points of view "empower" us in quite a radical sense. They

---

[90]See Searle [103, 104] and Jackson [58].

[91]See, recently, Farkas [30].

[92]First-person points of view are a peculiar sort of personal points of view. Also, they are a sort of reflective points of view. We can say that they are personal points of view adopting a reflective move over the personal character of some of our personal points of view.

[93]Moreover, following the structure of an argument suggested before, if to identify natural kinds requires being a person (at least, to share our criteria of similarity, relevance, rationality, etc.), then the peculiar kind of points of view supporting our status as persons has to constitute a "natural kind".

continuously redefine the spaces of possibilities that are relevant for a subject. Points of view open spaces of possibilities that would not be available if the subject were not to take them, or if the subject were to abandon them.

References

1. Apel, H.-O. (1988). *Understanding and explanation. A transcendental pragmatic perspective.* Cambridge: MIT Press.
2. Baier, K. (1958). *The moral point of view.* Ithaca: Cornell University Press.
3. Bermúdez, J. (1998). *The paradox of consciousness.* Cambridge: MIT Press.
4. Boghossian, P. (2008). *Content and justification.* Oxford: Oxford University Press.
5. Brandom, R. (1982). Points of view and practical reasoning. *Canadian Journal of Philosophy, 12,* 321–333.
6. Brandom, R. (1994). *Making it explicit.* Cambridge: Harvard Univ. Press.
7. Brandom, R. (Ed.). (2000). *Rorty and his critics.* Oxford: Blackwell.
8. Brandom, R. (2002). *Articulating reasons.* Cambridge: Harvard University Press.
9. Burge, T. (2007). *Foundations of mind.* Oxford: Oxford Univ. Press.
10. Burnyeat, M. (1976). Protagoras and Self-Refutation in Plato's. *Theaetetus Philosophical Review, 85,* 172–195.
11. Carnap, R. (1950). Empiricism, semantics, and ontology. *Revue Internationale de Philosophie, 4,* 20–40.
12. Cavell, S. (1979). *The claim of reason. Wittgenstein, scepticism, morality and tragedy.* Oxford: Oxford University Press.
13. Charro, F., & Colomina, J. J. (2014). Points of view beyond Models: towards a formal approach to points of view as access to the world. *Foundations of Science, 19*(2), 137–151.
14. Chisholm, R. (1966). *Theory of knowledge* (3rd Ed., 1989). Englewood Cliffs: Prentice Hall.
15. Chisholm, R. (1976). *Person and object.* Chicago: Open Court.
16. Clark, M. (1990). *Nietzsche on truth and philosophy.* Cambridge: Cambridge University Press.
17. Cooper, D. (1999). *Existentialism.* Oxford: Blackwell.
18. Crane, T. (Ed.). (1992). *The contents of experience.* Cambridge: Cambridge University Press.
19. Crane, T. (1992). The nonconceptual content of experience. In T. Crane (Ed.), *The contents of experience* (pp. 136–157). Cambridge: Cambridge Univ. Press.
20. Davidson, D. (1974). On the very idea of a conceptual scheme. *Proceedings and Addresses of the American Philosophical Association, 47,* 5–20. [Also in Davidson (1984)].
21. Davidson, D. (1984). *Inquiries into truth and interpretation.* Oxford: Clarendon Press.
22. Davidson, D. (2001). *Subjective, intersubjective, objective.* Oxford: Oxford University Press.
23. de Unamuno, M. (1913). *El sentido trágico de la vida* [*Tragic Sense of Life,* 1954. New York: Dover Publications].
24. Dennett, D. (1987). *The intentional stance.* Cambridge: MIT Press.
25. Dennett, D. (1990). The Interpretation of texts, people, and other artifacts. *Philosophy and Phenomenological Research (Supplement), 50,* 177–194.
26. Dennett, D. (1991). *Consciousness explained.* Boston: Little Brown.
27. deVries, W., & Triplett, T. (2000). *Knowledge, mind, and the given.* Indianapolis: Hackett.
28. Diamon, C. (1991). *The Realistic Spirit: Wittgenstein Philosophy and the Mind.* Cambridge: Bradford Books.
29. Earnshaw, S. (2006). *Existentialism: A Guide for the perplexed.* London: Continuum.
30. Farkas, K. (2008). *The subject's point of view.* Oxford: Oxford University Press.
31. Flynn, T. (2006). *Existentialism: A very short introduction.* Oxford: Oxford University Press.

32. Foucault, M. (1966). *The order of things: An archaeology of the human sciences* (p. 2002). London: Routledge.
33. Gadamer, H.-G. (1960). *Truth and method*. New York: Continuum. (1994).
34. Gadamer, H.-G. (1976). *Philosophical hermeneutics*. Berkeley: University of California Press.
35. García-Carpintero, M., & Kölbel, M. (Eds.). (2009). *Relative truth*. Oxford: Oxford University Press.
36. Gärdenfors, P. (2000). *Conceptual spaces: On the geometry of thought*. Cambridge: MIT Press.
37. Goodman, R. (Ed.). (2005). *Pragmatism: critical concepts in philosophy*. London: Routledge.
38. Guignon, C. (2003). *The existentialists: Critical essays on Kierkegaard, Nietzsche, Heidegger, and Sartre*. New York: Rowman and Littlefield.
39. Gunther, Y. (Ed.). (2003). *Essays on nonconceptual content*. Cambridge: MIT Press.
40. Haack, S. (1996). Reflections on relativism: From momentous tautology to seductive contradiction. *Philosophical Perspectives, 10*, 297–315.
41. Haack, S. (Ed.). (2006). *Pragmatism. Old and new*. Amherst: Prometeus.
42. Habermas, J. (1988). *On the logic of social science*. Cambridge: Polity Press.
43. Hager, P. (1994). *Continuity and change in the development of Russell's philosophy*. Berlin: Kluwer.
44. Hales, S. (2000). Review of Adrian Moore's book *points of view*. *Mind, 109*(433), 166–169.
45. Hales, S. (2006). *Relativism and the foundations of philosophy*. Cambridge: MIT Press.
46. Hales, S. (Ed.). (2011). *A companion to relativism*. Oxford: Wiley-Blackwell.
47. Hales, S., & Welshon, R. (2000). *Nietzsche's perspectivism*. Illinois: Univ. of Illinois Press.
48. Harman, G. (1990). The intrinsic quality of experience. *Philosophical Perspectives, 4*, 31–52.
49. Hautamäki, A. (1983). The logic of viewpoints. *Studia Logica, 42*(2/3), 187–196.
50. Hautamäki, A. (1986). Points of view and their logical analysis. *Acta Philosophica Fennica, 41*.
51. Hautamäki, A. (1992). A conceptual semantics approach to semantic networks. *Computers & Mathematics with Applications, 23*(6–9), 517–525.
52. Heck, R. (2000). Non-conceptual content and the 'space of reason'. *Philosophical Review, 109*(4), 483–523.
53. Heidegger, M. (1927). *Being and time* (p. 1996). New York: State University of New York Press.
54. Heidegger, M. (1950). On the origin of the work of art. In D. Farrell (Ed., 2008), *Basic writings. 1st Harper perennial modern thought edition* (pp. 143–212). New York: Harper Collins.
55. Horgan, T., & Tienson, J. (2002). The intentionality of phenomenology and the phenomenology of intentionality. In D. Chalmers (Ed.), *Philosophy of mind: Classical and contemporary readings* (pp. 520–532). Oxford: Oxford University Press.
56. Horgan, T., Tienson, J., & Graham, G. (2004). Phenomenal Intentionality and the brain in a vat. In R. Schantz (Ed.), *The externalist challenge: new studies on cognition and intentionality* (pp. 297–319). Berlin: Walter de Gruyter.
57. Hurley, S. (1998). *Consciousness in Action*. Cambridge: Harvard Univ. Press.
58. Jackson, F. (1986). What Mary didn't know. *The Journal of Philosophy, 83*, 291–295.
59. James, W. (1907). *Pragmatism: A new name for some old ways of thinking*. Indianapolis: Hackett.
60. Kaipainen, M., & Hautamäki, A. (2011). Epistemic Pluralism and multi-perspective knowledge organization. Explorative conceptualization of topical content domain. *Knowledge Organization, 38*(6), 503–514.
61. Kripke, S. (1982). *Wittgenstein on rules and private language*. Cambridge: Harvard University Press.
62. Kuhn, T. (1970). *The structure of scientific revolutions* (2nd Ed.). Chicago: Chicago University Press.

63. Linhares-Días, R. (2006). *How to show things with words. A study on logic, language, and literature*. Berlin: Mouton de Gruyter.
64. Liz, M. (2013). Models and points of view. The analysis of the notion of point of view. In L. Magnani (Ed.) *Model-based reasoning in science and technology. Studies in applied philosophy* (pp. 109–128). Heidelberg: Springer.
65. Lloid, G. (1993). *Being in Time: Selves and Narrators in philosophy and literature*. London: Routledge.
66. Loar, B. (2003). Phenomenal intentionality as the basis of mental contents. In M. Hahn & B. Ramberg (Ed.), *Reflections and replies. Essays on the philosophy of Tyler Burge* (pp. 229–258). Cambridge: MIT Press.
67. McDowell, J. (1994). *Mind and world*. Cambridge: Harvard University Press.
68. McDowell, J. (1994). The content of perceptual experience. *Philosophical Quarterly, 44*, 190–205.
69. McGinn, C. (1983). *The subjective view. Secondary qualities and indexical thoughts*. Oxford: Clarendon Press.
70. Moline, J. (1968). On points of view. *American Philosophical Quarterly, 5*, 191–198.
71. Moore, A. (1987). Points of view. *Philosophical Quarterly, 37*, 488–491.
72. Moore, A. (1997). *Points of view*. Oxford: Oxford Univ. Press.
73. Nagel, T. (1974). What is it like to be a vat? *Philosophical Review, 83*, 435–450 [Also in T. Nagel. (1979). *Mortal questions*. Cambridge: Cambridge University Press].
74. Nagel, T. (1986). *The view from nowhere*. Oxford: Oxford University Press.
75. Needham, P. (1975). *Temporal perspective. A Logical Analysis of Temporal Reference in English*, Uppsala: Univ. of Uppsala Press.
76. Noë, A. (2002). Is perspectival self-consciousness nonconceptual? *The Philosophical Quarterly, 52*(207), 185–194.
77. Noë, A. (2004). *Action in perception*. Cambridge: MIT Press.
78. Ortega y Gasset, J. (2004–2010). *Obras Completas [Philosophical Papers]. 10 vols*. Madrid: Fundación Ortega y Gasset - Taurus/Santillana.
79. Peacocke, Ch. (1983). *Sense and content*. Oxford: Oxford Univ. Press.
80. Peacocke, Ch. (1986). Analogue content. *Proceedings of the Aristotelian Society. Supplementary, 15*, 1–17.
81. Peacocke, Ch. (1989). Perceptual content. In J. Almog, J. Perry, & H. Wettstein (Eds.), *Themes from Kaplan* (pp. 297–392). Oxford: Oxford University Press.
82. Peacocke, Ch. (1992). Scenarios, concepts, and perception. In T. Crane (Ed.), *The contents of experience* (pp. 105–136). Cambridge: Cambridge University Press.
83. Perry, J. (1979). The problem of the essential indexical. *Noûs, 13*, 3–21 [Also in Perry, J. (1993). *The Problem of the Essential Indexical and other Essays*, (pp. 33–50). Oxford: Oxford University Press].
84. Pitt, D. (2004). The phenomenology of cognition or what is it like to think that P? *Philosophy and Phenomenological Research, 69*, 1–36.
85. Popper, K. (1945). *The open society and its enemies*. London: Routledge.
86. Popper, K. (1957). *The poverty of historicism*. London: Routledge.
87. Putnam, H. (1981). *Reason, truth, and history*. Cambridge: Cambridge University Press.
88. Putnam, H. (1987). *The many faces of realism*. La Salle: Open Court.
89. Putnam, H. (1992). *Renewing philosophy*. Cambridge: Harvard University Press.
90. Putnam, H. (1994). *Words and life*. Cambridge: Cambridge University Press.
91. Putnam, H. (1994). *Pragmatism*. Oxford: Blackwell.
92. Putnam, H. (1994). *Realism with a human face*. Cambridge: Harvard University Press.
93. Putnam, H. (1999). *The threefold cord mind, body and world*. New York: Columbia University Press.
94. Putnam, H. (2002). *Collapse of the fact/value dichotomy and other essays*. Cambridge: Harvard Univ. Press.
95. Rescher, N. (1973). *The Primacy of Practice*. Oxford: Basil Blackwell.

96. Ricoeur, P. (1974). *The conflict of interpretations: Essays in hermeneutics*. Illinois: Northwestern Univ. Press.
97. Rorty, R. (1982). *Consequences of pragmatism*. Minneapolis: University of Minnesota Press.
98. Rorty, R. (1991). *Objectivity, relativism, and truth*. Cambridge: Cambridge Univ. Press.
99. Rorty, R. (1998). *Truth and progress*. Cambridge: Cambridge University Press.
100. Russell, B. (1912). *The problems of philosophy*. London: Williams and Norgate.
101. Russell, B. (1914). *Our knowledge of the external world, as a field for scientific method in philosophy*. Chicago: Open Court.
102. Russell, B. (1918). The relation of sense-data to Physics. In *Mysticism and logic, and other essays* (pp. 113–140). London: George Allen & Unwin.
103. Searle, J. (1980). Minds, brains, and programs. *Behavioral and Brain Sciences, 3*, 417–457.
104. Searle, J. (1985). *Minds, brains and science*. Cambridge: Harvard University Press.
105. Searle, J. (1995). *The construction of social reality*. New York: Free Press.
106. Segerberg, K. (1973). Two dimensional model logic. *Journal of Philosophical Logic, 2*, 77–96.
107. Sellars, W. (1963). *Philosophy and the scientific image of the man perception science and reality*. Atascadero: Ridgeview Publishing Company.
108. Sellars, W. (1967). *Science and metaphysics*. London: Routledge and Kegan Paul.
109. Shoemaker, S. (1996). *The first-person perspective and other essays*. Cambridge: Cambridge Univ. Press.
110. Siegel, H. (1987). *Relativism refuted: A critique of contemporary epistemological relativism*. Dordrecht: Reidel.
111. Siewert, Ch. (1998). *The significance of consciousness*. Princeton: Princeton University Press.
112. Solomon, R. (1996). Nietzsche ad hominen: Perspectivism, personality, and ressentiment. In B. Magnus & K. Higgins (Eds.), *The Cambridge companion to Nietzsche* (pp. 180–222). Cambridge: Cambridge University Press.
113. Sosa, E. (1991). *Knowledge in perspective*. Cambridge: Cambridge University Press.
114. Sosa, E. (1997). Mythology of the given. *The History of Philosophy Quarterly, 14*, 275–286.
115. Sosa, E. (2007). *A virtue epistemology*. Oxford: Oxford University Press.
116. Sosa, E. (2009). *Reflective knowledge*. Oxford: Oxford University Press.
117. Strawson, G. (1994). *Mental reality*. Cambridge: MIT Press.
118. Tuomela, R. (1995). *The importance of us. A philosophical study of basic social notions*. Stanford: Stanford Univ. Press.
119. Tuomela, R. (2006). Joint intention. We-mode and I-mode. *Midwest Studies in Philosophy, 30*(1), 35–58.
120. Turing, A. (1950). Computing machinery and intelligence. *Mind, 59*, 433–460.
121. Vázquez, M., & Liz, M. (2011). Models as points of view. The case of system dynamics. *Foundations of Science, 16*(4), 383–391.
122. Vázquez, M., & Liz, M. (2013). Simulations models of complex social systems. A constructivist and expressivist interpretation. In L. Magnani (Ed.), *Model-based reasoning in science and technology, studies in applied philosophy*, (pp. 563–582). Heidelberg: Springer.
123. Vázquez, M., Liz, M., & Aracil, J. (1996). Knowledge and reality: Some conceptual issues in system dynamics modelling. *The System Dynamics Review, 12*(1), 21–37.
124. von Wright, G. (1971). *Explanation and understanding*. New York: Cornell University Press.
125. Williams, B. (1978). *Descartes: The project of pure enquiry*. Harmondsworth: Penguin Books.
126. Wittgenstein, L. (1921). *Tractatus Logico-Philosophicus* (p. 1961). London: Routledge & Kegan Paul.
127. Wittgenstein, L. (1953). *Philosophical investigations*. Londres: Blackwell.
128. Wittgenstein, L. (1969). *On certainty*. Oxford: Blackwell.

Chapter 2
Subjective and Objective Aspects of Points of View

Antonio Manuel Liz Gutiérrez and Margarita Vázquez Campos

Abstract One of the most puzzling features of points of view is their bipolarity between the subjective and the objective. First, we will distinguish in a precise way subjective points of view from objective ones. Both of them have a subject as their bearer, so the distinction between subjective and objective points of view will have to be made over the peculiar explicit contents of the points of view involved. After doing that distinction, we will define other connected notions as those of inter-subjective points of view and private points of view. Finally, we will consider in detail the positions of relativism and perspectivism. This will offer, so to speak, a panoramic view from the subjective side of points of view. From the objective side, we will analyse the notions of independence from a perspective, absolute points of view, and transcendental points of view. Also, we will distinguish between independence from all perspectives and independence from any particular perspective. The second notion will be crucial for a certain way of understanding objectivity.

1 Subjective and Objective Points of View

Points of view have both subjective and objective aspects. The subjective aspects derive from the relationships between the bearer of the point of view and its explicit contents. Some of the strongest forms of relativism are rooted in those subjective

This work has been granted by Spanish Government, "Ministerio de Economía y Competividad", Research Projects FFI2008-01205 (*Points of View. A Philosophical Investigation*), FFI2011-24549 (*Points of View and Temporal Structures*), and FFI2014-57409-R (*Points of View, Dispositons, and Time. Perspectives in a World of Dispositions*).

A.M. Liz Gutiérrez (✉) · M. Vázquez Campos
University of La Laguna, Tenerife, Canary Islands, Spain
e-mail: manuliz@ull.es

M. Vázquez Campos
e-mail: mvazquez@ull.es

© Springer International Publishing Switzerland 2015
M. Vázquez Campos and A.M. Liz Gutiérrez (eds.), *Temporal Points of View*,
Studies in Applied Philosophy, Epistemology and Rational Ethics 23,
DOI 10.1007/978-3-319-19815-6_2

aspects. But the objective aspects are no less important. It is by adopting some peculiar points of view that we are able to construe an objective scientific image of the world and of ourselves. We will begin by asking what the difference is between subjective and objective points of view.

We have to distinguish between the notion of a point of view with a subject as its bearer and the notion of a subjective point of view. Both subjective and objective points of view can have a subject as their bearer, and a set of attitudes connecting these subjects with the explicit contents of the points of view. Hence, the distinction subjective/objective points of view cannot depend on that. It has to be made in relation to how this can affect the peculiar explicit contents of the points of view involved.

The contrast subjective/objective points of view does not depend directly on the subjects which are the bearers of the points of view. It depends on the effects of the relationships that those bearers do in fact maintain with the explicit contents that the points of view have. More precisely, the explicit contents of a point of view can be said to be subjective in two main senses:

1. Subjective impregnation from attitudes: Non-conceptual and conceptual explicit contents can be subjective in the sense of being "subjectively impregnated" with the experiential, or qualitative, or phenomenal, non-conceptual contents linked to the psychological attitudes maintained by the subject, in such a way that these implicit non-conceptual contents determine the explicit contents. We will not define with more precision the notion of "subjective impregnation". However, the idea is clear: some implicit non-conceptual contents determine the explicit, either non-conceptual or conceptual, contents. Some experiential, or qualitative, or phenomenal, features are projected onto the explicit contents of our points of view. Colours, sounds, smells, textures, etc., all the so called "secondary quali-ties", many times are said to be subjective in that sense. Also, when a "non-cognitivist" stance about an area of discourse is maintained, what is gen-erally claimed is that the intended conceptual contents belonging to that area are subjective in the sense that they only "express" our attitudes: desires, emotions, feelings, etc. If there is subjective impregnation, then our attitudes determine the contents. And that determination will produce cases of subjective points of view.

2. Subjective relativisation to a certain position: There is another way in which a point of view can be said to be subjective. This time, it is a way involving only conceptual contents. A point of view can be subjective when the conceptual contents explicitly contained in the point of view cannot be semantically eval-uated solely in confrontation with the world, and some knowledge is required about how the bearer of the point of view is "placed" in the world. We can say that the conceptual contents are "subjectively relativized to a certain position or emplacement defined by some subject". All indexical thoughts (demonstrative, temporal, self-referential, etc.) are subjective points of view in that sense. Also, when I claim "such-and-such, in my view", and "in my view" is not redundant, i.e., when it does not mean simply that it is me who is claiming that, the conceptual content expressed by that such-and-such becomes subjectively

relativized in that sense. The truth or falsity of the proposition that such-and-such is relativized to the perspective, or point of view, from which I am having (maintaining, stating, considering, etc.) that content.

There is an important remark to make in relation to the second sense of "subjective points of view". Only points of view with a subjective bearer, paradigmatically personal points of view, can be subjective in the first sense. In that case, the explicit contents, either non-conceptual or conceptual, of the point of view can be determined by a "subjective impregnation" coming from the implicit non-conceptual contents linked to attitudes. In contrast, any point of view could be subjective in the second sense. The conceptual contents of any point of view can be relativized to how the bearer of the point of view is placed in the world, having in perspective the contents in question. A special case of this, involving points of view not having a subject as their bearer, would happen when we attribute a certain propositional content to a particular state of an instrument; for instance when we are measuring something, and we make corrections according to how the instrument is placed in the circumstances.

However, the subjective relativisation in those cases is strongly dependent on "our" attributions of points of view with some explicit conceptual contents. Moreover, it is arguable that the relevant subject of relativisation is not the entity to which the point of view is attributed but the subject that is making those attributions. We will not discuss that issue here.

In general, we can make a distinction between subjective and objective points of view as follows:

A subjective point of view is a point of view having explicit contents which are subjective in at least one of the above two senses: either through a subjective impregnation coming from attitudes or through a relativisation to a certain subjective position.

An objective point of view is a point of view having explicit contents which are not subjective in either of those senses.

Now, it is clear that not all points of view with a subject as their bearer have to be subjective points of view. They can be completely objective. If objective points of view are called "impersonal", we can also say that points of view with a subject as their bearer can be completely impersonal.

Traditionally, it has been assumed that most of the explicit non-conceptual contents of our personal points of view are always impregnated with subjectivity (i.e., with the subjectivity coming from the bearer's attitudes). Concerning them, our points of view could not but be subjective. The classical distinction between "secondary" and "primary" qualities is based on that assumption. In contrast, it has been also generally assumed that our conceptual contents can be completely objective (i.e., that it would be possible to eliminate from them any subjective relativisation), at least in principle. Both assumptions can be questioned.

Subjectivism and objectivism are two philosophical positions in conflict. They can be defined as follows:

Subjectivism claims that all points of view are necessarily subjective.
Objectivism claims that objective points of view are possible.

The possibility of objectivity is enough in order to be objectivist. In other words, the philosophical position of subjectivism seems to be much more demanding that the philosophical potition of objectivism. Because of that, it looks also much less plausible.

Here, we can introduce the notion of "epistemic independence". It can be defined through the notions of subjective impregnation from the attitudes and subjective relativisation to a position.

The content of a point of view is epistemically independent if that content has no subjective impregnation from the attitudes of the subject which is the bearer of the point of view and it is not subjectively relativised to the position of that bearer either.

For subjectivism, the explicit contents of all points of view are necessarily epistemically dependent. For objectivism, some points of view can have explicit contents epistemically independent.

Epistemic dependence/independence can be applied both to non-conceptual and conceptual contents. And we can say that whereas the contents of objective points of view are epistemically independent, the contents of subjective points of view are epistemically dependent.

It makes sense to say that some contents have more subjective impregnation coming from the attitudes of the subject which is the bearer of the point of view than other ones. Also, it makes sense to say that some contents have more subjective relativisation to the position of the bearer of the point of view than other ones. The subjective or objective character of a point of view is a matter of degree. And epistemic dependence/independence is also a matter of degree.

Both objectivity and truth are connected with knowledge and science. They are, however, very different aims. Furthermore, It is generally supposed that only scientific knowledge is capable of achieving points of view at the same time increasingly objective and increasingly true. This is another assumption that must be questioned. Objectivity and truth can follow distinct ways. Falsity can be as objective as truth. And it makes perfect sense to speak of true contents which cannot be evaluated except in a subjective way.

2 Intersubjective Points of View

The notion of an "intersubjective" point of view is closely connected with the notion of an objective point of view. It is arguable that objectivity and intersubjectivity go in parallel, and that we can find both intersubjectivity in our search for objectivity and objectivity in our search for intersubjectivity.[1] Moreover, even if the contrary of being "intersubjective" is to be "private", and not to be "subjective", it is arguable that it is much easier for an objective point of view to be intersubjective than it is for a subjective point of view to be so.

Intersubjective points of view cannot be simply contrasted with subjective points of view. This is not a good contrast. By themselves, intersubjective points of view can be very subjective. There is no reason why some strongly subjective points of view, in the senses above introduced, cannot be "shared" by a number of different subjects.

We cannot confuse intersubjective points of view with collective ones, either. In principle, there could be collective points of view which are not intersubjective ones. A collective point of view requires a collective subject, not a number of different subjects.

The notion of intersubjective points of view can be defined as follows:

An intersubjective point of view is a point of view that is taken by more than one subject, individual or collective.

Intersubjective points of view are points of view "shared" by a variety of subjects, individual or collective. So, a point of view can be collective without being intersubjective, and it can be intersubjective without being collective.

The opposites of intersubjective points of view are "private points of view". And not only individual points of view can be private. A point of view can be at the same time collective and private, in the sense that it can be a very "idiosyncratic" point of view of a certain collective subject.

In any case, points of view can be more or less intersubjective. Alternatively, they can be more or less idiosyncratic ones. Private points of view would be the limiting case of idiosyncratic points of view. Private points of view can be defined in the following way:

Private points of view are points of view maximally idiosyncratic.

Private points of view could not be intersubjective because of their idiosyncratic character. The more idiosyncratic a point of view is, the less intersubjective it can be.

Both intersubjective and private points of view have been philosophically relevant for many reasons. On the one hand, intersubjectivity has been repeatedly considered a necessary condition for things like meaning, communication, language, collective agency, society, normativity, rationality, etc. On the other hand, privacy also has been considered a necessary condition for things like mentality, personhood, freedom, morality, etc. From a Cartesian perspective, the existence of private points of view is one of the "marks" of the mental. For Kant, morality is crucially a private business. In contrast, from the perspectives of authors like the

---

[1]About that, see Davidson [33].

pragmatists, or Habermas, there could not be normativity without intersubjectivity.[2]

There is another important role for intersubjectivity. Even assuming all the nuances previously noted, intersubjective points of view also could have a crucial role in the "interplay" between the subjective and the objective. This is a very classical idea. And it has been recently emphasised by Donald Davidson.[3] According to him, intersubjective points of view would help to "triangulate" our subjective points of view in the search for objectivity. We can go from the subjective to the objective only through the mediation of the intersubjective.

3 Relativism

Any analysis of the notion of points of view has to deal with relativism. However, the relationships between points of view and relativism are complex. On the one hand, the distance from perspectivism to relativism is very short. On the other hand, not every kind of relativism depends on the notion of points of view, and it is possible to make a deep philosophical use of the notion of points of view without embracing any relativistic conclusion.

In order to identify with precision how the notion of points of view is connected with relativism, we will begin by introducing some important distinctions. Then, we will analyse the conditions in which the notion of point of view can lead to relativism. We will continue by exploring some of the most relevant fields where relativism has been proposed. Finally, we will consider the widespread relativist attitude that can be found in recent philosophy under the banner of "postmodernism".

3.1 Absolutism, Relativism, and Perspectivism as Philosophical Programs

The notion of points of view is relevant in many contexts. Absolutism, relativism, and perspectivism offer different philosophical accounts of that fact. These positions permeate all the history of philosophy. A precise characterisation would be the following one:

---

[2]In that line, see more recently Putnam [111, 113, 114, 115], Rorty [130, 131] and Brandom [12, 13]. One crucial difference between the Wittgenstein of the *Tractatus* [171] and the Wittgenstein post-*Tractatus* lies precisely in the contrast between private and intersubjective points of view. The solipsist option of the *Tractatus*, a private point of view which cannot but be the only correct one, is completely discarded as a serious option by the Wittgenstein of the *Investigations* [173], especially in relation to the problematic of "following a rule". And it is discarded too, although for different reasons, by the Wittgenstein of *On certainty* [172].

[3]Davidson [33].

<u>Absolutism</u> claims that there is a stable way in which things are in themselves, with independence from any point of view, so that any other way in which things can be in relation to a point of view is reducible, at least in principle, to that epistemically independent stable way of being.

<u>Relativism</u> rejects the claim of absolutism, maintaining that there is no stable way in which things are in themselves with independence from any point of view.

<u>Perspectivism</u> claims that there are some stable ways in which things are in themselves, with independence from any point of view, and that there are also non-reducible ways in which things are the way they are only in relation to some points of view.

Absolutism includes a positive thesis about reality in itself and, as a consequence, a reductivist thesis about reality in relation to perspectives. The positive thesis is that reality in itself has an epistemically independent stability. The reductivist thesis is that reality in relation to perspectives, i.e., reality from a certain point of view, is reducible to that epistemically independent stable way of being.

It is very important to distinguish the notion of objective points of view, and objectivism as the philosophical position claiming that objective points of view are possible, from the notion of absolute points of view and absolutism as a philosophical position. In other sections, we will define with precision absolute points of view. Now, we will focus on the philosophical positions of absolutism, relativism and perspectivism.

Absolutism only makes sense if it is possible to adopt objective points of view. However, to be capable of adopting objective points of view, i.e., points of view which are not subjective in any of the two senses above defined, subjective impregnation and subjective relativisation, does not entail to embrace absolutism. The existence of an epistemically independent reality is not enough for the truth of absolutism. That epistemically independent reality has to have a minimally stable way of being. And everything else has to be reducible to it.

Absolutism, relativism, and perspectivism can be interpreted as giving place to three very different kinds of "philosophical programs". Absolutism is adopted by reductive physicalism and by eliminativist physicalism. Points of view would have to be reduced to other more basic realities, or they would have to be ontologically eliminated. In any case, the world in itself would not contain points of view. That program has close links with the notion of an "absolute conception of the world".[4] The notion of an absolute conception of the world is the notion of a true and objective conception of reality independent of our points of view. A complete absolute conception of reality would show that points of view are either reducible or eliminable.

Now, let us consider relativism. It is the philosophical program adopted by many forms of idealism. The world without points of view is rejected as the basic reality. The basic reality is constituted by a number of points of view, or by a privileged point of view. And the world without points of view, a world independent of points

---

[4]See Williams [166], Moore [97] and Putnam [112].

of view, is understood as a construction, or projection, or postulation, made from those points of view.

We can find that program in Nietzsche also, and we can find it in many varieties of radical constructivism in the contexts of both continental and analytical philosophy. In this second context, it is a program explicitly adopted by Nelson Goodman, and by many other antirealists.[5]

The program of perspectivism is not as easy to introduce as the other two. It is suggested by claims like that of Putnam that, "the mind and the world jointly make up the mind and the world".[6] The crucial idea is that we cannot either reduce all points of view to a supposedly more basic reality, or eliminate them completely from our ontology. One of the main reasons for that impossibility, either of reduction or of elimination, is that any supposedly more basic reality could only be identified from some point of view. Moreover, the very distinction between "to see something, or to think of something, as being in a certain way" and "to be really in that way" also seems to depend on the adoption of a certain point of view.

Perspectivism entails a great amount of indeterminacy. From some points of view, we assume a reality independent of our points of view. From other points of view, we assume that at least some points of view are not reducible or eliminable. However, if the programs of absolutism and relativism are rejected, then our only option is to make that lack of determinacy acceptable.

3.2 Relativism, Skepticism, and Subjectivism

Relativism has to be distinguished from skepticism and from subjectivism. On the one hand, whereas relativism contains a positive claim about reality, namely, that there is no stable way in which things are in themselves with independence from points of view, skepticism does not contain any positive claim. Skepticism about a certain area is the rejection that we have, or that we can have, any knowledge about that area.

On the other hand, subjectivism can be a variety of relativism. A subjectivist relativism about a certain area would be that kind of relativism according to which all things inside that area are relativized to a certain subject (in any of the senses above introduced). However, there are other non-subjectivist varieties of relativism. In principle, relativism can be of a subjectivist kind and of a non-subjectivist kind.

We can distinguish two main ways of being relativist. One of them is the Protagorean way. According to Protagoras, the human being is the measure of all things. There is no place for a reality independent of human points of view. Usually, this is not meant to be equivalent to solipsism. It is supposed that there are a variety of subjects constituting reality. From Plato on, the consistency of that position has been an open problem. In any case, another way of being relativist is the

---

[5]See Goodman [51, 50].

[6]Putnam [110:xi]. Also, it comes close to other projects like Dennett [37]'s "heterophenomenology".

Heraclitean way. According to Heraclitus, there is no stable reality. Everything is in flux, like a "river". Everything is relative to the particular position in which it is placed in a fluent reality. This would be an important sort of relativism of a non-subjectivist kind.

Heraclitean relativism rejects absolutism. However, it assumes the existence of objective points of view. There are points of view having explicit contents which are not subjective in either of the two senses above introduced: subjective impregnation from attitudes and subjective relativisation to a certain position. Moreover, it is by adopting an objective point of view that Heraclitean relativism rejects the absolutist requirement of stability.

The two ways of being relativist, the Protagorean and the Heraclitean, can be found in many guises, both inside and outside philosophy. And both of them can be found in the works of Nietzsche. However, they are different. The sort of Heraclitean relativism we have described does not make any relevant use of the notion of point of view. Only the first one, the relativism rooted in Protagoras, depends clearly on that notion.[7]

3.3   Relativism Requires Constitution and Plurality

We have to note two very important features of relativism. The first one comes from the need to distinguish between a relational thesis and a relativist thesis. Being relational is not enough for being relative, or it is so only in a very weak sense. It cannot be accepted that every relation entails a relativisation. For something to be relative in a stronger sense, that relational character has to be "constitutive".

Many properties manifest a relational character. However, only if that relational character is constitutive of the properties in question, in the sense that changes in the relations entail changes in the properties themselves, can it be properly said that the properties are relative. Only in that case would the properties be relative in a strong sense. In other cases, we could vary all the relevant relational parameters without any variation in the properties. There is a well known relation, for instance, between some gas having a certain temperature and the gas having a certain volume and exerting a certain pressure over its container. However, the property of having that temperature is not constituted by the gas having that volume and exerting that pressure. It is constituted by the kinetic energy of the molecules of the gas. Relativism requires constitution. And the property of having a certain temperature, even though it is lawfully related to the properties of having a certain volume and exerting a certain pressure, is not relative to them.

What relativism has traditionally maintained is that some relevant properties are relative to certain factors in the strong, constitutive sense. This is what we find both

---

[7]A detailed analysis of Nietzsche's positions can be found in Conant [31, 32]. See also Hales and Welshon [59].

in Protagorean relativism and in Heraclitean relativism, with the difference that only in the first sort of relativism is there a relativisation to the points of view of a number of subjects. In Protagorean relativism the factor of relativisation is "inside" the points of view of some subjects, whereas in Heraclitean relativism the factor of relativisation is in the reality "outside" their points of view. The classical refutations of relativism also focus on that constitutive sense. Plato's claimed that the very existence of thought and language requires some stable reality, a certain way things are in themselves. This was a claim against both Protagorean relativism and Heraclitean relativism in the strong sense, not in the weak one.[8]

Another important feature of relativism, such as it has been traditionally maintained, is that it needs to claim 1) that there are many ways in which things can be strongly relative, or in any case more than one way; and 2) that those ways exclude each other. If there could be but "only one way" in which things are strongly relative, then relativism would not make sense. If there could be more than one way but "without exclusion", then relativism would not make sense either. Relativism requires a non-reducible plurality of mutually exclusive alternatives.

That feature of relativism is also present both in Protagorean forms of relativism and in Heraclitean forms. A Protagorean sort of relativism, involving crucially the notion of points of view, needs to claim that there are many possible points of view, or in any case more than one, and that those points of view exclude each other. A Heraclitean form of relativism would not appeal to the notion of points of view. However, it needs to admit that reality can flow in more than one possible way, and that those possible ways exclude each other.

This second feature is no less important than the first one. Even if subjectivism can be a variety of relativism, there are also subjectivist positions which are not relativist. And solipsism is one of them. According to solipsism, the whole of reality is necessarily determined by an individual subject. Solipsism is a form of subjectivism, but not a relevant form of relativism.

In exactly the same sense, it would not be enough for Protagorean relativism to say that from a certain point of view it is "as if there were" other points of view. In that case, those other points of view would exist only, so to speak, "inside" the first point of view. Relativism in its full sense needs a "real plurality", or at least a "really possible plurality", of points of view in conflict.

3.4   From Points of View to Relativism

The distinctions and commentaries we have made have a crucial importance regarding how the notion of points of view can lead to relativism. We can summarize them as follows.

---

[8]In particular, see *Cratilo*, *Teethetus*, and *Republic*.

1. The sort of relativism connected with the notion of points of view is not Heraclitean relativism, but Protagorean relativism.
2. Protagorean relativism, as any other relativism, requires that the relations between points of view and reality have a "constitutive" power. The mere existence of relations between points of view and reality would not be enough. If they were enough, because points of view have a relational nature, the need to adopt a point of view in any effort to know reality, and to know ourselves, would directly entail the truth of relativism.
3. Protagorean relativism would be a sort of subjectivism. However, as relativism, it cannot collapse into a subjectivism of a solipsistic variety. It has to assume the "real existence", or at least the "really possible existence", of a number of points of view in conflict.

In order to articulate a relativist position about a certain area from the notion of points of view, we would have to argue 1) that there are a number of different points of view about that area, 2) that they have a constitutive power over the properties that give structure to the phenomena in that area, 3) that those different points of view are in conflict, and 4) that there is no stable reality in that area that can remain out of that conflict.[9]

3.5 Two Dimensions in Relativism

Relativism can be projected into two dimensions: "scope" and "modal force". Relativism can have more or less scope. It can be only local, affecting particular areas or fields of phenomena. Or it can have a maximal generality. In the last case, its scope is global.

Relativism also can have a more or less strong modal force. Even though the relativist relations need to be constitutive, it can be maintained that their modal force has limits. This would mean that at some modal level those relations could be not so constitutive. Alternatively, it can be maintained that the modal force of the relevant constitutive relations is maximal, and that they are completely unavoidable. In other words, the constitutive relations can be understood as contingent at some modal level, or as something completely necessary at every modal level.

The degree of generality defines the scope of the constitutive relations. Their degree of contingency or necessity defines its modal force. This allows us to distinguish the following four kinds of relativism:

---

[9]A recent rejection of relativism based on what would be entailed by the identification and interpretation of "other" conceptual schemes, is Davidson [35]. In close connection with some ideas of the Wittgenstein of the *Investigations*, Putnam [118] is also very interesting. Other analyses and refutations of relativism can be found in Siegel [143]. About relativism in general, see again Clark [29], Haack [56], Hales [57] and Hales (ed.) [58].

Relativism-1   Relativism with both the maximum of generality and the maximum
               of necessity
Relativism-2   Relativism with the maximum of generality but with a contingent
               modal force
Relativism-3   Relativism with only a local scope but with the maximum of
               necessity
Relativism-4   Relativism with only a local scope and with a contingent modal
               force

It is arguable that Relativism-1 is inconsistent. Relativism-1 would be necessarily false because it would have to be false even if it is true. If the relativist position is claimed as something necessary, then it cannot be maximally generalised. And if it is maximally generalised, then it cannot be understood as something necessary. Traditional
self-refutations of relativism always have made use of these ideas.[10] However, many times it has also been maintained that, even if it is self-refuting "to claim" that kind of relativism, or "to believe" it, nevertheless it can reflect, or represent, our true situation. That move would go in parallel to similar moves that can be made, and that traditionally have been made, for protecting radical scepticism from a direct self-refutation.[11]

There would be only three consistent possibilities for relativism: Relativism-2. Relativism-3, and Relativism-4. How plausible are they? In Relativism-2, the relativist constitutive relations are not seen as something necessary at every modal level. There is at least one modal level in which they do not apply. Hence, at that modal level, the content of that relativism, what it says, could be consistently claimed, or believed, as being true.

The problem with Relativism-2 is twofold. On the one hand, it does not seem to be true. Reality shows many aspects that do not seem to be so strongly relative to our points of view. At least, they seem to be independent of any particular point of view. In principle, any point of view could be enriched with those aspects of reality by "simple addition".[12] On the other hand, it is difficult to see how its contingent character can be combined with its maximal generality. If it is assumed that "in fact", or "at some modal level", there are relativist constitutive relations over absolutely every field of reality, why not claim also that those relativist constitutive relations are necessary?

The second consistent possibility is Relativism-3, a local relativism with a modal force of necessity. The third possibility is Relativism-4, a local relativism without the modal force of necessity. We can consistently maintain both positions. Here, we would have relativist constitutive relations only with a local scope, and with a more or less strong modal force.

---

[10]See, for instance, Putnam [118].

[11]See Stroud [156].

[12]About that idea, see Moore [97].

3.6 Fields for Protagorean Relativism

Some kinds of contents are prone to be considered a relativist matter in a Protagorean sense. They can be called "fields for Protagorean relativism". The following list establishes a ranking of them:

1. Sensorial taste
2. Aesthetic taste
3. Social institutions
4. Moral norms and values
5. Meaning
6. Knowledge
7. Rationality

The list goes from fields, in the top, with a great propensity of being considered relative to a subject, individual or collective, in a strong sense, to fields that tend to be outside of relativist considerations.

Nevertheless, there are very important relativist positions about meaning, knowledge and rationality. As we have noted, we can find in Nietzsche a radical relativist approach about those matters. Sociology of knowledge,[13] and the more recent so called "Strong Program",[14] offer no less radical relativist conclusions. In combination with philosophies inspired by marxism and psychoanalysis, the influence of Foucault has been crucial also, mainly in Continental Philosophy.[15] Furthermore, current Postmodernism maintains very strong relativist positions.[16]

As we have said, it can be argued that such extreme relativist positions cannot be properly understood as "making a claim". Radical relativist claims of a Protagorean sort are self-refuting. In order to restore consistency, they have to be understood in other ways. In fact, this is accepted by many radical relativists. And very often radical relativism is interpreted more as an attitude than as a claim.

In any case, there is a general tendency to consider that even if it may be adequate to see meaning, knowledge, and rationality as contextual phenomena, and even if a certain perspectivism about them can be reasonably maintained, they cannot be simply relativised to things like psychology, social relations, culture, etc., without losing their "normative" functions. Here, relativism would be in the same boat with psychologism and naturalisation. The old reasons of Husserl, Frege, and Russell against psychologism and naturalisation would be also reasons against relativism. Meaning, knowledge and rationality have an anti-relativist "conceptual behaviour". In other words, to say that they are radically relative appears to be the same as to say that there is no meaning, no knowledge, and no rationality at all.

---

[13]See Berger and Luckmann [7].

[14]See Bloor [8].

[15]See Foucault [42, 41].

[16]See, for instance, Lyotard [85]. For a critical view of relativism, see Boghossian [10].

On the other pole of our ranking, things like sensorial taste, aesthetic taste, and social institutions (including here all of our "natural languages") appear to be highly relative. Even if we suppose objective properties in reality capable of being relevantly connected with them, the contribution of the points of view of the subjects, individual or collective, always seems to be determinant.

In the case of sensorial taste, the subjective equipment of the bearer of the point of view is the decisive factor. As we have noted, the classical distinction between "secondary" sensible qualities and "primary" ones puts the emphasis in that point.[17] In the case of aesthetic taste, the standards of taste, for a certain social group, in a certain context, would play a role similar to that of sensorial equipment. In the case of our social institution (for instance, "natural languages"), that role is played by our intentions, decisions and conventions.[18]

The field of moral norms and values occupies a very unstable position in between these poles. Sometimes, they have been considered at the same level as secondary qualities.[19] At other times, they have been considered to be social institutions. And there are also many approaches claiming a more objective status for moral norms and values.[20]

3.7   Relativism and Postmodernism

Nowadays, relativism is a very influential cultural perspective. Curiously, many times, natural science is appealed to "in support" of such relativism. This has been so especially in the case of relativity theory and quantum mechanics. Also, evolutionism and genetics are repeatedly mentioned as giving support to the idea that all our mental life is biologically determined. However, in spite of these appeals to natural science, the main sources of recent relativism are the social sciences and the humanities.

The movement known as Postmodernism maintains explicitly relativist theses. Postmodernism is an epigone of French Philosophy. It rejects all the rationalist and empiricist philosophical projects rooted in the Enlightenment, maintaining also an attitude of suspicion towards marxism, psychoanalysis and structuralism understood as "big theories".

Very often, as in the case of Nietzsche and others, that relativism is preserved from inconsistency by being presented not as a set of claims, i.e., as something we would have to evaluate as true or false, but as something expressing an attitude, or having a rhetorical status.

---

[17]See McGinn [93].

[18]About that, see Searle [139] and Tuomela [159, 160].

[19]See for instance, McDowell [92].

[20]With respect to relativist approaches to norms and values, see Harman [64], Honderich (ed.) [65], Krausz and Meiland (eds.) [70] and Mackie [90].

Postmodernism has received a harsh answer from more classical intellectual attitudes. The result is what has been called "The science wars".[21] The traditional conflict between the perspective of the natural sciences and the perspective of the social sciences and humanities reaches its highest intensity here. It is not only a methodological conflict,[22] or a conflict between two cultures,[23] or a conflict between two images of the world and of the human being in the world,[24] but also, or even mainly, a conflict of "interests" and "cultural power".

An important battle in the context of that war is "The Sokal's case". We can interpret it as a confrontation between those who believe in the possibility of absolutism and those who reject it, claiming some local kinds of relativist theses and adopting a generalised relativist attitude.[25]

The point of view of gender also has led to relativist approaches, sometimes of quite a radical sort. And very often, this has been in connection with Postmodernism.[26]

4 Perspectivism

The notion of point of view is deeply involved in our conceptions of the world and of ourselves. And the three philosophical reactions to that fact are absolutism, relativism, and perspectivism. We have defined them. Absolutism claims that there is a stable way in which reality is in itself, with independence from our points of view, and that everything else is either reducible, at least in principle, to that way of being or eliminable. Relativism claims that there is no such stable reality independent of our points of view. Perspectivism tries to place itself "between" absolutism and relativism.

Like absolutism, perspectivism assumes that there are some stable ways in which reality is in itself. However, like relativism, it also assumes that there are other non-reducible ways in which reality is dependent on our points of view. Perspectivism draws something from absolutism and something from relativism.

Another equivalent way to define perspectivism would be by maintaining the thesis that absolutism and relativism, even if they are stated with a maximum of

---

[21]See Ashman and Barringer (eds.) [4, 14], Callon [18], Gross and Levitt [54], Labinger and Collins (eds.) [75], Parsons (ed.) [106], Sokal [147] and Sokal and Bricmont [148].

[22]See Davidson [34] and von Wright [174].

[23]See Show [146].

[24]See Sellars [141].

[25]See Sokal and Bricmont [148] and Sokal [147]. Other authors with relevant contributions to all of these debates are Boghossian [10], Frankfurt [43, 44], Nagel [99], Searle [139] and Williams [168]. From different perspectives, all of them argue against relativism and defend the value of things like truth, reality, objectivity and rationality.

[26]As an example of that kind of gender relativism, see Hardin [61, 62, 63].

modal force, can only have a "local" sense. Perspectivism looks like a reasonable position. However, it is very difficult to articulate it in a fully elaborated way.

We said that the program of perspectivism entails a great amount of "indeterminacy". From some points of view, we assume a stable reality independent of our points of view. From other points of view, we assume that at least some points of view are not reducible or eliminable. In any case, perspectivism only is coherent if the following two conditions are fulfilled:

1. the scientific descriptions we have of physical, chemical and biological phenomena are not complete in the sense of exhausting every aspect of reality, and
2. the points of view we have about reality do not entail by themselves any relativist position.

The two conditions involve realist compromises. If condition 1 is not satisfied, then the claim that the world "really" contains points of view could not make sense. If condition 2 is not satisfied, then the claim that we can really "know" from some points of view that the world really contains points of view could not make sense either.

The most important problem for perspectivism is to distinguish those aspects of reality that are stable and independent of points of view from those aspects of reality that are not.

4.1 Contemporary Perspectivism in Philosophy

Many philosophical positions have adopted perspectivist positions concerning a certain area of phenomena. We will dedicate this section to offering a little guide about contemporary perspectivism in various philosophical disciplines.

4.1.1 Epistemology

There are very strong tendencies toward perspectivism and relativism in epistemology. Moreover, many times it is very difficult here to identify clearly the differences between each position.

The fact that things can be seen with different colours and shades, with different shapes, etc., from different perspectives, or by different subjects, or by the same subject in different conditions, etc., always has constituted one of the main motivations for perspectivism and relativism. The same point would hold regarding any other sensorial modality. Perception has a very "circumstantial", or "situated", character.

There is a very common argument that goes from that circumstantial character to the conclusion that none of the things we perceive can be objective. The argument is that there are so many different perceptual aspects in any object that none of them can be assumed to be its "objective" or "real" aspect, an aspect the object has independently from points of view. Any object can be seen, for instance, with so many different colours, even with so many different shades of a certain colour, that none of them can be said to be its "objective", or "real", colour. Perceptual contents would be merely subjective.[27]

The distinction between "primary" and "secondary" qualities puts a boundary on the above move. In contrast with secondary qualities (properties like colour, sound, smell, texture, etc.), primary qualities (properties like form, quantity, etc.) can be considered "objective", or "real", properties of the objects. That distinction has been a disputed topic throughout all the history of philosophy, and it continues to be so.[28]

Going from perception to belief, there are also very strong tendencies toward perspectivism and relativism. Coherentism, for example, conceives justification and knowledge in ways that make it very difficult to avoid the possibility of alternative systems of beliefs that are maximally coherent and comprehensive.[29]

Pragmatism is another example of an epistemological approach that makes justification and knowledge at least partially dependent on other things apart from the way things can be in themselves. Practical value is dependent on the subjects and their points of view.[30]

Things are more implicit with the epistemological position known as confiabi-lism. Confiabilism seems to be a position that tries to do justice to the notion of objective truth. It defines justification and knowledge in close connection to it. However, in order to deal with real situations of knowledge, confiabilism always needs to include contextual references to concrete subjects and circumstances, and this entails a certain amount of perspectivism.[31]

Another approach that has had a crucial role in recent epistemological debates is the one called "virtue epistemology". In virtue epistemology, the contextual aspects of justification and knowledge are very important. What can be an epistemic virtue for a subject does not have to be an epistemic virtue for other subjects, and what is an epistemic virtue in one context does not have to be an epistemic virtue in other contexts.[32]

Reflective points of view about our own points of view also are very important in virtue epistemology. In some cases, coherence would not be enough to get justification and knowledge, nor would it be enough to fulfil all sorts of practical requirements. And we can say the same of the reliability of our representational states. Sometimes, justification and knowledge require an epistemic ascent: to take

---

[27]A paradigmatic presentation of that argument can be found in Russell [133].

[28]See Hamlyn [60], McGinn [93] and Stroud [157].

[29]See Bender [6], Bonjour [11], Davidson [35], Lehrer [80, 81], Rescher [124, 125] and Sosa [149].

[30]Two recent and very important approaches in that sense are Rorty [129] and Stich [153]. Among classical pragmatists, James [68] constitutes the most explicit assumption of perspectivism.

[31]See Armstrong [3], Goldman [47] and Nozick [102].

[32]See Sosa [150, 151, 152]. See also Greco (ed.) [52] and Greco [53].

an adequate epistemic perspective over our own epistemic states, and their sources.[33]

The perspective offered by reflection has been very important in contemporary epistemology in another sense also. Nelson Goodman proposed a way to understand the relationships between inductive practices and inductive rules that has been called "reflective equilibrium".[34] Inductive practices are corrected when they do not follow sound inductive rules, and inductive rules are changed when they are not in accordance with persistent inductive practices. Induction has a dynamics grounded in that reflective equilibrium. The same strategy has been applied to other areas such as the theory of justice and conceptual analysis.[35] Reflective equilibrium seems to be at the very core of rationality.

All the epistemological approaches we have examined are anti-foundationalist. Some of them are closer to relativism than others. Coherentism and pragmatism are very close to relativism, whereas confiablilism is not. In any case, all of them suggest some kind of perspectivism.

There is also an important kind of perspectivism in many foundationalist epistemologies. Descartes's epistemology is a classical example of foundationalism. But, it is also a classical example of the first-person point of view. In Descartes, there is a peculiar blend of foundationalism and perspectivism. Descartes's foundationalism is grounded in the first-person point of view.[36] It has been defended that it is possible to separate the two ingredients, foundationalism and the first-person point of view, in Descartes. According to some authors, whereas foundationalism would not be an adequate epistemology, the first-person point of view, such as it is elaborated by Descartes, defines the very nature of the mind.[37]

Chisholm is another example of foundationalist epistemology. This time, the first-person point of view is taken to be essential to any process of assessment and justification of our beliefs in order to achieve knowledge. All our knowledge would be justified by certain "evidences", and to be or not to be evident is a subjective matter. It depends on a certain perspective that only can be achieved from a first-person viewpoint.[38]

We have said that coherentism and pragmatism are quite close to relativism. Other epistemological positions assume an explicit relativism. The possibility of alternative conceptual frames, or alternative conceptual schemes, or situations of theoretic incommensurability, has been maintained, or suggested, by many authors in contemporary philosophy.[39]

---

[33]About that requirement, see specially Sosa [150, 152].

[34]See Goodman [49].

[35]With respect to justice, see Rawls [122]; with respect to conceptual analysis, see Sosa [150].

[36]About that, see Farkas [39], Quinton [121], Williams [166] and Williamson [169].

[37]This is argued in Farkas [39].

[38]See Chisholm [26].

[39]We can mention Feyerabend [40], Foucault [42], Goodman [50], Kuhn [71], Quine [119, 120], Putnam [110, 111, 113, 114, 115]; and in a very radical way Rorty [128, 129, 130, 131].

In one way or another, the possibility of alternative conceptual frameworks has a very strong Kantian inspiration. That possibility was also considered by Carnap, in connection with his crucial distinction between "internal" and "external questions". Internal questions make sense only inside a certain conceptual framework. External questions are questions about the frameworks themselves. The last questions do not have answers that can be true or false. Conceptual frameworks are simply chosen.[40]

4.1.2 Philosophy of Language

If the circumstantial, or situated, character of perception has been the main motivation for perspectivism and relativism about non-conceptual contents, the circumstantial and situated character of language has been the main motivation for perspectivism and relativism in relation to conceptual content.

The circumstantial and situated character of language has many faces. All of them suggest a certain perspectivism, and sometimes also relativist positions. One such face, with a long history, has to do with the quite simple and obvious fact that there are "many" natural languages.

That linguistic pluralism has sometimes been transformed into a linguistic perspectivism, or even into a linguistic relativism. This is the case with the so called Sapir-Whorf's relativist hypothesis. According to that hypothesis, natural languages shape different ways of conceptualising the world, even different ways of perceiving it. Moreover, those configurations are alternative in quite a radical sense.[41] This position has been highly influential. Interestingly enough, that linguistic relativism is usually grounded in empirical studies comparing very different languages, as for instance Hopi language and English with respect to temporal concepts.[42]

The close relations between languages and conceptual frameworks, or conceptual schemes, means that many of the authors who maintain an epistemological relativism also can be considered as maintaining a linguistic relativism, and vice versa.[43]

A second sort of perspectivist approach connected with the circumstantial and situated character of language involves Quine's theses about the "inscrutability of reference", the "indeterminacy of translation" and "ontological relativity".[44] Those expressions suggest an explicit alignment with relativism. However, it is not easy to interpret Quine's claims. Certainly, they can be interpreted as being very close to relativism. Reality would be dependent on language and conceptual framework.

---

[40]Carnap [20].

[41]See Whorf [165] and Gumperz and Levinson (eds.) [55].

[42]For a reconstruction and criticisms of these relativist ideas, see Malotki [91].

[43]In particular, this is so with Feyerabend [40], Davidson [35], Goodman [49], Kuhn [71], Quine [119, 120], Putnam [110, 111, 113, 114, 115] and Rorty [128, 129, 130, 131].

[44]See Quine [119, 120].

However, Quine can be also interpreted as undermining, or undercutting, the very possibility of philosophical relativism. This would be so when the conclusion of Quine's theses, closely tied to his rejection of the analytical/synthetical distinction, is intended to be that neither absolutism nor relativism make sense.

Quine's approach has been very influential. Davidson's ideas about language, closely connected to notions such as "translation", "radical interpretation", "charity principle", "rationality", etc., have their roots in Quine. Davidson always emphasises the need to rationalise, and this entails adopting a very peculiar point of view, different from the points of view of the natural sciences. Also important is Davidson's anti-relativist thesis about the incoherence of the idea of a conceptual scheme completely different from our own conceptual scheme.[45]

A third very important perspectivist face of language, derived from its circumstantial and situated character, is "context dependence". There are many kinds and subkinds of contextual dependence. And they can affect syntax, semantics, and pragmatics.[46]

We will mention two other perspectivist issues connected with the circumstantial and situated character of language. One of them has to do with the peculiarities of indexicality in the first-person case. The indexicality of "I" is very special. Many other indexicals could be defined from it. Furthermore, it is not at all clear the sort of meaning, and the sort of knowledge, that are involved in the use of the indexical "I".[47]

The other one has to do with the recent discussion of "faultless disagreements". There are cases of disagreements (for instance, I say "No doubt, avocados are tasty"; you say "Absolutely false, they are not tasty") where the truth of what is said seems to be ultimately dependent on some social standards, or simply dependent on the peculiar taste of the subjects involved. There seems to be a genuine disagreement concerning some truth, but the conflict cannot be solved with more information. The final result is a kind of "perspectival truth". In the extreme case, it can be a kind of non-reducible "relative truth".[48]

Sometimes, the phenomenon of faultless disagreements is rejected because its incompatibility with an absolutist position. At other times, it is interpreted as a confirmation of relativism. Indeed, if faultless disagreements were the rule, then

---

[45]See Davidson [35].

[46]Recanati [123] offers a very clear and useful classification of the main forms of "context dependence". He distinguishes between pre-semantic context dependences and semantic context dependences. Among the first ones, the most relevant cases are language-relativity, syntactic ambiguity and lexical ambiguity. Among the second ones, the most relevant cases are circumstance-relativity, indexical token-reflexivity, indexical semantic under-specification and modulation. About contextualism in general, see Preyer and Peter [108]. About the relationships between contextualism and relativism, see Richard [127].

[47]About that, see Perry [107].

[48]Among the vast literature concerning this topic, see García-Carpintero and Kölbel (eds.) [45], Kölbel [72, 73, 74]; Lasersohn [77, 78], MacFarlane [87, 88, 89]; Preyer and Peter [108], Recanati [123], Richard [127], Williamson [169], Cappelen and Hawthorne [19], Stojanovic [154] and López de Sa [86].

large parts of our use of language, and large parts of our thought, would lead to "relative truths". However, between the extremes of absolutism and relativism, faultless disagreements also could receive a perspectival interpretation.

How to obtain such a perspectival interpretation? It can be argued that faultless disagreements constitute very unstable situations. Sometimes, we are inclined to say that there is some fault in the disagreement, other times we are inclined to say that there is no disagreement at all. It would depend on the "perspective" adopted. Under that diagnosis, faultless disagreements would always have to be understood in a dynamic context.

4.1.3 Philosophy of Mind

The contrast between the first-person point of view and the third-person point of view is crucial in current philosophy of mind. At a personal level, a subject having a mental state always has a first-person perspective about it. The subject has a direct, empathic access to his, or her, own personal mental states. And all other subjects only have an indirect access to them. This "asymmetry" between the point of view of the first-person and the point of view of the third-person is a constant source of problems.[49]

On the one hand, the relevance of the third-person point of view has been maintained, not only for a scientific study of the mind but in any context. In recent years, Daniel Dennett has been one of the leading authors in that sense. For him, the mind is no more than the result of an attribution made from an intentional stance. Alan Türing maintained that the capacity to manipulate symbols from a third-person point of view, in particular the capacity to simulate a conversation, is the only adequate perspective for attributing mentality. This idea was very influential at the beginning of artificial intelligence. Behaviourism in psychology and philosophy, the rejection of "the ghost in the machine", the critique of the existence of "private languages", etc., also entail a passionate defence of the third-person point of view. All sorts of reductionist and eliminativist approaches coming from neurology also maintain the prevalence of a third-person point of view in order to know adequately the nature of the mind.[50]

The thesis that mentalistic concepts are theoretical concepts, introduced for predicting and giving an explanatory account of the behaviour of some complex entities, assumes quite directly a third-person perspective. That thesis comes from Wilfrid Sellars, and has had a tremendous impact on many recent developments.[51]

---

[49]See Tye [162, 163, 164]. See also Levine [82], who coined the expression "explanatory gap" to emphasise the differences between the first-person point of view and scientific third-person points of view.

[50]See Dennett [36, 37], Ryle [135], Türing [161], Patricia Churchland [27] and Wittgenstein [173].

[51]See Sellars [140] and Paul Churchland [28].

On the other hand, it has been also maintained that the first-person point of view has some privileges that cannot be obviated. And that this fact puts serious limits to the possibility of "construing machines" able to have properly a mental life. Sometimes, the first-person point of view is connected with a certain way of sensing the world and ourselves, a "what-is-it-like" producing a peculiar qualitative, phenomenal content. Other times, it is connected with a certain "know-how", some abilities or competences not reducible to propositions or rules.[52]

The classic defender of the privileges of the first-person view is Descartes. His "cogito" can be seen in clear contrast with the materialist third-person point of view of authors like Hobbes. The contents of the "cogito" only can be accessed from a first-person perspective.

The debate between "externalism" and "internalism" about mental content and mental states is one of the areas in philosophy of mind where the contrast between the first-person and the third-person points of view is more explicit. According to internalism, to have a thought with a certain content has to entail knowing that we have that thought with that content. According to externalism, the mental contents of our thoughts, and those very thoughts, are determined by environmental or social factors that can be completely outside our epistemic horizon. Even thoughts about ourselves, i.e., self-knowledge, would be so determined.[53]

We have indicated some approaches giving relevance, in an exclusive way, either to the third-person point of view or to the first-person point of view. There are two other approaches that suggest some ways to overcome that tension. One of them consists in looking for a different perspective. And the "second-person" point of view is a perfect candidate. In situations of personal interaction, the mental attribution is reciprocal. And there is a dynamics of mutual attribution that determines the final result.[54]

The other approach breaking the dichotomy between the first-person and the third-person is the analysis of "multi-agents" contexts in artificial intelligence. The design of adequate models of interaction in those contexts requires taking into account at the same time both the first-person points of view and the third-person points of view, and their various sorts of interactions.

---

[52]See Cassam [21], Chalmers [23], Chisholm [25], Dreyfus [38], Farkas [39], Jackson [67], McGinn [93], Mellor [95], Nagel [100, 98], Searle [136, 137, 138], Lewis [83] and Shoemaker [142].

[53]See Putnam [117] and Burge [15, 16, 17]. See also Boghossian [9] and Liz [84]. The need for a perspectival self-consciousness is particularly demanding in the case of thoughts about oneself. The phenomenology of the "I" has been analysed by Chisholm [25] and Castañeda [22]. Its radical indexicality has been emphasised by Perry [107]. And the connections among perception, action, and self-consciousness have been stressed by Hurley [66]. Extending Hurley's ideas, Noë [104, 105] has defended the non-conceptuality of perspectival self-consciousness. Some of our analyses of points of view would have relevant implications here. Perhaps the proper space for "self-consciouness" and "self-knowledge" is that space which is internal to points of view without being internal to the subjects having those points of view.

[54]See Gomila [48].

4.1.4 Philosophy of Science

Philosophy of science has always lived between the poles of a dogmatic absolutism and of an extreme relativism.

On the one side, the analyses of the methodology of science, assuming a distinction between the "context of justification" and the "context of discovery", have favoured absolutist positions. This was the approach generally adopted by Logical Positivism and also by Popper. That approach gave place progressively to the idea that a better perspective is to treat scientific theories as interpretative frames or as conceptual systems.[55]

On the other side, researchers both in the history of science and in the social, economic and political aspects of science have favoured relativist positions.[56]

Against linearly progressive versions of the history of science, there is now a strong tendency to see the history of science as a dramatic "fight" between different paradigms: geocentrism versus heliocentrism, the old teleological concepts of premodern science versus modern biology, alchemy versus chemistry, etc. And recent science is also seen in that way, as a "fight" between different conceptions and interests where moments of peace are not the consequence of a rational victory but the result of the (academic) elimination of the enemy.

Inside basic science, things also have changed a lot in the past century. Many times, it has been said that two of the most important theories of that period, namely the Special Relativity Theory and Quantum Mechanics, involve a revision of the traditional notion of scientific objectivity and a vindication of the crucial role of the points of view of the observers.

Special Relativity rejects the existence of an absolute space-time. In particular, temporal relations between events depend on the inertial reference frame of the observer, other inertial frames being equally acceptable. According to Quantum Mechanics, the variables defining the state of a particle do not have specific values before the measurement process. The particle does not have a determinate state, but a superposition of different states, and it is only the operation of measurement that "fixes" some particular values. In Quantum Mechanics, at least under the standard interpretation, the state of a particle does not exist independently of the observer.

A very important consequence of those changes in basic science is that many of the fundamental concepts used in the description of reality, as for instance the concept of causality, are now seen as perspectival concepts.[57]

---

[55]For that change of perspective, see Lakatos and Musgrave (eds.) [76] and Toulmin [158].

[56]In the first field, Kuhn's notions of "paradigm" and "incommensurability" have had an enormous influence. See Kuhn (1996, 3 ed.). With respect to the second field, see Bloor [8], Barnes and Bloor [5] and Collins [30]. Holding a harder constructivism, see Latour and Wolgar [79] and Knorr-Cetina [69].

[57]About the perspectival character of causality, see Menzies and Price [96], Price [109] and Álvarez [1].

The relations between science and other cultural instances are also under the pressure of different points of view. The tension between the perspectives of the, so called, "manifest image" and of the "scientific image" is a perfect example.[58]

Some authors have proposed positions that try to maintain an equilibrium between "absolutism" and "relativism". A recent case is the "perspectival realism" maintained by R. Giere. He uses as an analogy the partial representations offered by maps. It is a very good analogy. Maps cannot be said to be true or false with independence from a perspective. But, unlike Kuhn paradigms, the different perspectives that can be adopted are not necessarily incommensurable.[59]

4.1.5 Perspectivism and Conceptions of Rationality

Many times, points of view are evaluated as being more or less rational. However, rationality can be understood in many senses. We can distinguish a theoretical rationality, in a more generic sense an epistemic rationality, from a practical one. Also, we can distinguish between merely formal conceptions of rationality and more substantive ones. We can distinguish between an instrumental rationality and a rationality including ends and values. We can also distinguish between "Humean" conceptions of rationality, in which beliefs are in the service of desires, and "Platonic" conceptions, in which desires have to be submitted to some rational control. Even if we do not want to adopt relativism, there is a wide room for perspectivism with respect to rationality.[60]

There is another contrast worthy of attention. It is the one between "ideal conceptions of rationality" and "conceptions of a bounded rationality". Two sorts of paradigmatic examples of ideal conceptions of rationality are the ones based on logic and the ones based on decision theory. Conceptions of a bounded rationality assume very directly a perspectivist approach. The origin of the notion of a "bounded rationality" is Herbert Simon's idea that many times it is more rational not to be as rational as an ideal conception of rationality would require.[61]

Bounded rationality defines very circumstantial and situated ways of selecting and organising our beliefs and actions. Rationality is not defined by conditions independent of how we are placed in the real world. Rationality is always relative to some particular subjects and some particular circumstances.[62]

There is a huge amount of evidence showing that ideal conceptions of rationality do not have a clear "descriptive" value. The open question is whether they can be said to have a certain "normative" value, and how they can have it.

---

[58]See Sellars [141] and Rosenberg [132].

[59]See Giere [46].

[60]About all those distinctions, and many others, see Wilson [170] and Rescher [126].

[61]See Simon [144, 145].

[62]About "bounded rationality", and its contrast with "ideal conceptions of rationality", see Cherniak [24].

In any case, one thing is to adopt, or to take, a point of view which can be more or less rational, and another different thing to evaluate a point of view as being more or less rational. The second situation, but not necessarily the first one, entails taking a normative point of view. And this seems to require having a certain conception of rationality, and applying it. The fact that our conceptions of rationality change, and have changed, offers another very sound argument for perspectivism in this field.

4.2 Perspectivism Without Relativism

Now, let us compare more closely the positions of relativism and perspectivism. As we have said, perspectivism would claim that there are some stable ways in which things are in themselves, with independence from points of view, and that there are also other ways in which things are the way they are only in relation to some points of view. Perspectivism assumes both absolutism and relativism in a local sense.

We distinguished between a relativism of a Heraclitean sort and a relativism of a Protagorean sort. Only the second one involves in a crucial way the notion of points of view. What was the diagnostic for Protagorean relativism? On the one hand, Protagorean relativism with a general scope and a maximal modal force is inconsistent. Its non-relative truth cannot be stated. This is the classical refutation of that sort of relativism. On the other hand, Protagorean relativism with a general scope but with a contingent modal force is not a stable position. In principle, there would not be any sound reason to consider it in that general but only contingent way.

Protagorean relativism is an option only in a local sense. In that local sense, even with a maximal modal force, Protagorean relativism is an innocuous position. In any case, the truth of such relativism would depend on its peculiar scope. So understood, relativism becomes an "objective" claim that can be tested and controlled. Here, the positions of relativism and perspectivism come very close. Perspectivism can be locally relativist in this sense.[63]

Also, we distinguished between Protagorean relativism and subjectivism. Solipsism would be a subjectivist position, but not a relativist one. Even a Protagorean relativism with a general scope and a maximal modal force would have to assume the real existence of "other points of view". There have to be other alternative points of view which are assumed as real. There cannot be only other points of view "from my own perspective". Also in that sense, a consistent relativism assuming the real existence of "other points of view" offers a suggestive way towards perspectivism. Such relativism makes it possible to analyse what is entailed by such "objective" acknowledgement of other points of view. Again, the positions of relativism and perspectivism are very close here.

----

[63]About that, see Anderson [2].

Perspectivism can accept local Protagorean relativism. Also, it can accept any relativism assuming seriously the reality of other points of view. The open problem for perspectivism would be to determine in detail the "objective" contents of those relativist positions. Perhaps that cannot be done for Heraclitean reasons. Here, the other way of being relativist, the Heraclitean way, might have a relevant role to play. In any case, now, it would be a role played "inside a perspectivist frame".

5   The Objective Side of Points of View

Let us summarise briefly some of our results. Objective points of view are points of view which are not subjective. And subjective points of view are points of view with explicit contents having a subjective impregnation from the attitudes or having a subjective relativisation to a certain position. Both objective and subjective points of view can have a psychological subject as their bearer. The difference between them is a difference of content.

We have distinguished between objective points of view, objectivism, and absolutism. The contents of objective points of view are epistemically independent. Objectivism claims that such points of view may exist. And absolutism claims that some parts of reality have an epistemically independent stability.

Objectivism is a philosophical position in contrast with subjectivism, and absolutism is a philosophical position in contrast with relativism. The opposition between objectivism and subjectivism is exclusive. Each one has been defined as the negation of the other one. However, the opposition between absolutism and relativism is not exclusive. It is possible to be absolutist with respect to certain parts of reality and relativist with respect to other ones. This combination of absolutism and relativism defines perspectivism.

Objectivism and subjectivism are philosophical positions about points of view, more concretely about the nature of the explicit contents of points of view. Absolutism and relativism are philosophical positions about reality. It is very important to introduce exclusivism with respect to the first contrast and not to introduce it with respect to the second contrast. That way, we can obtain the following relevant conclusion:

Objectivism is entailed by absolutism, by perspectivism, and also by all the non subjectivist varieties of relativism. Only subjectivist relativism entails subjectivism.

Our intuitive, ordinary conceptual framework tends to objectivism. Many perspectivist positions, even many relativist positions, require a certain amount of objectivity. Our philosophical analyses have to reflect that feature.

According to absolutism, there is objectively a stable reality. According to Heraclitean relativism, there is objectively no stable reality. Also, it is important to make room for the possibility of rejecting absolutism from an objectivist stance. Heracliten relativism is the prototype of a very relevant variety of relativism different from Protagorean relativism. It is characterised by the aim of claiming objectively that beyond our epistemic contributions there is no stable reality.

In previous sections, we have focused on the subjective side of points of view. The analysis of subjective points of view, of relativism, and of perspectivism has offered, so to speak, a panoramic view of points of view from their subjective side. From their objective side, we have paid attention to the notions of objective points of view and to the philosophical positions of objectivism and absolutism. Now, let us consider more closely other aspects of that objective side.

In general, the objective side of points of view is manifest in six sorts of features:

1. The possibility of having objective points of view, with an objective content, and not only subjective ones.
2. The existence in points of view of ingredients which are internal to the points of view but external to the subjects having those points of view.
3. The non-eliminable and non-reducible character of points of view, particularly their non-reducibility to psychology.
4. The possibility of reflective moves producing objectivity.
5. The search for contents that are independent from perspective.
6. The aspiration to absolute and transcendental points of view.

In other places, we have discussed 1, 2, and 3.[64] Now, let us focus on 4, 5, and 6.

5.1 Reflection and Objectivity

Reflection can be understood in various ways. One way is to understand it as the achievement of some sort of "objective" point of view about our own points of view. In that case, the contents of reflection would have to be objective, and they could be also "intersubjective". Moreover, perhaps those contents could not be objective unless they can be also intersubjective.

Many ideas of Wittgenstein's *Philosophical Investigations*, would be relevant here. His arguments against "private languages" and the nonsense of "following rules in a private way" are closely connected with that point. We can say that reflection about the rules we are following only makes sense if it is possible to get some "objectivity" about them. Moreover, according to Wittgenstein, the existence of such objectivity is possible only if the existence of an "intersubjectivity" with respect to those rules is also possible.

We have defined objective points of view in a very general way. Obviously, science tries to obtain points of view which are objective. However, we can ask, is science the "only place" from which such objectivity can be obtained? There are strong reasons for claiming it is not. The crucial point is that it is very difficult to

---

[64] With respect to 1, see the preceding sections. With respect to 2 and 3, see the previous chapter of this book.

reject that there may be "also" objectivity in ordinary knowledge (about chairs, tables, mountains, etc.), in practical knowledge (for instance, about the best and worst ways to open a can), in personal knowledge (for instance, in self-knowledge), in inter-personal knowledge (for instance, in knowledge by testimony), etc.

Furthermore, is the objectivity obtained in reflection a "scientific" objectivity? Do the objective points of view achieved in reflection have to be always a "scientific" subject matter? Can they be so at all?

There are many important problems involved in these questions. One of them, really crucial, has to do with the requirement that in reflection there has to be a true and objective conceptual content about the identity, in our case a "personal identity", between the subject who is making the reflective movement and the subject the reflection is about.

Even thought it is a conceptual content, that content is a, let us say, highly "personal" content. And it has also a very strong "indexical" character. If scientific descriptions alone cannot contain that kind of personal content, and if they cannot show that kind of indexical character, then the objective points of view achieved in reflection could not be, in the last term, a scientific subject matter.

This would be a very important conclusion. Any scientific result has to be assumed reflectively. And through that reflective move, we aim to obtain also certain objectivity. However, the objectivity we can obtain in reflection is of a kind very different from the kind of objectivity we can obtain by means of scientific knowledge.

5.2   Independence from Points of View

According to absolutism, reality has a stable way of being which is independent from points of view. According to Heraclitean relativism, there is no such stable reality. According to Protagorean relativism, there is no reality independent from points of view. Both absolutism and Heraclitean relativism are objectivist. And their definitions make use of the notion of "independence from points of view". What does that notion mean?

The notion of "independent from points of view" is equivalent to the notion of "epistemic independence" defined in other sections. In both cases, the crucial features are subjective impregnation from the attitudes of the subjects which the bearer of the point of view and subjective relativisation to a certain position of that bearer. A content is independent from points of view to the extent it has no subjective impregnation and it is not subjectively relativised.

In both cases, there is also a serious ambiguity. We will present it using the first notion, but the other case would be equivalent. Something like "independent from points of view" can be interpreted in two main senses:

1. It can mean "independent of, or apart from, all points of view".
2. Alternatively, it can mean "independent of, or apart from, any particular point of view".

The quantification is very different in 1 and 2. The second sense is entailed by the first one, but it is weaker than it. Absolutism seems to require the first sense, the stronger one. Absolutism claims that there is a stable reality independent from all points of view. The sense of independence that is rejected by Protagorean relativism seems to be also the first one. The claim that something, or even everything, is dependent on points of view is not merely a rejection of independence from points of view in the second sense.

However, the second sense of independence, the weaker one present in 2, has a great importance. Perhaps, this second sense is all we can have when we are looking for things like objectivity, intersubjectivity, and perspectival invariance. Moreover, beyond first appearances, perhaps the second sense also is enough when our aim is to obtain an absolute perspective about reality, or a transcendental perspective about our place in it.

Most of the difficulties in making full sense of absolutism and transcendentalism derive from the implausibility of adopting a perspective which can be independent from "all points of view". To adopt one such perspective amounts to something like intending to think without thinking, or like intending to say something without speaking, or like intending to do something without acting. Sometimes this is just what is suggested by some transcendentalists. And "mystic transcendentalism" would feel comfortable with such formulations. However, it is not easy to make sense of statements like these.[65]

The second sense of independence is weaker than the first one. But, as we have indicated, perhaps it is the only one that can be available for us. It is possible to consider that second sense of independence as the limiting case of the notion of "perspectival invariance". The notion of "perspectival invariance" is crucial in many contexts. We can define it as follows:

Something has perspectival invariance if it has a way of being that is invariant under changes of points of view or perspective.

Perspectival invariance comes in degrees. Some things have more perspectival invariance than others. In any case, objectivity entails perspectival invariance. And perspectival invariance "suggests" objectivity. Points of view with explicit contents having a relevant high degree of perspectival invariance can be taken (more precisely, they can be taken in some contexts) as objective points of view.[66]

Perspectival invariance has effects over our points of view about other points of view. We claimed that relativism needs the "reality" of a plurality of points of view, and not only other points of view that "seem" to exist from a particular point of

---

[65]For a discussion of this subject, see Nagel [98, 99]. See, also, the discussions of Moore [97] of the "ineffable" character of some theses of absolutism, relativism, and transcendentalism.

[66]About the relations between "invariance" and "objectivity", see Nozick [103].

view. The notion of perspectival invariance offers a sense in which it is plausible to suppose that this need is satisfied. We can say that the other points of view can be assumed as real when they are capable of displaying a relevant high degree of perspectival invariance with respect to our points of view.

In other words, perspectival invariance could make us capable of "transcending" our points of view without adopting any either absolute or transcendental position. This is a very important idea that can be used in relation to many problems.

5.3   Absolute Points of View

We have said that the notion of perspectival invariance makes room for transcendence without absolutism, in the last term it makes room for transcendence without transcendentalism. We need to define with more accuracy what is it to adopt an absolute point of view, and what is it to adopt a transcendental point of view.

The number of authors defending relativist positions in contemporary philosophy is really impressive. However, there are also many authors defending positions completely contrary to relativism. These positions can be characterised as absolutist. Absolutism entails that there is a non-perspectival stable way in which reality is in itself, and that everything else has to be eliminated or reduced, at least in principle, to that way of being.

Nowadays, it is very common to identify absolutism with "scientific realism". Science would have access to how reality is "in itself", with stability and independence from all perspectives. Moreover, any perspective or point of view would have to be eliminated or reduced to some combination of features of such reality "in itself". Curiously, the modern notion of an "absolute conception" of the world, and of ourselves, began closely linked to the Cartesian search for certainty from a reflective point of view centred in the "subject". And from such a subjectively centred reflective point of view, the crucial questions were transcendental ones: Who am I? How am I epistemically related with reality? How can I know something about the world and about myself? What do I know of the world and of myself?[67]

In order to illuminate the issue, we will introduce the following distinctions:

Absolutism/Absolute points of view
Absolute points of view/Transcendental points of view
Transcendental points of view/Transcendental, but non-conceptual, points of view
Transcendental points of view/Transcendentalism

To begin with, we can define the notion of an absolute point of view as follows:

---

[67]About "absolute" points of view, see Williams [166] and Moore [97]. About "transcendental" points of view, see Moore [97].

An <u>absolute point of view</u> is an objective point of view with the conceptual content that, independently from all points of view, reality is in a certain way.

According to our definition, absolute points of view would be objective points of view of a "conceptual" kind. Their aim is to contain a certain amount of true conceptual content. We can also say that absolute points of view would constitute the extreme case of conceptual objectivity.[68]

The sense of "independence from points of view" involved in absolute points of view has to be the strongest one. Perhaps this is not possible, but the aim of absolute points of view is to obtain epistemic independence in the strongest sense.

Absolute points of view can have a "local" scope. Hence, there may be a number of different absolute points of view, all of them having a local character. It is arguable that, in relation to them, truth can increase by "simple addition", or "direct integration".[69]

We need to distinguish between "absolutism" and "absolute points of view". Absolute points of view are a kind of objective point of view, perhaps an empty one. Absolutism is an ontological thesis about reality, perhaps a false one.

As we have said, absolutism would maintain that there is a stable way in which reality is with independence from all points of view. According to Protagorean relativism, there is no reality independent from points of view. Protagorean relativism rejects both absolutism and the existence of absolute points of view. However, Heraclitean relativism assumes the existence of absolute points of view.

Heraclitean relativism rejects the existence of a stable reality. However, this is a fact that is intended to be stated from an "absolute point of view". It is a fact that is intended to be independent from all points of view. Reality in itself is that way. Heraclitean relativism intends to be an objective claim, independent from all points of view, about how reality is. It is important to appreciate this. Even if Heraclitean relativism rejects absolutism, it does so for different reasons than Protagorean relativism. Because of that, it can be committed to the existence of a certain absolute point of view.

If we do not recognize any difference between defending absolutism and defending the existence of absolute points of view, we could not appreciate what is peculiar in a position like that of Heraclitus. Absolutism needs an absolute point of view. But, it is possible to maintain the existence of absolute points of view and, at the same time, to reject absolutism.

---

[68]As we said, the notion of an "absolute conception" of the world, and of ourselves as part of it, comes from Williams [166, 167], and has one on its main sources in Descartes. It has been recently analysed and vindicated by Moore (1987, [97], and criticised by Nielsen [101] and Putnam [110, 111, 113, 114, 115, 116].

[69]See again Moore [97]. We can also say that, in those cases, truth works as an "extensive" measure. In non-absolute (relative) points of view, truth would work as an "intensive" measure.

5.4   Transcendental Points of View

Are absolute points of view possible? Are some of them true? Even if it is not possible to obtain true conceptual contents such as those required for absolute points of view, the structure of absolute points of view is important. In many cases, they involve a peculiar transcendental move. They include, or at least they are associated with, a certain point of view about how we are, in the last instance, i.e., with independence from all points of view, epistemically connected to reality.

That combination can be called the "transcendental mood".

The transcendental mood is a combination of 1) the aim to obtain an absolute point of view involving true conceptual contents with 2) a transcendental point of view about how we are epistemically connected to reality with independence from all points of view.

It is possible to adopt the transcendental mood not only with respect to our epistemic connections to reality, but in many other fields. For instance, we can adopt the transcendental mood with respect to how we are, with independence from all points of view, practically connected to reality, or morally connected to reality, or aesthetically connected to reality, or religiously connected to reality, etc.

According to our characterisation, absolute points of view have a conceptual character. Many times, the transcendental move involved in them also is conceptual. However, whereas the target of absolute points of view is the whole of reality, the target of transcendental points of view is only certain parts of reality. More precisely, we can define transcendental points of view in the following way:

A transcendental point of view is an absolute point of view about that part of reality constituted by how we are epistemically (or practically, or morally, or aesthetically, or religiously, etc.) connected to reality.

Also, we can characterise transcendentalism as follows:

Transcendentalism maintains absolutism with respect to how we are epistemically (or practically, or morally, or aesthetically, or religiously, etc.) connected to reality.

Transcendentalism is an ontological thesis about some peculiar parts of reality. Again, as in absolutism, the sense of the perspectival independence involved is in principle the strongest one. It do not seem to be enough to interpret that independence as "independence from any particular point of view".

The distinction between transcendentalism and transcendental points of view runs in parallel with the distinction between absolutism and absolute points of view. Transcendentalism is an ontological thesis about our epistemic position (or practical position, or moral position, etc.), perhaps a false thesis. Transcendental points of view are a kind of absolute, and so objective, points of view, perhaps an empty one. To adopt absolutism entails the adoption of an absolute point of view, and to adopt transcendentalism entails the adoption of a transcendental point of view. Moreover, in the same Heraclitean sense in which one can reject absolutism while adopting an absolute point of view, one could reject transcendentalism while adopting a transcendental point of view.[70]

There is a modification we need to introduce. Even if it does not make clear sense to claim that there may be absolute points of view of a non-conceptual character (perhaps it can have only a "mystical" sense), it makes sense to claim that there may be transcendental "but non-conceptual" points of view about certain peculiar parts of reality. Reality as a whole is not a possible non-conceptual content. However, our epistemic relations with reality (also our practical relations, our moral relations, etc.), or a part of them, can be non-conceptual contents of some points of view. Simply, we can experience them.

Let us introduce the following notion:

A transcendental, but non-conceptual, point of view is an objective point of view having as non-conceptual content the way we are epistemically (or practically, or morally, etc.) related with reality independently from all points of view.

Some transcendental points of view can be understood as non-conceptual ones, and they have been so understood many times. According to that, we can consider that transcendentalism entails the adoption of some transcendental points of view, either conceptual or non-conceptual. And we have to admit also the possibility of a rejection of transcendentalism from a transcendental but non-conceptual point of view.

The blend of absolute and transcendental points of view that we have called the "transcendental mood" also can involve transcendental but non-conceptual points of view. This is not so when the transcendental mood comes from "scientific knowledge and naturalization".[71] But it is typically so in many cases in which the transcendental mood is purely philosophical.

In absolute points of view, there is always a strong emphasis on conceptual truth. Certain semantically evaluable conceptual contents are intended to be true. Moreover, it is a widespread idea that among absolute points of view, truth would increase by "simple addition", or by "direct integration". This feature suggests a strong "conceptual" character for those points of view. This is especially clear when science is taken as the paradigmatic example of an absolute conception.[72] In contrast, many transcendental points of view do not look for the truth in that conceptual sense. They try to "grasp" our epistemic relations with reality in other ways. Sometimes, they look for something like a very special "intuition", or

---

[70]The classical locus for what we are calling the "transcendental mood" in epistemology is Plato's criticism of the Sophists rejection that things have a way of being in themselves. Such transcendentalism offers an absolute, non-perspectival ontological position about our epistemic relation with reality.

[71]Our definitions of absolute points of view, transcendental points of view (of a conceptual kind), absolutism and transcendentalism fit very well with an exclusivist scientific realism involving projects for naturalizing epistemology, ethics, etc. The peculiar transcendental mood that we find here can be called "scientificism".

[72]About that, see again Williams [166], Moore [97] and Nozick [103].

"vision", or "grasping". Other times, they look for something that only can be "shown", but not "said", or for something only manifest `in our practices'. In any case, many times they look for something non-conceptual.[73]

5.5 To Transcend Our Points of View Without Adopting Any Transcendental Mood

Let us discuss briefly a last, but very important, question. It has been mentioned in several places of other sections. Relativism needs a real plurality of points of view in conflict. Perspectivism also needs to affirm the real existence of different points of view. Both positions entail that there is more than just our own perspective. To claim the real existence of different perspectives entails to "transcend" our points of view. But, how to transcend our "own" points of view without being involved in the transcendental mood?

We will answer by emphasising a very important claim: we can do so in one of the two senses distinguished above for the expression "with independence from points of view". The second and weaker sense, i.e., independence from any particular point of view, offers a way in which we can transcend our own points of view without assuming a transcendental point of view.

How to obtain such independence from any particular point of view? The notion of "perspectival invariance" is here crucial. We have said that the notion of perspectival invariance makes room for transcendence without absolutism. In the last term, that notion makes room for transcendence without transcendentalism. Perspectival invariance could make us capable of transcending our particular points of view without embracing any transcendental mood.

It is plausible to argue that, in a strict sense, we cannot take a transcendental point of view of any sort. It is difficult to see how the contents of transcendental points of view could be independent from all points of view. Also, it is doubtful that transcendentalism can make sense. In general, it is much more easy to understand reality apart from us than our relations with reality.[74] However, to transcend our particular points of view by trying to obtain "perspectival invariance" has a very clear and useful sense.

---

[73]Plato, for instance in his *Cratilus*, develops the first option. The second one is one of he main topics of Wittgenstein in his *Tractatus*, one of the more important works in the transcendentalist tradition.

[74]About that, see Stroud [155]. According to him, the main claim of radical scepticism is that a full understanding of the whole of reality is simply non possible.

6 Subjective and Objective Aspects of Time

We can apply our analyses to the problem of understanding time. The place of time in reality has been shown to be highly unstable. It does not seem to be completely adequate to place time in reality "in itself", nor does it seem to be completely adequate to place time in our "pure subjectivity".

Let us begin by distinguishing between "time" and "temporal points of view". Time is intended to be something of reality, perhaps a feature, more or less constitutive, or a dimension, or a frame, etc. Temporal points of view are a kind of points of view. The real existence of time has been repeatedly questioned, in particular the real existence of a "fluent time" involving a past, a present, and a future.[75] However, it is not so easy to question the existence of temporal points of view. We adopt them "time after time". This suggests that, perhaps, the real existence of a fluent time is closely linked to the real existence of temporal points of view (in the last term, to the real existence of points of view).

Let us say understand temporal points of view as follows:

A temporal point of view is a point of view identifying some differences in non-conceptual contents (qualitative, phenomenal, experiential contents) as "changes" of content.

The identification can be either conceptual or not conceptual. This is a very important point. Subjects without conceptual capacities could be capable of adopting temporal points of view. In any case, in a temporal point of view certain differences in non-conceptual content count as a "change": something future becoming present, or something present becoming past.

The idea behind that characterisation of temporal points of view is very simple. Temporal points of view take some differences in the non-conceptual contents of experience as being temporal differences entailing a "change". This is the crucial point.

Of course, temporal points of view also can take a content as having a "permanence" in time, i.e., as something continuing from the past to the present, or from the present to the future. However, we can consider that the identification of permanences in time is dependent on the possibility of identifying changes. In other words, to identify a permanence is to identify possible but not actual changes.

Now, let us explore briefly how temporal points of view so understood can be projected onto the analyses and distinctions we have been making in previous sections.

We began by distinguishing between "subjective" and "objective" points of view. Our distinction was relative to the explicit contents of the points of view. The

---

[75]Apart from McTaggart, we have in Mellor [94], one of the most elaborated rejections of the real existence of a "fluent time". In other chapters of this book, these issues will be discussed in depth.

explicit contents of subjective points of view are such that either 1) there is a determinant subjective impregnation coming from the non-conceptual, qualitative, phenomenal, experiential features of the attitudes of the subject of the point of view, or 2) the conceptual contents are constitutively relative to the position or emplacement of the bearer of the point of view. Are temporal points of view, such as we are understanding them, subjective in either of these two senses?

The subjective relativisation in the conceptualisation of something either as being in the future, or as being in the present, or as being in the past, seems to be unavoidable. Some reference, either direct or indirect, to the bearer of the temporal point of view is always necessary. What about subjective impregnation? Even though it can be claimed that there is not always a determinant subjective impregnation of the temporal contents involved, so that the different temporal positions are independent from the qualitative features of our attitudes, in many cases there is in fact such a subjective impregnation.

We can leave open the question whether there may be temporal points of view without a subjective relativisation and without a subjective impregnation. Perhaps this can be possible in some temporal points of view. In any case, let us suppose that temporal points of view are subjective points of view. Even in that case, this would not entail that the "fluent time" we can find in temporal points of view is merely internal to the subjects having those temporal points of view. Even if fluent time is "internal to temporal points of view", it can be "external to the subjects" that are the bearers of these temporal points of view.[76]

This is a very important result. The contents of subjective points of view are subjective in the sense of having a certain amount of subjective impregnation or being to a certain extent subjectively relative. However, nothing of that entails that those contents are internal to the subjects which are the bearers of those points of view. Those contents are internal to the points of view, but are not internal to the subjects adopting those points of view. Subjective points of view are not internal to the subjects having them. They are no more internal to the subjects than objective points of view are. With respect to this issue, it does not matter whether temporal points of view are objective or subjective. In any case, the "fluent time" we can find in them is not reducible to properties and conditions merely internal to the subject individualistically considered.

In our analyses, we also distinguished between "private" and "intersubjective" points of view. As we have said, the contents of temporal points of view are not internal to the subjects having those temporal points of view. But, would the subjective character of temporal points of view mean that they are private points of view, i.e., completely "idiosyncratic" ones? It does not either. If they were private,

---

[76]In Russell [134], we can find a clear case of a relational time which is internal to a construed "space of perspectives" without being merely internal to the subjects from which that space of perspectives is construed. The construction of a Russellian space of perspectives is explained in other chapters of this book.

then no "rational control" over their contents would be possible. However, as a matter of fact, we are capable of having rational control over the contents of our temporal points of view. We use all sorts of clocks to "coordinate" our behaviours.[77]

A simple explanation of that fact would be that temporal points of view can be "intersubjective". Even though temporal points of view are subjective, they are not merely internal to the subjects. Moreover, they can be intersubjective ones. This entails that a number of "different subjects" can make the same temporal identifications, or at least that they can make similar temporal identifications.

This would be also a very important result. It entails the existence of some "shared temporal contents". It is not only that the experienced fluent time is not something merely internal to the subjects that are adopting a certain temporal point of view. That fluent time can be a "shared fluent time". The fluent time I am experiencing may be subjective. However, it is not simply "inside me". And it can be "the same" fluent time than the fluent time you are experiencing, or very similar to it.

We established a sharp distinction among three ontological positions: "absolutism", "relativism", and "perspectivism". Which ontological position would be the adequate one with respect to the fluent time we can find in our temporal points of view? The existence of a fluent time seems to be something relative to the existence of temporal points of view. Fluent time is internal to some peculiar kinds of points of view, namely temporal points of view. That way, relativism appears to be the adequate position with respect to a fluent time. However, things are more complicated when we look to temporal points of view.

The existence of temporal points of view appears not to be relative to any point of view. Can it be argued that temporal points of view exist in reality in an absolute sense? At least, they do not exist only "inside other points of view". This would lead to a regression. It would be a particular case of the general kind of regression that we would have to face if we claim that points of view only exist from some point of view. From this perspective, absolutism would seem to be the adequate position to adopt. We can conclude saying that Temporal points of view seem to be subjective points of view existing in an absolute way.

According to their contents, temporal points of view seem to be subjective points of view. Even though there were no determination coming from a subjective impregnation of the qualitative features of attitudes, their contents would always be strongly relative to the positions or emplacements of the bearers of the temporal points of view in non-reducible ways. However, the existence of temporal points of view does not seem to be relative to further points of view. Temporal points of view seem to exist in a completely absolute sense.

Moreover, when temporal points of view become contents of other points of view, for instance when we think reflectively about them, these second points of view seem to be capable of being fully objective exactly in the same sense in which

---

[77]And this is so with independence of all the problems about the possibility of "simultaneity" in relation to physical time,.

we can adopt a fully objective point of view about, let us suppose, our subjective points of view in matters of taste. We can convert, for instance, the subjective point of view of subject S with the content

1. Avocados are delicious

into an objective point of view with the content

2. For S, avocados are delicious.

So, there are reasons for relativism with respect to fluent time. And there are some reasons for absolutism with respect to temporal points of view. However, there are also reasons for relativism concerning temporal points of view themselves. Let us see with more detail this important issue.

A reflective point of view about temporal points of view would have to include, or to make reference to, the temporal contents of those temporal points of view. And it is doubtful that this can be done without inheriting the subjective relativisation of those temporal contents. The difference from the case of the avocados, and from other similar cases, is clear. Whereas it seems plausible to claim that we can make reference to the fact that S considers that avocados are delicious without any relativisation, even to describe that fact in fully objective terms, it is not easy to claim that we can do the same thing with respect to temporal contents such as, for instance, "It is raining now".

There is a deep tension here. To say 2, above, is very different from saying something like

3. For S, it is raining now

In 2, we can leave behind all the subjective relativity present in 1. We can think of many kinds of "objective relations" between S and the avocados capable of doing the work. But it is not clear that we can get the same in 3.

The crucial problem is that the content expressed in 3 cannot be completely identified without knowing "when" it is raining. And "when" it is raining cannot be known except by "sharing" the temporal point of view of S.

There are many ways of sharing a point of view: we can directly ask, or we can observe a certain behaviour, or we can have testimony from others, etc. In any case, it is one thing to "share" a point of view, and another very different thing to have an "objective" perspective over that point of view. And the problem with the contents of temporal points of view is that we can only "share" them.

How could we maintain the absolutist demands regarding the existence of temporal points of view? Would we have to maintain a transcendental point of view about the existence of temporal points of view? Would we have to maintain, at least, a transcendental but not conceptual, point of view? The answer is negative.

According to our definitions, absolute points of view would have to be objective points of view. This entails an obstacle in relation to temporal points of view. As we have said, perhaps we cannot maintain reflectively any objective conceptual point of view about our temporal points of view. However, something "close to absolutism" can be maintained from the inside of some reflective perspectives.

The notion of a reality being in a certain stable way, independently from "any particular point of view", has a very important role in many of our points of view. This is also applicable to the existence of temporal points of view. From the inside of some reflective perspectives, the existence of temporal points of view can be seen "very close to having an absolute sense".

The possibility that temporal points of view can be "intersubjective", so that we can have some "sharable temporal contents" in perspective, offers a clue.

It is plausible to think that if temporal points of view can be intersubjective, their contents can have a relevant perspectival invariance. At least, those temporal contents have to be invariant in relation to some idiosyncratic features of each particular temporal point of view. And perspectival invariance offers important reasons for absolutism. Even though the sense of independence from points of view involved in perspectival invariance is only the weak one (i.e., independence from any particular point of view), not the strong one (i.e., independence from all points of view), achieving that weak sense offers very sound reasons for thinking that temporal points of view are a feature of reality over which we can maintain an absolutist position.

Let us summarise our main results. With respect to temporal contents, a fluent course of things and events in time, we have claimed that even though they are internal to some temporal points of view, a peculiar kind of point of view, they are not merely internal to the subjects adopting those temporal points of view. Moreover, we have argued that temporal points of view can be intersubjective, and so temporal contents can be "sharable".

With respect to temporal points of view, our position has been placed very close to absolutism. Temporal points of view cannot be a merely subjective epiphenomenon. They have to be included in a complete ontological picture of reality such as it is with independence from any particular point of view. At least in that sense of "independence from points of view", temporal points of view are not ontologically dispensable.

All of that would be to "transcend" our particular temporal contents and our own temporal points of view. But it would be to transcend our temporal condition without adopting any "transcendental point of view".

Perhaps we are capable of having transcendental points of view of a "non-conceptual kind" about the existence of a fluent time, independent of all our points of view, and about the existence of temporal points of view. Perhaps this is possible through a very special sort of "intuition" ("vision", "grasping", etc.) of our temporal relations with reality, or through something that only can be "shown" but not "said", or through something having to do with "agency" and action". However, we can leave this question open. We can get a great amount of intersubjectivity for temporal contents and a great amount of objectivity for the existence of temporal points of view even though we reject that possibility.

Our strategy would make it possible to transcend both our particular temporal contents and our own temporal points of view without adopting any "transcendental mood". Intersubjective temporal points of view, displaying certain sorts of relevant perspectival invariances, make possible to share a fluent time (more precisely, a fluent world of things and events in the past, the present, and the future). Also, they point at absolutism about the existence of temporal points of view. But they do all of that without embracing any "claim", or any "intuition", about the non-perspectival existence of fluent time, or about the non-perspectival place of temporal points of view in reality.

In conclusion, we could defend both relativism about the existence of fluent time and something "very close to absolutism" about the existence of temporal points of view. Fluent time would be internal to some temporal points of view, without being internal to the subjects having those temporal points of view, and these temporal points of view could exist in reality in quite an absolute sense, or at least in a very objective sense.

This means that, in general, we could be "perspectivists" with respect to time: we can be relativists in relation to some temporal phenomena (the existence of a "fluent time") and we can be absolutists, or something close enough, in relation to other temporal phenomena (the existence of "temporal points of view").

That temporal perspectivism would not be but a particular case of the general kind of perspectivism we can hold in many other fields. Furthermore, many of the perspectivist ideas we have maintained can be found in Russell's notion of a "space of perspectives".[78] According to Russell, the contents of our experience (he calls them "sensibilia") have a place in a space construed through their relations: similarities, differences, groupings in classes, existence of series with a limit, etc. These relations make possible to define ordinary things, physical objects, matter, time, etc. In that space, our perspectives occupy also a position. They are like "points" in that space. And each perspective defines a different "space of experience". From these spaces of experience, we construe a space of perspectives. In a certain sense, that space of perspectives is only a "virtual" space. However, it can also have a very important sort of "intersubjectivity" and "objectivity". For Russell, there is integration, not opposition, between the spaces of our experiences and the space of perspectives. The physical world is construed out of that space.[79]

Any scientific result has to be assumed reflectively. The same is true of any intuitive or ordinary statement. Through all these reflective moves, we aim to obtain some intersubjectivity and objectivity. However, the intersubjectivity and objectivity we obtain reflectively is always very different from the kind of intersubjectivity

---

[78]See Russell [134]. A very important reference for Russell's approach was Leibniz's *Monadology*. Leibniz was also an important reference for the perspectivist position of the Spanish philosopher Ortega y Gasset.

[79]The details of Russell's construction of a "space of perspectives" are explained in other chapters of this book. Russell claims that the spaces of experience are "private". This is consistent with his insistence in that the constructions offered (of a space of perspectives, of a physical space and a physical time, of ordinary things and physical objects, of matter, etc.) could be made from a solipsist basis. According to our definitions, however, a "private point of view" could not be intersubjective. So, to the extent that spaces of experience can be intersubjective, and this can put us on the track of objectivity, they could not be "purely private".

and objectivity we obtain by means of scientific knowledge. The intersubjectivity we have claimed for temporal contents and the objectivity we have claimed for the existence of temporal points of view have its sources in reflection.

References

1. Álvarez, S. (2014). Causation and the Agent's Point of View. *International Journal for Theory, History and Foundations of Science, 29*(79), 133–47.
2. Anderson, R. (1998). Truth and objectivity in perspectivism. *Synthese, 115*(1), 1–32.
3. Armstrong, D. (1973). *Belief, truth, and knowledge.* Cambridge: Cambridge University Press.
4. Ashman, K., & Barringer, Ph (Eds.). (2001). *After the science wars.* London: Routledge.
5. Barnes, B., & Bloor, D. (1982). Relativism, rationalism, and the sociology of knowledge. In M. Hollis & S. Lukes (Eds.), *Rationality and relativism* (pp. 21–47). Cambridge: MIT Press.
6. Bender, J. (1989). *The current state of the coherence theory.* Dordrecht: Kluwer.
7. Berger, P., & Luckmann, T. (1966). *The social construction of reality: A treatise in the sociology of knowledge.* New York: Anchor Books.
8. Bloor, D. (1992). *Knowledge and social imaginary* (2nd ed.). Chicago: University of Chicago Press.
9. Boghossian, P. (1989). Content and self-knowledge in philosophy of mind. *Philosophical Topics, 17*(1), 5–26.
10. Boghossian, P. (2006). *Fear of knowledge. Against relativism and constructivism.* Oxford: Clarendon Press.
11. Bonjour, L. (1985). *The structure of empirical knowledge.* Cambridge: Harvard University Press.
12. Brandom, R. (1994). *Making it explicit.* Cambridge: Harvard University Press.
13. Brandom, R. (2002). *Articulating reasons.* Cambridge: Harvard University Press.
14. Brown, J. (2001). *Who rules in science? An opinionated guide to the wars.* Cambridge: Harvard University Press.
15. Burge, T. (1979). Individualism and the mental. *Midwest Studies in Philosophy, 4*(1), 73–121.
16. Burge, T. (1988). Individualism and selfknowledge. *The Journal of Philosophy, 85,* 649–663.
17. Burge, T. (1989). Individualism and psychology. *Philosophical Studies, 40,* 39–75.
18. Callon, M. (1999). Whose impostures? Physicists at war with the third person. *Social Studies of Science, 29*(2), 261–286.
19. Cappelen, H., & Hawthorne, J. (2009). *Relativism and monadic truth.* Oxford: Oxford University Press.
20. Carnap, R. (1950). Empiricism, semantics, and ontology. *Revue Internationale de Philosophie, 4,* 20–40.
21. Cassam, Q. (1989). Reductionism and first-person thinking. In D. Charles & K. Lennon (Eds.), *Reductionism, explanation and realism* (pp. 361–379). Oxford: Clarendon Press (1992).
22. Castañeda, H. (1999). *The phenomeno-logic of the I. Essays in self-consciousness.* In J. Hart & T. Kapitan (Eds.) Bloomington: Indiana University Press.
23. Chalmers, D. (1996). *The conscious mind. In search of a fundamental theory.* New York: Oxford University Press.
24. Cherniak, Ch. (1992). *Minimal rationality.* Cambridge: MIT Press.
25. Chisholm, R. (1976). *Person and object.* Chicago: Open Court.
26. Chisholm, R. (1966). *Theory of Knowledge.* Englewoods Cliffs: Prentice Hall [3rd. ed., 1988].

23. Churchland, Patricia. (1986). *Neurophilosophy: Toward a unified science of the mind/brain*. Cambridge: MIT Press.
28. Churchland, Paul. (1988). *Matter and consciousness*. Cambridge: MIT Press.
29. Clark, M. (1990). *Nietzsche on truth and philosophy*. Cambridge: Cambridge University Press.
30. Collins, H. (1981). Stages in the empirical program of relativism. *Social Studies of Science, 11*, 3–10.
31. Conant, J. (2005). The dialectic of perspectivism, 1. *Nordic Journal of Philosophy, 6*, 5–50.
32. Conant, J. (2006). The dialectic of perspectivism, 2. *Nordic Journal of Philosophy, 7*, 6–57.
33. Davidson, D. (2001). *Subjective, intersubjective, objective*. Oxford: Oxford University Press.
34. Davidson, D. (1963). Actions, reasons and causes. *Journal of Philosophy* 60: 685–700 [Also in Davidson, D. (1980). *Essays on actions and events*. Oxford: Clarendon Press].
35. Davidson, D. (1974). On the very idea of a conceptual scheme. *Proceedings and Addresses of the American Philosophical Association* 47: 5–20 [Also in Davidson, D. (1984). *Inquiries into truth and interpretation*. Oxford: Clarendon Press].
36. Dennett, D. (1987). *The intentional stance*. Cambridge: MIT Press.
37. Dennett, D. (1991). *Consciousness explained*. Boston: Little Brown.
38. Dreyfus, H. (1972). *What computers (still) can't do: A critique of artificial reason* (6th ed.). Cambridge: MIT Press (1999).
39. Farkas, K. (2008). *The subject's point of view*. Oxford: Oxford University Press.
40. Feyerabend, P. (1975). *Against method: Outline of an anarchist theory of knowledge, (1993* (3rd ed.). London: Verso.
41. Foucault, M. (1980). *Power/knowledge: Selected interviews and other writings, 1972–1977*. New York: Pantheon Books.
42. Foucault, M. (1969). *The archeology of knowledge*. New York: Pantheon Books (1972).
43. Frankfurt, H. (2005). *On bullshit*. Princeton: Princeon University Press.
44. Frankfurt, H. (2005). *On truth*. Nueva York: Alfred A. Knopf.
45. García-Carpintero, M., & Kölbel, M. (Eds.). (2009). *Relative truth*. Oxford: Oxford University Press.
46. Giere, R. (2006). *Scientific perspectivism*. Chicago: University of Chicago Press.
47. Goldman, A. (1986). *Epistemology and cognition*. Cambridge: Cambridge University Press.
48. Gomila, A. (2010). Musical expression and the second person perspective. In I. Álvarez, F. Pérez-Carreño, & H. Pérez (Eds.), *The expression of subjectivity in the performing arts*. Cambridge: Cambridge Scholars Publishing.
49. Goodman, N. (1954). *Fact, fiction and forecast*. Indianapolis: Hackett.
50. Goodman, N. (1978). *Ways of worldmaking*. Indianapolis: Hackett.
51. Goodman, N. (1960). The way the world is. *Review of Metaphysics, 14*, 48–56 [Also in Goodman, N. (1972). *Problems and projects*. New York: Bobbs-Merrill].
52. Greco, J. (Ed.). (2004). *Ernst sosa and his critics*. London: Blackwell.
53. Greco, J. (Ed.). (2011). *Virtue epistemology. Stanford encyclopedia of philosophy*. http://plato.stanford.edu.
54. Gross, P., & Levitt, N. (1994). *Higher supersticion: The academic left and its quarrels with science*. Baltimore: Johns Hopkins University.
55. Gumperz, J., & Levinson, S. (Eds.). (1996). *Rethinking linguistic relativity*. Cambridge: Cambridge University Press.
56. Haack, S. (1996). Reflections on relativism: From momentous tautology to seductive contradiction. *Philosophical Perspectives, 10*, 297–315.
57. Hales, S. (2006). *Relativism and the foundations of philosophy*. Cambridge: MIT Press.
58. Hales, S. (Ed.). (2011). *A companion to relativism*. Oxford: Wiley-Blackwell.
59. Hales, S., & Welshon, R. (2000). *Nietzsche's perspectivism*. Illinois: University of Illinois Press.
60. Hamlyn, D. (1961). *Sensation and perception. A history of the philosophy of perception*. London: Routledge.
61. Hardin, S. (1986). *The science question in feminism*. Milton Keynes: Open University Press.

62. Hardin, S. (1991). *Whose science? Whose knowledge? Thinking from women's lives*. Milton Keynes: Open University Press.
63. Harding, S. (1998). *Is science multicultural? Postcolonialism, feminism and epistemology*. Bloomington: Indiana University Press.
64. Harman, G. (1975). Moral relativism defended. *Philosophical Review, 84*, 3–22.
65. Honderich, T. (Ed.). (1985). *Morality and objectivity*. Londres: Routledge and Kegan Paul.
66. Hurley, S. (1998). *Consciousness in action*. Cambridge: Harvard University Press.
67. Jackson, F. (1986). What mary didn't know. *The Journal of Philosophy, 83*, 291–295.
68. James, W. (1907). *Pragmatism: A new name for some old ways of thinking*. Indianapolis: Hackett.
69. Knorr-Cetina, K. (1983). The ethnographic study of scientific work: Towards a constructivist interpretation of science. In K. D. Knorr-Cetina & M. Mulkay (Eds.), *Science observed* (pp. 189–204). Hollywood: Sage.
70. Krausz, M., & Meiland, J. (Eds.). (1982). *Relativism: Cognitive and moral*. Notre Dame: University of Notre Dame Press.
71. Kuhn, T. (1970) *The Structure of Scientific Revolutions*.Chicago: University of Chicago Press.
72. Kölbel, M. (2002). *Truth without objectivity*. London: Routledge.
73. Kölbel, M. (2003). Faultless disagreement. *Proceedings of the Aristotelian Society, 104*, 53–73.
74. Kölbel, M. (2004). Indexical relativism versus genuine relativism. *International Journal of Philosophical Studies, 12*(3), 297–313.
75. Labinger, J., & Collins, H. (Eds.). (2001). *The one culture? A conversation about science*. Chicago: University of Chicago Press.
76. Lakatos, I., & Musgrave, A. (Eds.). (1970). *Criticism and the growth of knowledge*. Cambridge: Cambridge University Press.
77. Lasersohn, P. (2005). Context dependence, disagreement, and predicates of personal taste. *Linguistic and Philosophy, 28*, 643–686.
78. Lasersohn, P. (2009). Relative truth, speaker commitment, and control of implicit arguments. *Synthese, 166*, 359–374.
79. Latour, B., & Woolgar, S. (1979). *Laboratory life: The social construction of scientific facts*. Beverly Hill: Sage.
80. Lehrer, K. (1974). *Knowledge*. Oxford: Clarendon.
81. Lehrer, K. (1986). The coherence theory of knowledge. *Philosophical Topics, 14*(1), 5–25.
82. Levine, J. (1983). Materialism and qualia: The explanatory gap. *Pacific Philosophical Quarterly, 64*, 354–361.
83. Lewis, D. (1979). Attitudes De Dicto and De Se. *Philosophical Review*, (88), 513-43.
84. Liz, M. (2003). Intentional states: Individuation, explanation, and supervenience. In M. J. Fapolli & E. Romero (Eds.), *Meaning, basic knowledge, and mind* (pp. 129–223). Chicago: University of Chicago Press.
85. Lyotard, J. (1979). *The postmodern condition. A report on knowledge*. Minneapolis: Minnesota University Press.
86. López de Sa, D. (2009). Presuppositions of commonality: An indexical relativist aconceptual contentsount of disagreement. In M. García-Carpintero & M. Kölbel (Eds.), *Relative truth* (pp. 297–310). Oxford: Oxford University Press.
87. MacFarlane, J. (2003). Future contingents and relative truth. *The Philosophical Quarterly, 53*, 321–336.
88. MacFarlane, J. (2005). Making sense of relative truth. *Proceedings of the Aristotelian Society, 105*, 321–336.
89. MacFarlane, J. (2007). Relativism and disagreement. *Philosophical Studies, 132*, 17–31.
90. Mackie, J. (1977). *Ethics: Inventing right and wrong*. Harmondsworth: Penguin.
91. Malotki, E. (1983). *Hopi time: A linguistic analysis of the temporal concepts in the hopi language*. Berlin: Mouton.
92. McDowell, J. (1998). *Mind, Value, and Reality*. Cambridge: Harvard University Press.

93. McGinn, C. (1983). *The subjective view: Secondary qualities and indexical thoughts.* Oxford: Oxford University Press.
94. Mellor, D. (1998). *Real time II.* London: Routledge.
95. Mellor, D. (1991). I and now. In *Matters of metaphysics* (pp. 17–29). Cambridge: Cambridge University Press.
96. Menzies, P., & Price, H. (1993). Causation as a secondary quality. *The British Journal for the Philosophy of Science, 44,* 187–203.
97. Moore, A. (1997). *Points of view.* Oxford: Oxford University Press.
98. Nagel, T. (1986). *The view from nowhere.* Oxford: Oxford University Press.
99. Nagel, T. (1997). *The last word.* Oxford: Oxford University Press.
100. Nagel, T. (1974). What is it like to be a vat? *Philosophical Review, 83,* 435–450 [Also in Nagel, T. (1979). *Mortal questions.* Cambridge: Cambridge University Press].
101. Nielsen, K. (1993). Perspectivism and the absolute conception of the world. *Crítica XXV, 74,* 105–116.
102. Nozick, R. (1981). *Philosophical explanations.* Cambidge: Harvard University Press.
103. Nozick, R. (2001). *Invariances. The structure of the objective world.* Cambridge: Harvard University Press.
104. Noë, A. (2002). Is perspectival self-consciousness nonconceptual? *The Philosophical Quarterly, 52*(207), 185–194.
105. Noë, A. (2004). *Action in perception.* Cambridge: MIT Press.
106. Parsons, K. (Ed.). (2003). *The science wars: Debating scientific knowledge and technology.* New York: Prometheus Books.
107. Perry, J. (1979). The problem of the essential indexical. *Noûs, 13,* 3–21 [Also in *The Problem of the Essential Indexical and other Essays,* 1993: 33–50. Oxford: Oxford University Press].
108. Preyer, G., & Peter, G. (Eds.). (2005). *Contextualism in philosophy: Knowledge, meaning and truth.* Oxford: Oxford University Press.
109. Price, H. (2007). Causal perspectivalism. In H. Price & R. Corry (Eds.), *Causation, physics and the constitution of reality,* (pp. 225–292). Oxford: Clarendon Press.
110. Putnam, H. (1981). *Reason, truth and history.* Cambridge: Cambridge University Press.
111. Putnam, H. (1987). *The many faces of realism.* La Salle: Open Court.
112. Putnam, H. (1992). *Renewing philosophy.* Cambridge: Harvard University Press.
113. Putnam, H. (1994). *Words and life.* Cambridge: Cambridge University Press.
114. Putnam, H. (1994). *Pragmatism.* Oxford: Blackwell.
115. Putnam, H. (1994). *Realism with a human face.* Cambridge: Harvard University Press.
116. Putnam, H. (1999). *The threefold cord mind, body and world.* Nueva York: Columbia University Press.
117. Putnam, H. (1975). The meaning of "meaning". In K. Gunderson (Ed.), *Language, mind, and knowledge* (pp. 131–193). Minnesota: University of Minnesota Press.
118. Putnam, H. (1982). Why reason can't be naturalized. *Synthese 51* [Also in *Realism and Reason* (pp. 3–24). Cambridge: Cambridge University Press].
119. Quine, W. (1960). *Word and object.* Cambridge: MIT Press.
120. Quine, W. (1969). *Ontological relativity and other essays.* Nueva York: Columbia University Press.
121. Quinton, A. (1966). The foundations of knowledge. In B. Williams & A. Montefiore (Eds.), *British analytical philosophy.* London: Routledge.
122. Rawls, J. (1971). *A theory of justice.* Cambridge: Harvard University Press.
123. Recanati F. (2007). *Perspectival thoughts. A plea for (Moderate) relativism.* Oxford: Oxford University Press.
124. Rescher, N. (1973). *The coherence theory of truth.* Oxford: Clarendon Press.
125. Rescher, N. (1973). *The primacy of practice.* Oxford: Basil Blackwell.
126. Rescher, N. (1988). *Rationality.* Oxford: Oxford University Press.
127. Richard, M. (2004). Contextualism and relativism. *Philosophical Studies, 119,* 215–242.
128. Rorty, R. (1979). *Philosophy and the mirror of nature.* Princeton: Princeton University Press.

129. Rorty, R. (1982). *Consequences of pragmatism*. Minneapolis: University of Minnesota Press.
130. Rorty, R. (1989). *Contingency, irony, and solidarity*. Cambridge: Cambrdge University Press.
131. Rorty, R. (1991). *Objectivity, relativism, and truth*. Cambridge: Cambridge University Press.
132. Rosenberg, J. (2007). *Wilfrid sellars: Fusing the images*. Oxford: Oxford University Press.
133. Russell, B. (1912). *Problems of philosophy* (p. 1968). London: Oxford University Press.
134. Russell, B. (1918). The relation of sense data to physics. In *Mysticism and logic, and other essays* (pp. 113–140). London: George Allen & Unwin.
135. Ryle, G. (1949). *The concept of mind*. Chicago: Chicago University Press (1963).
136. Searle, J. (1980). Minds, brains, and programs. *Behavioral and Brain Sciences, 3*, 417–457.
137. Searle, J. (1985). *Minds, brains and science*. Cambridge: Harvard University Press.
138. Searle, J. (1992). *The rediscovery of mind*. Cambridge: Cambridge University Press.
139. Searle, J. (1995). *The construction of social reality*. New York: Free Press.
140. Sellars, W. (1956). Empiricism and the philosophy of the mental. In *Empiricism and the philosophy of the mental (introduction by Richard Rorty and study guide by Robert Brandom)*. Cambridge: Harvard University Press (1997).
141. Sellars, W. (1962). Philosophy and the scientific image of the man. In *Science, perception and reality* (pp. 1–40). Atascadero: Ridgeview Publishing Company (1963).
142. Shoemaker, S. (1996). *The first-person perspective and other essays*. Cambridge: Cambridge University Press.
143. Siegel, H. (1987). *Relativism refuted: A critique of contemporary epistemological relativism*. Dordrecht: Reidel.
144. Simon, H. (1957). *Models of man*. New York: Wiley.
145. Simon, H. (1969). *The sciences of the artificial*. Cambridge: MIT Press.
146. Snow, C. P. (2007). *The two cultures (the 1959's work, the 1964's suplement, and later revisions)*. Cambridge: Cambridge University Press.
147. Sokal, A. (2008). *Beyond the hoax: Science, philosophy and culture*. Oxford: Oxford University Press.
148. Sokal, A., & Bricmont, J. (1998). *Intellectual impostures*. London: Profile Books.
149. Sosa, E. (1980). The raft and the pyramid. *Midwest Studies in Philosophy, 5*, 3–26.
150. Sosa, E. (1991). *Knowledge in perspective*. Cambridge: Cambridge University Press.
151. Sosa, E. (2007). *A virtue epistemology*. Oxford: Oxford University Press.
152. Sosa, E. (2009). *Reflective knowledge*. Oxford: Oxford University Press.
153. Stich, S. (1990). *The fragmentation of reason. Preface to a pragmatic theory of cognitive evaluation*. Cambridge: MIT Press.
154. Stojanovic, I. (2008). Talking about taste: Disagreement, implicit arguments, and relative truth. *Linguistics and Philosophy, 30*, 691–706.
155. Stroud, B. (1968). Transcendental arguments. *Journal of Philosophy, 65*, 241–256.
156. Stroud, B. (1984). *The significance of philosophical scepticism*. Oxford: Clarendon Press.
157. Stroud, B. (2000). *The quest for reality. Subjectivism and the metaphysics of colour*. Oxford: Oxford University Press.
158. Toulmin, S. (1972). *Human understanding. The collective use and evolution of concepts*. Princeton: Princeton University Press.
159. Tuomela, R. (1995). *The importance of Us. A philosophical study of basic social notions*. Stanford: Stanford University Press.
160. Tuomela, R. (2006). Joint intention. We-mode and I-mode. *Midwest Studies in Philosophy, 30*(1), 35–58.
161. Turing, A. (1950). Computing machinery and intelligence. *Mind, 59*, 433–460.
162. Tye, M. (1986). The subjective qualities of experience. *Mind, 95*, 1–17.
163. Tye, M. (1989). *The metaphysics of mind*. Cambridge: Cambridge University Press.
164. Tye, M. (1995). *Ten problems of consciousness*. Cambridge: MIT Press.
165. Whorf, B. (1956). *Language, thought and reality*. Cambridge: MIT Press.
166. Williams, B. (1978). *Descartes: The project of pure enquiry*. Harmondsworth: Penguin Books.

167. Williams, B. (1985). *Ethics and the limits of philosophy*. Cambridge: Harvard University Press.
168. Williams, B. (2002). *Truth and truthfulness*. Princeton: Princeton University Press.
169. Williamson, T. (2005). Knowledge, context, and the agent's point of view. In G. Preyer & G. Peter (Eds.), *Contextualism in philosophy*. Oxford: Clarendon Press.
170. Wilson, B. (1970). *Rationality*. New York: Harper and Row.
171. Wittgenstein, L. (1922). *Tractatus logico-philosophicus* (p. 1961). London: Routledge & Kegan Paul.
172. Wittgenstein, L. (1969). *On certainty*. Oxford: Blackwell.
173. Wittgenstein, L. 1953. *Philosophical investigations*. Londres: Blackwell (1974).
174. von Wright, G. (1971). *Explanation and understanding*. New York: Cornell University Press.

Chapter 3
Temporal Aspects of Points of View

Antonio Manuel Liz Gutiérrez and Margarita Vázquez Campos

Abstract Time has a highly unstable place between the objective and the subjective. On the one side, there are very well known philosophical arguments trying to show that time has only a subjective reality, even that it is merely a subjective epiphenomenon. On the other side, we are compelled to take points of view as non dispensable elements of reality, at least of a reality capable of containing beings like us. And points of view offer a world of temporal entities existing in an objective way. Moreover, points of view themselves appear to be temporal entities among other temporal entities. We analyse both aspects of time. Our main focus will be McTaggart's arguments against the reality of a fluent time, what he called temporal series of kind A. We will distinguish three very different arguments in McTaggart works. We analyse them in detail. And we reject their conclusive character. Our final target is to maintain that there is a room for fluent time in what is internal to points of view but external to the subjects adopting those points of view.

Is time a merely subjective epiphenomenon? Are there conclusive reasons against the objective reality of time? What is it to adopt a temporal point of view? Are points of view, themselves, temporal, or tensed, entities? If so, how to characterise their peculiar dynamics? We will try to offer some answers to these questions. And the first thing we will do is to face directly McTaggart's well known argumentative strategies against the reality of time.[1]

This work has been granted by Spanish Government, "Ministerio de Economía y Competividad", Research Projects FFI2008-01205 (*Points of View. A Philosophical Investigation*), FFI2011-24549 (*Points of View and Temporal Structures*), and FFI2014-57409-R (*Points of View, Dispositons, and Time. Perspectives in a World of Dispositions*).

---

[1]We will focus on McTaggart [13, 14].

A.M. Liz Gutiérrez (✉) · M. Vázquez Campos
University of La Laguna, Tenerife, Canary Islands, Spain
e-mail: manuliz@ull.es

M. Vázquez Campos
e-mail: mvazquez@ull.es

© Springer International Publishing Switzerland 2015
M. Vázquez Campos and A.M. Liz Gutiérrez (eds.), *Temporal Points of View*,
Studies in Applied Philosophy, Epistemology and Rational Ethics 23,
DOI 10.1007/978-3-319-19815-6_3

1   McTaggart's Arguments

In McTaggart's approach to time, we can distinguish three different arguments against the reality of temporal A-series, the ones constituted by the application of the characteristics "to be past", "to be present" and "to be future". One argument is merely negative. It tries to show that there is no reason for believing that temporal A-series are real. According to McTaggart, the A-series necessarily require some external reference outside the series themselves, but it is very difficult to imagine what that external reference can be. The other two arguments are positive. They offer reasons for believing that the A-series cannot be real. One of these positive arguments is based on the thesis that A-series are themselves contradictory. The other positive argument puts the emphasis on the fact that temporal determinations in the A-series are circular or regressive. We will maintain that none of these three arguments is conclusive. That being so, the doors would be open for an ontological analysis of the ways in which an A-theory of time, i.e., a theory maintaining the reality of temporal A-series, could be formulated and defended.

1.1   What Is the Issue?

Time is a central topic in McTaggart's philosophy. And the rejection of the reality of time is a constant thesis in the various stages of his thought.[2] At first sight, that rejection of the reality of time is surprising when we consider the strong dependence that McTaggart's approach has on Hegel's philosophy. McTaggart explicitly says that Hegel, together with Spinoza and Leibniz, maintained that time is not real. However, this is in sharp contrast with the standard interpretation of Hegel. According to that interpretation, the notion of time is crucial in order to understand Hegel's system. Heidegger, for instance, criticised Hegel for having over-conceptualised temporality. According to Heidegger, the problem was not that Hegel maintained that time is unreal, but that his conception of time was abstract and not personal.

A central part of McTaggart's approach to time is the claim that time implies change, and that change only is possible if things take temporal positions with respect to a distinction between future, present, and past. Applications of these temporal characteristics constitute the temporal series of kind A. McTaggart argued that relations like "earlier-than", or "later-than", together with "simultaneous to", would not make enough room for change. There would not be any change in an event with "temporal parts" placed earlier than others, and hence the last ones being placed later than the first ones. According to McTaggart, temporal positions with respect to relations like "earlier-than", or "later-than", and "simultaneous to", constitute temporal series of kind B. And the B-series depend on the A-series.

---

[2]This is so, at least, from McTaggart [13] to McTaggart [14].

Properly, the B-series only are "temporal" series thanks to their dependence on a certain A-series. Hence, there cannot be time without change. And there cannot be change but with respect to A-series.

Many authors have focussed their attention on that part of McTaggart's approach to time, the dependence of a B-series on an A-series. For some of them, the B-series would not depend on any series of kind A. Also, it has been claimed that change only requires different positions in a B-series. Usually, both claims are maintained by the same people. To have a B-theory of time consists in that. We will not take part in this discussion.

We will not address McTaggart's conception of events as "substances" that could be placed in one position or another in the A-series and B-series either. This is a very strange way of understanding the notion of an event. On an ordinary reading, "event" and "change" are nearly synonymous. "To be an event" means "to suffer a change". On more sophisticated readings, there also could be events consisting in "resisting change through a certain period of time". In any case, the notion of event and the notion of change are strongly interconnected.

Nowadays, we have two main theories of events: Davidson's theory of events and Kim-Goldman's theory of events. In both of these theories, events are temporal, or tensed, entities. Davidsonian events are primitive entities with no structure. But they have a temporal nature, they are tensed entities. According to Kim-Goldman, events are objects instantiating a property, or relation, in a certain time or period of time. Here, again, events are tensed entities. So, events would have an "essential" tensed character in our two most important theories of events. But, if events are tensed entities, then they have to have by themselves a position in time, both in the A-series and in the B-series. In the B-series, they have a stable position. In the A-series, their position is not stable. In any case, events do not seem to be substances, in the sense of being the "substrata of change", or the "subject of temporal determinations". In other words, if they are substances, they seem to be essentially "tensed substances".

The issue is not only terminological. On the one hand, it would be to beg the question of the "unreality of time" to say, without argument, that what is placed in the A-series and B-series are timeless substances having an unproblematic real existence. That way, the A-series and B-series could be no more than two families of predicates we can attribute to a timeless reality. This would convert the A-series and B-series into a mere epistemological, or descriptive, recourse. On the other hand, to say that what is placed in the A-series and B-series are tensed events would be to beg the question in the opposite direction. It would entail that some positions in the A-series and B-series have a direct ontological value. There is, however, a crucial difference between these two options. In the first option, it is very difficult to understand why there are A-series and B-series. In other words, what could it be the point of having them? This is not a problem for the second option. Simply, reality is itself tensed. And through A-series and B-series we would try to conceptualise the temporal nature of reality.

In the frame of the second option, we need to give more ontological weight to A-series and B-series. We need to claim that events are essentially tensed. They have, in themselves, some ontological positions in A-series and in B-series. Moreover, whereas their positions in the ontological B-series are stable, their positions in the ontological A-series are not so stable. These events can be future events, present events, or past events according to their own, let us say, "internal temporality". That way, to try to place events into some other A-series, or into some other B-series, would always be to make an epistemic guess about some very special sort of "coordination" of their internal temporality, which is essential to them, with other attributed temporalities, perhaps with our own temporality.

We can conclude this preliminary discussion by saying that McTaggart's conception of events seems to be not only unclear, but deeply misleading. However, as we have said, this is not the issue we want to be concerned with here.

There are other problems we want to avoid. We will avoid, for instance, the problem posed by temporal appearances. Even if we come to have very strong reasons against the reality of time, we will continue perceiving and feeling time and change. We acquire new beliefs, and we abandon others. There seem to be histories and news. And we seem to communicate to each other, and to share, those perceptions and feelings, those beliefs, histories (trying to distinguish between histories and stories), and news (for instance, in newspapers). We adopt an impressive variety of intersubjective points of view about a fluent time. Moreover, to communicate something, and to make assertions and arguments, takes time. We will not discuss in depth here, either, the status of all such appearances. But they have an enormous weight.

McTaggart tries to make sense of temporal appearances through another way of ordering events, the C-series.[3] An alphabetic order would be an example of such C-ordering. McTaggart uses as an example an order like M,N,S,T. By themselves, C-series of events do not constitute a temporal series. In particular, they lack a determinate direction. The previously mentioned order, for instance, is symmetrical. It can also be seen as T,S,N,M. Again, it is only together with an A-series that a C-series can determine a temporal B-series. However, C-series can exist objectively. And McTaggart claims that they can give an "objective support" to our temporal appearances. That way, our temporal appearances could be "well founded".[4]

Indeed, McTaggart's treatment of time is complex. His arguments against the reality of time are entangled with a huge number of other claims. And we cannot fully understand and assess McTaggart's arguments in isolation from other aspects of his philosophy.[5]

---

[3]Mainly, he does it in book VI of his *The Nature of Existence* (1927).

[4]The notion of a "well founded apparent relation", a *"bene fundata* appearance", would come from Leibniz. This topic will be discussed in other chapters of the book.

[5]About that complexity, see Nyiri [17]. She also offers a very interesting overview about the different reactions to McTaggart's arguments in the last hundred years, in particular their connections to Einstein-Minkowski's conception of space-time. We will not address any of these topics here.

Having said all of that, what then is the issue we are going to discuss? The issue we want to discuss is whether McTaggart's arguments against the ontological reality of temporal A-series are conclusive. We will argue that they are not.

We would not be in a comfortable position if we could admit only the C-series as real. Nor we would be in a comfortable position if we could only say that even though no temporal A-series can be real, change does not need any such temporal series, but only some B-series. Neither the C-series nor only a B-theory of time is what we expected to have in order to understand time. C-series are completely independent of time. And B-series are completely independent of what can be taken as the paradigm of change: our experience of a fluent time.[6]

We would not be in a comfortable position either if we considered that both past and future are unreal, or inexistent, and that only the present has reality or existence in a full sense; or that past and future are just as real as the present; or that only the past, a "growing past", is real. Nor would we be in a comfortable position if we simply considered that the problems posed by McTaggart, about the reality of B-series and A-series, and about the reality of change and time, will be alive "for ever".

The three most important philosophical theories of time rejecting the reality of the A-series are Presentism, Eternism, and the Growing Block Universe Conception. Roughly, they can be characterised as follows

Presentism claims that only exists the present.
Eternism claims that past and future are just as real as the present.
The Growing Block Universe Conception claims that only is real the past, a growing past.

What is extremely puzzling with these three philosophical theories is that they are usually explained in terms of the A-series (more or less, in the way we have done). Only Eternism can preserve a certain sense in terms of the B-series.[7] Moreover, if these three theories are understood as theories of the physical time, they will have serious problems concerning the identification of the "present", and its distinction from the "past" and from the "future". The Special Theory of relativity entails that the identification of the present is relative to the place where we

---

[6]That A-series are essential to our experience of time, and that an adequate account of time needs them, are the two main claims about time of Lynne Rudder Baker ([2], Chap. 7). In her own words, "[...] both the B-series (that orders time in terms of unchanging relations like "earlier than") and the A-series (that orders time in terms of changing properties like "being past", "being present", and "being future") are needed for an adequate account of time. Neither series is dispensable, and neither by itself is a sufficient account of time. [...] it is a deep fact about time that it can be experienced only as transient." (pp. 155–156).

[7]Suppose this situation: either I am singing, or I have sung, or I will sing. What is it ultimately real of that situation? For Presentism only that I am singing at the present time is real. For Eternism, that I have sung and that I will sing would be as much real as that I am singing at the present time. For the Growing Block Universe Conception, only that I have sung would be real. It is very difficult to reformulate this example exclusively in terms of B-series! Only Eternism seems not to depend on a sharp distinction between the "past", the "present", and the "future".

are doing the identifications. So, the correct answer to the question "What is it real in the universe?" would depend on the "place" or "position" where we were answering it.[8] Again, only Eternism seems to be capable of having a chance to deal with this problem. Simply, for eternism all temporal positions would be real, without any further qualification.

In any case, Eternism, Presentism and the Growing Block Universe Conception would reject the reality of the A-series, i.e., the reality of a fluent time. So, our experience of a fluent time would be only a "mere illusion". Indeed, this is not a comfortable position. Moreover, it cannot be a comfortable position if we do not have any good explanation of why we come to have that temporal illusion.

Fortunately, we can do something better. We can resist McTaggart's arguments. In particular, we can resist McTaggart's arguments against the reality of temporal series of kind A.

1.2   Three Arguments

As we have said, we can distinguish in McTaggart three different arguments against the reality of temporal series of kind A. One of them is a negative argument offering reasons for "not assuming" their reality. The other two arguments are positive arguments offering reasons for "rejecting" that reality.

The negative argument is based on the need to appeal to a certain element X, external to the A-series, in order to construe the series. According to McTaggart, it is very difficult to say what that element X could be. And this would offer a negative reason against the A-series. We would have a negative argument in the sense that it is not an argument for rejecting the A-series, but only an argument for not assuming it. One of the positive arguments for rejecting the reality of an A-series is based on the existence of an internal contradiction in the concepts involved in it. The other positive argument is based on the existence of circles or regressive situations when we try to place something in an A-series. Usually, the negative argument has been ignored in the literature, and the two positive arguments have been considered to be one and the same. This is a mistake.

In the most famous passage of his book *The Nature of Existence* (1927), Chap. 33, McTaggart himself combines the two positive arguments. Let us see the full scene:

1. McTaggart argues that the only genuine source of time we can get is the one involving change. That is, there is no time without change.

---

[8]Very often, the "spacialisation of time" in Special Theory of Relativity is assumed without taking into account its metaphysical consequences in relation to Presentism, Eternism, and the Growing Block Universe Conception. For Presentism and the Growing Block Universe Conception, what is real would be relative to our place or position. With respect to the physical world, only Eternism seems to have clear advantages.

2. Then, he argues that "the only change we can get is from future to present, and from present to past" (#329). That is, a change in the temporal positions of some event according to an A-series.

3. Then, he affirms that "being past", "being present" and "being future" are incompatible determinations or characteristics. If something is past then it is not present or future, if something is present then it is not past or future, etc.

4. After that, he claims that "every event has them all" (#329). That is, every event would have to be past, present, and future. All the three characteristics would belong to each event (or at least two of them, if we consider the first and the last elements of an A-series).

5. There is a direct contradiction between 3 and 4. The intended conclusion is: "The reality of the A-series, then, leads to a contradiction and must be rejected" (#333).

6. McTaggart considers the most obvious way of trying to escape from that contradiction: "The characteristics are only incompatible when they are simultaneous, and there is no contradiction to this in the fact that each term has all of them successively" (#329).

7. But he rejects that strategy arguing that we cannot make sense of that "successive character" except in a viciously circular or regressive way. He says: "Thus, our first statement about [an event] M—that it is present, will be past, and has been future—means that M is present at a moment of present time, past at some moment of future time, and future at some moment of past time. But every moment, like every event, is both past, present, and future. And so a similar difficulty arises" (#331).

8. The consequence of rejecting the above strategy would again be that, "The reality of the A-series, then, leads to a contradiction and must be rejected" (#333).

9. The final conclusion is that, "Nothing is really present, past, or future. Nothing is really earlier or later than anything else or temporally simultaneous with it. Nothing really changes. And nothing is really in time" (#333).

In that argumentation, we can distinguish two different positive arguments against the reality of temporal series of kind A. There is a first argument constituted by 3, 4 and 5. And there is a second argument constituted by 6, 7 and 8. As is stated in 9, the conclusion of both arguments would be the same. However, it is important to appreciate that they are very different arguments. The second one is an argument against a certain way of trying to resist the first one.

We will argue that the first argument 3-4-5, based on the existence of an internal contradiction in the very notion of an A-series, depends crucially on other theses of McTaggart. And that these theses are far from being acceptable. Also, we will argue that the second argument 6-7-8, based on the existence of a circle or regress when we try to place something in an A-series, does not necessarily lead to the conclusion that the time defined by such A-series is not real. Once the two positive arguments are rejected, McTaggart's approach only has the support of the negative argument. We will argue that this is a very weak support.

1.3   The First Positive Argument: The Existence of an Internal Contradiction in the Very Notion of an A-Series

The most contentious claim of the argument 3-4-5 is 4. It is the claim that every event would have to be past, present, and future. Why accept that claim?

This is one of the clearest cases where it is necessary to place McTaggart's arguments against the reality of time in a broader context. As it stands, claim 4 is simply unacceptable. Moreover, in the text mentioned, McTaggart does not offer any clear reason why we should have to accept it. The only loose explanations McTaggart gives of that claim appear in fragments like the following one:

> The characteristics, therefore, are incompatible. But every event has them all. If M is past, it has been present and future. If it is future, it will be present and past. If it is present, it has been future and will be past. Thus all the three characteristics belong to each event. How is this consistent with their being incompatible? (#329)

Our question remains open. In what sense would all the three temporal characteristics of the A-series have to belong to each event? It is in the context of the second argument, 6-7-8, that we find a clue:

> But what is meant by 'has been' and 'will be'? And what is meant by 'is', when, as here, it is used with a temporal meaning, and not simply for predication? (#331)

There is here a very important distinction between "temporal meaning" and "predicative meaning". It is in the predicative meaning that all the three temporal characteristics of the A-series would have to belong to each event.

In the predicative meaning, we would have to accept conditionals such as the following ones, let us call them <u>conditionals T</u>:

- If M has been future (with a temporal meaning), then M is future (with a predicative meaning).
- If M will be past (with a temporal meaning), then M is past (with a predicative meaning).
- If M is present (with a temporal meaning), then M is present (with a predicative meaning).

The predicative meaning is an "absolute" and "tenseless" meaning. And it is in that sense in which every event M would have all the three temporal characteristics of the A-series (we are leaving apart the first and the last events in time).

So, 4 is true "only" in a predicative sense. This point is crucial. However, not all commentators have noted this aspect of McTaggart's arguments. Generally, the emphasis is placed on in the second positive argument. Paul Horwich is an exception. He says[9]:

---

[9]Horwich [9], Section "McTaggart's argument for the unreality of time".

If events are located in a real A-series, then each event acquires the absolute properties past, now and future. A real A-series entails that for every event such as E, there is a fact, included in the totality of facts that constitutes the universe, consisting of E's having the quality of *presentness*, that is,
E is (or, E is now)
but also the universe must contain the facts
E will be (or, E is future)
and
E was (or, E is past)
Given what is meant by 'a real A-series', such facts are not relations between events and times. They are not, in other words, the exemplifications of merely *relative* properties, which can both apply and fail to apply to the same event relative to different frames of reference. Rather, such facts consist in the exemplification by events of absolute properties.

Hence, in a predicative (absolute, timeless) sense, every event would have to be past, present, and future. One of the best analogies for understanding the predicative sense of temporal predicates, and so the absolute character that temporal characteristics can have, is to think of reality as a movie in a box (or in a CD, or DVD, or any other format). In a certain sense, in the predicative sense intended by McTaggart, all the events of the movie really are in the box.

Another good analogy would be offered by the music contained in a score, in comparison with a particular performance of the score, and with the A-series created by such a performance. Also, we can think of some written text, for instance a book, in comparison with a particular reading of the text, beginning with some parts, ending in other parts, and with the A-series created by such a reading. Anyway, let us continue using the analogy of "the movie in the box".

The movie in the box constitutes a C series. And it would be a temporal B-series only in relation to the movie being displayed in a certain way, i.e. in relation to a certain A-series. If the movie were to be displayed in some "non-standard" way, for instance beginning at the end, we would obtain a different B-series from exactly the same C series. For McTaggart, some C series (some movies in their respective boxes) would constitute the ontologically most basic, and epistemologically most objective, structure of reality. McTaggart is a pluralist (in clear contrast, for instance, to Bradley). Reality is not Parmenidean, but plural. And that plurality is organised into a complex set of different series of kind C.

In any case, we can ask, why does the predicative meaning have to be the only relevant meaning in our discussion? More precisely, why does the predicative meaning have to be the only "ontologically" relevant meaning? Why cannot the temporal meaning be the basic one?

Before answering these questions, it will be relevant to comment on an important point made by Dummett. He sees a crucial difference between "time" and things like "space" or "personality".[10] The case of personality is less clear (and surely very close to the case of time). So, let us consider only the case of space.

In relation to space, we can also identify positions both according to perspectival properties like "here" and "there" (a kind of let us say, "spatial A-series") and

---

[10]Dummett (1978, v.o. 1960) [5].

according to non-perspectival, or absolute, properties like "near to" and "far from" (a kind of let us say, "spatial B-series"). Let us focus on expressions like "here" and "there". They are token-reflexive expressions. When a token-reflexive expression occurs in a sentence, the sentence can have different truth values according to the circumstances of its utterance. In relation to space, every position can be described both with the help of expressions like "here" and "there" and with the help of expressions like "near to" or "far from". Dummett agrees with McTaggart in that A-series are essential to time. And this is what establishes a sharp contrast with space. For whereas, as we have said, the use of token-reflexive expression is not essential to our descriptions of objects as being in space, it seems to be essential to our descriptions of objects as being in time.[11]

According to Dummett[12]:

> ... a description of events as taking place in time is impossible unless temporally token-reflexive expressions enter into it, that is, unless the description is given by someone who is himself in that time (1978: 354)

In fact, McTaggart rejected the reality of space and personality. They are not really such as they seem to be. And Dummett is right in that McTaggart does not reject them through the reasons he uses for rejecting the reality of time. The reasons for rejecting the reality of space and personality, such as they seem to be, are connected in McTaggart with the reasons for rejecting the existence of "matter". Anyway, the crucial question is the following: Why cannot the contrast Dummett is emphasising have a correspondence in reality? In other words, why cannot the essential character of token-reflexive expressions in our descriptions of "things being in time" be real but in the form of a psychologically epiphenomenal A-series having a pale ontological correlate in some C-series?

Dummett gives a very revealing answer to that question. His answer is that McTaggart's rejection of the reality of time ultimately rests on the assumption that there has to exist, at least in principle, a "complete description of reality in absolute terms". That would be the assumption that reality can be thought of as completely contained in "a set of movies in their respective boxes". Dummett says:

> I think the point is that McTaggart is taking for granted that reality must be something of which there exists in principle a complete description. [...] The description of what is really there, as it really is, must be independent on any particular point of view. Now, if time were real, then, since what is temporal cannot be completely described without the use of token-reflective expressions, there would be no such thing as the complete description of reality[13]

---

[11] According to Dummett, the case of personality would be similar to the case of space. However, it can be claimed that token-reflexive expressions are also essential for describing something as a particular person, for instance for describing something as being "me".

[12] Dummett [5].

[13] Dummett ([5]: 356).

We arrive at the core of the first positive argument of McTaggart. According to Dummett, we have two exclusive options:

1. The existence of complete absolute descriptions of reality entailing the unreality of time.
2. The reality of time entailing the non-existence of complete absolute descriptions of reality.

Faced with these two options, in the same text Dummett asks whether the thesis that what is in time cannot be fully described without token-reflexive expressions could not be taken

> ... rather as demonstrating the reality of time in a very strong sense, since it shows that time cannot be explained away or reduced to anything else?

The question, then, following Dummett, would be this: Why not adopt the option 2?

McTaggart's own answer can be found in the first two chapters of *The Nature of Existence*. There, McTaggart makes "reality" and "existence" equivalent. Furthermore, both notions are taken to be undefinable. If we try to define them, we become involved in circularities and regresses. But, even if the notions of reality and existence are not definable, McTaggart claims that we can identify the general sorts of things that are real, or existent. McTaggart argues that there is no other reality apart from the reality that exists in an absolute sense. In the last instance, for McTaggart there are not perspectival properties; there are not degrees of existence either; and there is no other possibility apart from actuality.

McTaggart rejects any non-actually existent reality. In particular, he rejects

1. the reality of propositions: semantically evaluable abstract objects of belief, desire, etc., that can have reality even when they are false and thus when there is nothing in reality corresponding to what is believed, desired, etc.
2. the reality of non-existent characteristics: properties that do not have actual instances,
3. the reality of non-existent facts: facts that are not actual facts, and
4. the reality of non-existent possibilities: real possibilities apart from what is actual.

In his own words[14]:

> It would seem, then, that there is nothing which compels us to believe in non-existent reality. There is nothing which makes it necessary for us to accept the reality of propositions, or of non-existent characteristics, facts, or possibilities. And these are, as far as I know, the only things which have been asserted to be real without existing.
> But are we entitled to go further, and conclude that there are reasons for positively rejecting non-existent reality? With regard to characteristics and possibilities, the course of our argument has justified us in asserting positively that they cannot be real without existing. For we saw, to begin with, that all characteristics were existent. And all statements of

---

[14]All the following fragments of McTaggart in this section come from ([14]: #35–36).

possibilities have been reduced either to statements about existent knowledge or to statements about the implications of characteristics, and are therefore statements about the
existent.

Let us focus on the case of possibilities. This would be of help in order to better
understand McTaggart's first positive argument against the reality of time.
McTaggart assumes that the notion of possibility is ambiguous. "It is possible
that…" can have two meanings: an epistemological meaning and an ontological
meaning. And he argues that, in either of these two senses, the notion of possibility
involves anything which is real but not existent.

In the epistemological sense, possibility would mean a "limitation of our
knowledge" in the following sense:

> Thus, if I say that it is possible that it may rain to-morrow, the most obvious sense of the
> words is that I do not know whether it will rain or not.

It is clear that this does not involve anything real but non-existent. As McTaggart
says,

> In this case, clearly, it is a statement, not about any non-existent reality, but about my
> existent knowledge

The ontological sense would be present when we say things like, for instance, "It
was possible that I should not have sneezed yesterday, although I did sneeze".
According to McTaggart,

> In this case the possibility means, I think, that there is nothing within some particular field
> of circumstances to ensure my sneezing. For example, it might have meant that the fact that
> I was alive on that day did not ensure my sneezing on it, as it did my breathing on it.

And McTaggart claims that

> … when possibility is taken in this sense, it is an assertion about the implication of one
> characteristic by another. And we have seen that the implication of one characteristic by
> another is always an existent fact. It is therefore no more necessary to accept the reality of
> anything non-existent when possibility is taken in this sense than when it is taken in the
> other.

Now, we can state the key point. Possibility in the ontological sense would mean
that there is "an implication between characteristics that do exist". Hence, because
the only objective correlate of an A-series would be an absolute and tenseless
ordering according to some C-series, the assertion of temporal possibilities would
have to mean that there are some sorts of implications between characteristics that
do exist in some particular C-series (as "the events of a movie" do exist in the box).

McTaggart's generalised use of the expression "unreality of time", instead of
using the expression "non-existence of time", is closely connected to that point.
McTaggart does not want to argue simply that time does not exist. His precise and
specific target is to argue that time is not "something that can be taken as real but in
some cases non-existent, in the sense of non actually existent".

McTaggart applies to "temporal" possibilities this general position about possibilities. Can we accept his proposal? Certainly, the intended meaning of our

assertions of temporal possibilities is not necessarily the epistemological one.[15] However, when it is not, it is plausible to argue that our assertions do not have the ontological meaning described by him either. Simply, our assertions of temporal possibilities do not seem to be assertions about implications between characteristics that do actually exist in some kind of C-series.

Why do we have to accept that McTaggart's ontological meaning is "the only adequate ontological meaning" that can be given to assertions of temporal possibilities? At this point, McTaggart's argumentation becomes badly circular. The predicative sense in which we would have to accept the above introduced conditionals T is taken to be the only ontologically relevant sense because temporal possibilities are considered to have "the same nature" as all the other possibilities. And with respect to all these other possibilities, McTaggart has claimed that, if our assertions of possibility do not mean limitations of our knowledge, then they have to mean "implications between characteristics that only exist in some kind of C-series". Assertions of possibility can only express either limitations of our knowledge or implications between characteristics.

However, the crucial feature of temporal possibilities is that they are "temporal". In contrast with other possibilities, when they have an ontological sense, they cannot be reduced to implications between characteristics that only exist in some sort of C-series. If we lose the temporal sense of temporal possibilities, then we lose them completely. If they are reduced to the nature of the other non-temporal possibilities, then their peculiar nature is eliminated. Hence, we cannot argue that temporal possibilities are no more than implications between characteristics "because" they are like all the other non-temporal possibilities. This would prejudge the issue.

In one of the texts above quoted, McTaggart says of propositions, of non-existent characteristics, of non-existent facts and of non-existent possibilities that they are, as far as he knows "the only things which have been asserted to be real without existing" (#36). The problem is right here. The problem is that temporal possibilities are "peculiar".

Hence, one can reject that temporal possibilities, in particular non-existent (always in the sense of being non-actually existent) temporal possibilities like "to be future", or "to be past", can be approached in the same way in which other possibilities are approached. It can be claimed that they do not only mean either limitations of our knowledge, or implications between characteristics that exist, in an absolute and tenseless sense, in some kind of C-series. In other words, it can be claimed that the temporal sense of possibilities like "to be future" or "to be past" cannot be reduced to any predicative sense. And it can be claimed that the temporal sense of "to be present" is not reducible to its predicative sense either!

---

[15]When, for instance, we say something about the future, we are not necessarily only expressing our ignorance. Moreover, the contrast between the past, the present, and the future (including here the asymmetry and directionality of a fluent time) cannot be reduced to a simple question of more or less knowledge.

Apart from his general metaphysical framework, there is nothing in McTaggart's arguments that excludes these claims. But, these claims would make conditionals T unacceptable. And they would put McTaggart's first positive argument against the reality of time in serious trouble.

We have arrived at a very important result. There are two ways of assuming it. One way of assuming it would be stronger than the other one. We can say that even if with respect to non-temporal possibilities the conditional

- If x is G in a non-temporal modal sense, then x is G in a predicative sense

were to be accepted, for any property G, one could reject the following T conditional:

- If x is G in a temporal modal sense, then x is G in a predicative sense

This would be the weak way of assuming our result. However, we could also say that because the "actualisation" of every non-temporal possibility always involves some temporal aspect, there is always something in non-temporal possibilities that cannot be "reduced" to a mere predicative meaning. It is easy to see that this second way of taking our result is very much stronger than the first one.

Let us conclude this section by saying that there is a crucial "change of meaning" in McTaggart's thesis that every event would have to be past, present, and future. In that thesis (4 in our reconstruction above), "past", "present", and "future" have a predicative, absolute, tenseless meaning. But this is not the temporal meaning that "past", "present", and "future" have in the A-series. Moreover, this is not the meaning that these words have when it is stated that each one of those characteristics is incompatible with the other ones (3 in our reconstruction).

The temporal meanings only entail the predicative meanings if we assume all the other metaphysical theses of McTaggart concerning the identity between reality and existence, and the implicit inclusion of temporal possibilities in his rejection of non-existent but real possibilities. However, there is much room for controversy with respect to all these matters. Therefore, there is no conclusive contradiction between 3 and 4, in the argumentative line above presented. And therefore, McTaggart's first positive argument against the reality of time is not conclusive.

1.4 The Second Positive Argument: The Existence of a Circle or Regress When Something Is Positioned in an A-Series

McTaggart's first positive argument was that the predicates "past", "present", and "future" involve a deep contradiction because, on the one hand, they are incompatible predicates and, on the other hand, all three apply to every event (for simplicity, we will follow McTaggart's use of the term "event"). As we have seen, a natural reply is that the predicates which apply are not simply "past", "present", and "future", but rather, for instance, "will be past in the future", "is present in the present", and "was future in the past", these new predicates being compatible. McTaggart's response to this reply is that it cannot offer any help. And that response constitutes his second argument, 6-7-8 in our reconstruction.

Dummett clarifies this point as follows[16]:

> Instead of three, we now have nine predicates, each of which still applies to every event and some of which are incompatible, for example, the predicates "was past" and "will be future". Admittedly the objector may again reply that the predicates which really apply to the same event are "is going to have been past" and "was going to be future", and that these are again compatible. But McTaggart can counter this move as before, and so on indefinitely

Dummett's conclusion with respect to McTaggart's reply is:

> If there is a contradiction connected with the predicates of the first level, the contradiction is not removed by ascending in the hierarchy [of temporal qualifications]

However, as we have said, there is no internal contradiction connected with the predicates of the first level. McTaggart's first positive argument is not conclusive. And if there is no such internal contradiction in the notion of an A-series, then the existence of a circle or regress when something is positioned in that series cannot lead to that contradiction either. The existence of such a circle or regress may constitute a hard problem, but it does not lead to the intended contradiction.

The majority of authors commenting on McTaggart's refusal of the reality of time have focussed on the supposed contradiction pointed out by the second argument. However, there is no such contradiction. There is only "a threat of circularity or regress". Once the conclusive character of the first positive argument against the reality of the temporal A-series is rejected, the second positive argument, i.e. the argument presented through 6-7-8, has to be reconsidered.

Nevertheless, the circles and regresses involved in the second argument pose an important problem. What is "that" problem? Let us introduce the main elements from which it arises.

We begin with a set of temporal predicates, or properties, or characteristics: "to be past", "to be present" and "to be future". And our task is to attribute some of these temporal characteristics to things that suffer a change. But, we cannot do it in an arbitrary way. There is the following "normative restriction" regulating our attributions:

(R) Nothing that changes can have in any of its temporal positions more than one different temporal characteristic, i.e., it has to be either "past", or "present", or "future"; and only the characteristic "to be in the future" could be had more than one time.

---

[16]In Dummett (1978, v.o. 1960) [5]. All the fragments of Dummett in this section come from here, pp. 351–352.

R is crucial. It establishes that nothing that change can have the same temporal position both in the past and in the present, nor both in the past and in the future, nor both in the present and in the future. And that only with respect to the future may there be more than one temporal position.[17]

The future is very peculiar. A thing that has changed only can have a position in the past. A thing that is changing only can have a position in the present. However, a thing that will change can be placed at different positions in the future. Not only because we can be ignorant of when it will change. To the extent that determinism "can be false", different positions in the future of a thing that will change are consistent with supposing a complete knowledge.[18]

R establishes a restriction that is not relativised to a particular A-series. Nothing that changes can be past according to a certain A-series and be present according to another A-series. Nothing that changes can be past according to a certain A-series and be future according to another A-series. And nothing that changes can be present according to a certain A-series and be future according to another A-series. In a literal sense, these things cannot occur "at the same real time". This is how our attributions of temporal characteristics work.

R does not exclude an "open future", i.e., different future possibilities. This can give sense to the asymmetry between, on the one hand, the past and the present and, on the other hand, the future. But R does not entail that the future is open either. Moreover, the same event could be placed more than one time in the future even though there were not but one only future. By themselves, our attributions of temporal characteristics according to R do not exclude "fatalism".[19]

Now, the whole problematic situation involving circles and regresses in McTaggart's second positive argument can be taken in two very different ways:

1. either as one in which the attributions of temporal characteristics to events are supposed to be "done", and we consider the results of those attributions, or
2. as a situation in which the attributions are something we are "doing", some sort of "work in progress".

Let us consider the characteristic "to be in the present". Let us call it Pr. That characteristic, Pr, has to have an extension E(Pr). E(Pr) is constituted by the class of all the things that are Pr, and only by those things. Even if E(Pr) is the null class, E(Pr) has to exist objectively as such a class. And it has to exist independently of the stable,

---

[17]Of course, persisting things could be placed "at the same time" in the past, the present and the future. R would not apply to them. However, we can think of A-series applied to persistent things something derivate from applications to the temporal positions of changing things.

[18]Fatalism can be defined as the thesis that determinism is "necessarily" true. The sort of distinction we are making between, on the one hand, the past and the present and, on the other hand, the future was a very important subject for Prior. See Prior [20, 21]. In fact, very often we think of future events as events that can happen in "one or another" point in the future.

[19]However, R would exclude other temporal scenarios. And it is important to note it. For instance, the possibility of having a perfect circularity of events in time: a circular time in which "absolutely identical events" (numerically identical events) would repeat again and again.

or unstable, character of the things that are Pr. Simply, if those things are not stable with respect to being Pr, then E(Pr) would not be stable either. The other characteristics would also have extensions E in that sense. Let us say that E(P) is the extension of "to be past", and E(F) the extension of "to be future". Again, with independence of the stable, or unstable, character of the things that are P, or F, the extensions E(P), and E(F), have to exist objectively. The only restriction, according to R, is that nothing can belong "at the same time" to more than one of such extensions (the last part of R would not be relevant here).

Also, let us introduce the notion of an "ostensive specification" of the above mentioned temporal characteristics. The ostensive specification of "to be present", let us call it O(Pr), would be constituted by our listing, or enumerating, the things that have the characteristic Pr. O(Pr) is a "doing". We can say that whereas E(Pr) has always a "closed texture", O(Pr) has always an "open texture". It is, we can also say, an "open doing".

E(Pr) is an objective class of things. E(Pr) is something "done", or the result of something "done". In contrast, O(Pr) is some sort of "work in progress". The other temporal characteristics of A-series also would have ostensive specifications. So, we can speak of O(P) and of O(F). Again, it does not matter whether these ostensive specifications have an unstable character or not. As we know, "time flies". Anyway, what continues being crucial is the normative restriction R. In our ostensive specifications, we cannot attribute more that one different temporal characteristic to the same things, and only the future can be attributed "more than one time" to the same things.

E and O are different things. But, there is a very important kind of dependence of extensions E on ostensions O. The determinations of the extensions E of temporal characteristics depend on their ostensive specifications O in the following sense:

We would only have a clear reason to believe that temporal characteristics have null extensions E if the ostensive specifications O were to be in some sense "defective".

In other words, even having a very unstable character, we think that non-null extensions E of temporal characteristics can exist to the extent that our ostensive specifications O are not defective.

Are temporal ostensive specifications O defective? Here is where McTaggart's second positive argument against the reality of time calls our attention to a very serious problem. However, the problem is not one of obtaining a contradiction, as in the first argument 3-4-5. The crucial problem is that there seems to be "no other way" of determining that the restriction R is fulfilled except by means of "some temporal ostensive specifications O".

Note that in order to follow R, and in order to know whether we are following R correctly, we have to assume for our temporal specifications themselves a temporal position with respect to the past, the present, and the future. In particular, we need to distinguish our present specifications from our past specifications and from our future specifications. This generates very directly a circular or regressive situation. And this situation can create the wrong feeling that temporal ostensive specifications O(P), O(Pr), and O(F), are "deeply defective".

Here is an example. Suppose that I have some doubts about whether an event *e* has to be ostensively specified as belonging to O(Pr). Perhaps, I guess, it was present in some very recent past, and now *e* has to be ostensively specified as belonging not to O(Pr) but to O(P). Or, perhaps, *e* is now only in the near future, and so it has to be ostensively specified as belonging to O(F). Very soon it will be present, but it is not present now. According to R, we can correctly attribute to *e* only one different temporal characteristic (and only the characteristic "to be in the future" could be had more than one time by the same thing). But, in order to attribute to *e* any temporal position, we have to ascend a Dummettian level. We have to attribute a temporal position to the very moment at which we are attributing temporal characteristics to *e*. Is that moment the present moment? Or, did that moment happened in a recent past? Or, will that moment happen in the near future?

Is there something wrong in that? The problem can be rephrased as follows: Are temporal ostensive specifications O(Pr), O(P), and O(F), defective because there is "no other way" of determining that R is correctly satisfied apart from making, in a circular or regressive way, some other temporal ostensive specifications? We said that only if temporal ostensive specifications were defective, we would have a clear reason for maintaining that the temporal characteristics of A-series have null extensions E. But, are they defective, moreover "deeply defective", simply because they involve the above circularity or regress?[20]

In the first chapters of *The Nature of Existence*, McTaggart considers that reality and existence are non-definable. According to him, they have to be taken as basic, or primitive, notions. He argues that they have to be so taken because when we try to define them, we can only use those notions in circular or regressive ways. McTaggart's approach has close connections with Moore's views about the un-definability of "good" and other moral characteristics. Anyway, the important point is this: Why does it have to be different with time? Why do circularity and regression have to entail non-definability, and a "basic, or primitive, ontological nature", in the case of reality and existence, and something "deeply defective" in the case of time?

The important thing is that to treat time in the same way in which reality and existence are treated would entail that our temporal ostensive specifications cannot be defective only because they involve circularity or regression. McTaggart's second positive argument for the unreality of time, the one based on the existence of a circle or regress when something is positioned in an A-series, is not conclusive.

---

[20]The point we are making is closely connected to the idea expressed by Dummett in one of the fragments previously quoted: "… a description of events as taking place in time is impossible unless temporally token-reflexive expressions enter into it, that is, unless the description is given by someone who is himself in that time" (Dummett 1978: 354). This involves circularity and regression, but not necessarily of a defective (or vicious) sort.

1.5 The Negative Argument: The Search of a Relational Element X

We have considered the two positive arguments that McTaggart offers against the reality of time. There is also a negative argument. McTaggart introduces this argument in the following way:[21]

> If, then, anything is to be rightly called past, present, or future, it must be because it is in relation to something else. And this something else to which it is in relation must be something outside the time-series.

His conclusion is this:

> We have come to the conclusion that an A-series depends on relations to a term outside the A-series. This term, then, could not itself be in time, and yet must be such that different relations to it determine the other terms of those relations, as being past, present, or future. To find such a term would not be easy, and yet such a term must be found, if the A-series is to be real.

This is the negative argument against the reality of A-series. To place something in a real A-series requires an external term. It requires something "outside" the A-series. But to find such a term, McTaggart claims, is not an easy task.

How to respond to the negative argument? The first thing would be to distinguish two senses in the condition that the entity X has to be placed "outside the A-series". Let us call it condition O. Condition O can have an epistemological sense and an ontological sense:

1. In the epistemological sense, the condition O would entail that the "correction of our specifications" of the changing relations between the past, the present, and the future is "independent of" our making those specifications.
2. In the ontological sense, the condition O would entail that "what fixes" the changing relations between the past, the present, and the future has an "existence independent" of the relations so fixed.

We can proceed according to those two senses. The way we have resisted McTaggart's "second positive argument" against the reality of A-series would offer a clue to deal with the epistemological sense in which the condition O would have to be satisfied. And the way we have resisted McTaggart's "first positive argument" would offer a clue to deal with the ontological sense in which the condition O would have to be satisfied.

Let us begin with the epistemological sense. In order to be correct, our ostensive temporal specifications have to satisfy restriction R. And they can be correct ones even though we have to make, again and again, other ostensive temporal specifications. So, our ostensive temporal specifications could be "correct specifications"

---

[21]McTaggart ([14], #327–328).

with independence of our making them. Circularity and regress do not pose any further problem here.

Now, let us turn to the ontological sense. For the condition O to be satisfied in the ontological sense, "what fixes" the changing relations between the past, the present, and the future needs to have an existence independent of the relations fixed. Is there something capable of doing this work?

In fact, we would have such a thing if what fixes the changing asymmetric relations between the past, the present, and the future is, by itself, something that, being real but not actually existent, is able to become something real and actually existent and, then, can become again something real but not actually existent. The "actualisation of possibilities" establishes a "before" and an "after" which is independent of any temporal determination. And that settlement of a distinction between a "before" and an "after" can be repeated again and again. So, the actualisation of some possibilities could fix asymmetric relations between the past, the present, and the future from the "outside" of those temporal determinations.

In the ontological sense of condition O, "what fixes" the changing relations between the past, the present, and the future needs to have an existence independent of the relations fixed. The actualisations of certain possibilities can do the work. The actualisations of some possibilities could fix the changing relations between the past, the present, and the future in such a way that the existence of those actualisations is independent of the relations that are so fixed. The actualisation of some possibilities would be the basic, or primitive, phenomenon able to constitute "the source of a fluent time". Condition O can be ontologically satisfied in that way.

We have offered some answers to the negative argument against the reality of A-series. We have made some proposals for giving content to the epistemological and ontological senses that the expression "outside the A-series" can have in condition O. However, our proposal in relation to the ontological sense invites consideration of a potential "plurality" of A-series. And this is a very important new problem.

If the source of a real A-series, i.e., the source of a fluent time, is the actualisation of some possibilities, then it makes sense to say that perhaps there are "more than one" real A-series. The actualisation of some possibilities would support A-series from the "outside" of any A-series. Those actualisations have an existence independent of the A-series. This generates the possibility of a "temporal pluralism". Simply, different actualisations could support different A-series.

It is possible to discard temporal pluralism through an "ad hoc" stipulation. We could claim that the actualisation of possibilities never will generate a pluralism of different A-series. However, this would have to be considered some sort of "last recourse". So, how to deal with the possibility of such temporal pluralism? How to avoid, for instance, the problem of "comparing temporally" a variety of A-series fixed by actualisations of different possibilities?

Let us guess at some answers. When we are placing "ourselves" in an A-series, it is easy to avoid those problems. In that case, when we are placing ourselves in an A-series, the ontological and epistemological senses of condition O could be one and the same. More precisely, the actualisations of certain possibilities fixing ontologically the changing temporal relations between the past, the present, and the future in our own case could be simply the various temporal specifications of our position in an A-series made from some temporal points of view able to produce correction. The "relevant actualisations of possibilities" would be a number of "correct temporal self-specifications".

When we are placing "ourselves" in the A-series, our making the temporal specifications we make (perhaps in a large part unconsciously) is the term "outside the A-series" on which the positions on the A-series depend. The correction of our making such temporal specifications is independent on our making them. That correction requires to satisfy R. And circularity and regress do not introduce necessarily any fatal problem. That way, condition O is satisfied in the epistemological sense. What about the ontological sense? We can say that our making temporal specifications in a correct way (surely, many of them unconsciously) entails the actualisation of relevant possibilities. These possibilities can be understood as some dispositions settled on us. But they may have an existence independent of the temporal relations fixed.

That hypothesis has a very important consequence. There has to be some relevant sort of temporal convergence in our own case. Our temporal self-specifications are self-correcting. This follows directly from condition R: nothing that changes can have more than one different temporal characteristic, and only the future can be had more than one time by the same things. When R is applied repeatedly to different temporal self-specifications, both actual and counterfactual, it leads to temporal convergence.

Each change I have suffered places me in the past; and this excludes that I have exactly that change in the present or in the future. Each change I am suffering places me in the present; and this excludes that I have exactly that change in the past or in the future. Each change I will suffer places me in the future, perhaps in more that one only place; and this excludes that I have exactly that change in the past or in the present.[22]

At the end, applying repeatedly the restriction R, there could not be more than one "correct" temporal specification of our own position in an A-series. At the end, there could not be any irreducible temporal pluralism with respect to our own temporal position.

Now, in order to understand the relationships between our correct temporal self-specifications and the temporal relations constitutive of "other" objects, events, processes, etc., different from ourselves, three options are open:

1. The first option consists in saying that a correct temporal specification of other objects, events, processes, etc., is simply a "good temporal measure" of them, in a purely operational sense. In other words, any other temporal specification would be a correct one if, from our temporal perspective, it is a useful way to

---

[22]Something can change "at the same time" with respect to more than one property. This does not pose any serious problem. We can say that X changes with respect to property F at the same time than it changes with respect to property G iff X changes with respect to H at that time, being H a certain combination of properties F and G.

describe those objects, events, processes, etc., or to predict them, or to control them, etc.

2. The second option consists in saying that the correct temporal specification of any other objects, events, processes, etc., is "reducible to the correct temporal specification of our own personal temporal relations". Another way to express this idea would be by saying that a subject making such temporal specification becomes a "temporally extended reality involving those other objects, events, processes, etc.".

3. The third option would be a blend of the other two. Perhaps with respect to some objects, events, processes, etc., even with respect to some objects, events, processes, etc., belonging to our bodies, or to our mental makeups, the first option is the most adequate; and with respect to other objects, events, processes, etc., the second option is the most adequate one.

The three options try to avoid temporal pluralism. They try to avoid the problem of the real existence of more than one A-series. Option 1 does it by means of an operationalist reduction of the meaning of "correct temporal specifications of other objects, events, processes, etc., different from ourselves". Option 2 does it by "reducing any such correct temporal specification to the case of our own temporal self-specifications". Option 3 does it in both ways.

The first option is very clear. There are normative contexts defining what can count, or cannot count, as a "good temporal measure" for many kinds of objects, events, processes, etc. Science provides us with a lot of such contexts. And so does ordinary knowledge. This option tries to avoid problems about the real existence of a plurality of A-series introducing a mere operational meaning for all the other temporal specifications apart from temporal self-specifications. This is quite a radical option. The sense in which we talk about our own temporal reality, the temporal reality of each one, and the sense in which we talk about the temporal reality of any other objects, events, processes, etc., would be "completely different". Moreover, they seem to be incommensurable.

The second option is no less radical than the first one. But its strategy is just the opposite of the strategy followed by the first option. The first option is operationalist. The second option is realist. It strongly suggests that there are other temporal, or tensed, entities apart from ourselves.

The realism of the second option calls our attention to a very important point. There is a sharp contrast between persons and other sorts of entities. That contrast has also a temporal face. And when the claim that there are other temporal, or tensed, entities apart from ourselves is combined with the possibility of a temporal pluralism, the result is very puzzling.

Those temporal, tensed entities would have parts. And those parts also would have to be temporal, or tensed, entities. Now, if we were to admit the possibility of a real existence of a plurality of A-series, then the entity itself could be temporally placed in different past, present, and future times than its parts. It is not that the entity can be "extended in time", or that it can have "temporal parts" extended in time, for instance some parts in the past and the present, other parts in the present, other ones in the future, etc., but that the entity and its parts could be in "different presents", in "different past times", and in "different future times".

In contrast with that, let us consider "persons". Strictly speaking, persons do not have parts. My hands are parts of my body, but they are not part of "me" as a person, as the person I am. I am no less a person if I lose my hands (or I am less a person only in a metaphorical sense). Hence, even if we admit the real existence of a plurality of A-series, that would not affect me as a person. I cannot be in a present (or past, or future) which is different from the present (or past, or future) of my parts simply because I do not have parts.[23]

With this contrast in mind, we can obtain a better understanding of the second option above introduced. According to it, our making temporal specification of some other objects, events, processes, etc., would entail for ourselves to become "extended realities" involving those other objects, events, processes, etc. They would be integrated, so to speak, into "my personal reality". And, in so becoming, there would not be any problem about a real plurality of A-series. This option is very suggestive. But it is also very radical.

Both option 1 and option 2 seem too radical. Perhaps a certain combination of the two might be not so radical. Option 3 would consist in a compromise between option 1 and option 2. What kind of compromise?

For persons, in the last analysis, there can be only one time, There is only one A-series. Moreover, interpersonal relations try to preserve that singular time. More concretely, what can be called "intersubjective temporal points of view" try to preserve that time. Other persons can become integrated into my personal reality. And I can become integrated into their personal reality. Perhaps the same can be said of other entities connected to us in some, let us say, "personal" ways. All of that would give a unifying sense to our "common history". In these cases, the option 2 seem to be completely acceptable. Moreover, it is full of important insights.

However, very often, when we try to specify the temporal relations of many other objects, events, processes, etc., what we try to obtain is simply a "good temporal measure" of those objects, events, processes, etc., in a purely operational sense. And in these cases, option 1 seems to be the best acceptable one. With

---

[23]The peculiarities of our personal experiences of time are emphasised by Russell [22]'s logical construction of time out of "sensibilia". Russell defends a relational, constructive (anti-Kantian and anti-Newtonian) Leibnizean theory of time. In his construction, time comes to be internal to a construed space of perspectives without being internal to the subjects from which that space of perspectives is construed. There is an asymmetry between determinations of temporal positions in the case of our own experiences and determinations of temporal positions in other cases. Whereas the first ones are direct, the second ones are indirect. And that indirect character entails the intervention of processes that, when they are projected over a physical space of perspectives, "take time". The construction of a Russellian space of perspectives, and of a physical space-time containing physical objects, matter, and perspectives, is explained in other chapters of this book.

respect to these cases, fluent time would be only a projection. A-series would be only a useful way of speaking.

1.6   Time Is not an Absolute Frame, nor a Kantian Scheme Either

Let us summarize our main results. None of the three arguments that McTaggart uses against the reality of temporal A-series is conclusive. The crucial point in the first positive argument is the rejection of any real but not actually existent possibility. We have found that this claim can be resisted. It depends on general metaphysical assumptions adopted by McTaggart. Real but non-existent temporal possibilities are assimilated to other modalities, and excluded without any clear justification. So, there is no internal contradiction in the reality of temporal A-series. Therefore, the second positive argument cannot lead to that intended contradiction either. However, the second positive argument poses a serious problem of another kind. And we have discussed it. The problem was that temporal specifications seem to be always circular or regressive. We have tried to show that this does not entail that they are defective. Such circularity or regress also can be taken as a symptom of undefinability.

Finally, we have addressed McTaggart's negative argument against the reality of time. According to it, to place something in a real A-series would require something outside the series. And it was not easy to say what that thing can be. We have distinguished two senses of "outside the A-series": an epistemological sense and an ontological one. In the epistemological sense, the correction of our temporal specifications in the A-series has to be independent of our making those specifications. In the ontological sense, what fixes the temporal relations of the A-series has to be independent on the relations fixed. We have claimed that the epistemological sense of the negative argument can be resisted in the same way in which we have rejected the conclusive character of the second positive argument. And that the ontological sense of the negative argument can be resisted in the same way in which we have rejected the conclusive character of the first positive argument. Following that strategy, we have faced the problem of "temporal pluralism": the problem posed by the possibility of a real existence of a plurality of A-series. In order to handle with that problem, we have considered two opposite and very radical options. And we have argued for a compromise between them.[24]

We have suggested that both reality and existence are tensed, at least in part. The past, the present, and the future are real. But the reality of both the past and the future depend in many ways on the reality of the present. The reality of the present is a present reality. Only the present actually exists. And only the present can be

---

[24]The BA-theory of time defended by Baker [2] would embrace the temporal duality present in the option 3.

known with accuracy. However, not only what actually exists is real. This is the common sense view of time. It is also the Aristotelian view. Sometimes, Aristotle seems to claim that time "is change". Other times, he says that time is simply something operational, "the measure of change". Both things can be true. In any case, change consists in something that is possible becoming actual, or in something that is actual becoming again only possible. And change, so understood, can be the source of time.

Perhaps the notion of time is as undefinable, as basic, as primitive, as the notions of reality and existence. Perhaps time is not detachable from reality and existence, or from certain parts of reality and existence. So understood, time could not be simply a Newtonian "absolute frame", or a transcendental "Kantian scheme", where real or existent things can be placed in one way or another. At the end of the day, McTaggart's rejection of the reality of time comes from understanding the A-series only as a kind of "absolute frame", or "Kantian scheme".

2 Temporal Points of View

Our points of view are full of indexical ingredients. Sometimes, that indexical character involves emplacement in space. Other times, it involves a relative position concerning some properties and relations instantiated by the subject and the environment. Other times, it involves emplacement in time.

Emplacement in time is especially important for subjects which are "persons". A person can become massively confused about her position in space, and about her relative position regarding many, perhaps all, of the properties and relations instantiated by herself and her environment, but she cannot become massively confused about her position in time. Being a person, at least a person like us, entails having a temporal perspective with a minimum of correction. Such correct temporal perspective, or temporal point of view, could be "internal" to our points of view. It could stand without any more "external", or more "objective", support. But, it has to exist.

Among the classical analyses of what it is to have a temporal perspective, or a temporal point of view, we have to make reference to Kant, Bergson, Husserl, McTaggart and Prior. Let us introduce very briefly some of their approaches.

For Kant, space and time would establish the conditions of possibility of having experiences of an "external world". We can say that there is in Kant a peculiar "transcendentalisation" of the Newtonian absolute concepts of space and time. Time also is crucial with respect to our "internal world". Without time, we could not have any "internal intuition", nor any kind of "self-intuition" of ourselves either.[25]

---

[25]See, in his *Critique of Pure Reason*, "Transcendental Aesthetic".

Bergson was very critical about conceiving of time in the same way as space. According to him, time has very peculiar features. Mainly, time is directional and it is not inert. The essence of time is "duration". Duration eludes any scientific approach. It can be only grasped through intuition. From a subjective point of view, the expression of time as duration is "memory".[26]

Husserl analysed in detail the phenomenological structure of temporal intentionality. According to him, "internal time" has a very complex structure. The present is never like a point. It is always some sort of "present continuous". It includes what has been just present, and also what is going to be present. Husserl's conception of time was very influential in the philosophy of the 20th century, mainly in Continental philosophy through Heidegger and Merleau-Ponty.[27]

We already know McTaggart's arguments against the reality of time. According to McTaggart, time is only a merely epiphenomenal subjective appearance. In contrast, Arthur Prior took very seriously temporal appearances.

Prior is the founder of modern temporal logic. Prior's central idea is that there is an internal representation of a "fluent time" in our language and thought, and that this internal time becomes crucial in the logical analyses of many inferences. Prior offered a huge variety of different logical systems defining temporal operators which are applied to propositions. The semantics for such systems are generally similar, with some extensions, to the ones for modal logic.[28]

All these authors and proposals insist on one idea: the essential role of our temporal points of view in order to constitute our identity as personal subjects capable of taking any other point of view.

2.1   Time and Temporal Points of View

We need to distinguish between

(A) The problem of understanding the existence and structure of time in reality; in particular, the problem of the real existence of a "fluent time" having the structure of McTaggart's A-series.

(B) The problem of understanding the existence and structure of temporal points of view.

Beyond all the discussions about problem A, it is plausible to claim that the existence of temporal points of view (hereafter, TPoV) cannot be denied. The existence of TPoV is something as manifest as the fact that you are "now" reading

---

[26]See, for instance, Bergson [3, 4].

[27]See Husserl [10], Heidegger [8], and Merleau Ponty [16].

[28]See Prior [19–21]. In the line of Prior, see Kamp [12], and more recently Øhrstrøm and Hasle [18], and Areces [1].

some words and phrases, perhaps "after" having read other ones, and "before" (we hope) reading still others.

As Prior argued, there are TPoV simply because our thoughts and our languages are tensed. To deny the existence of TPoV would be like denying that we have points of view involving the existence of an "external world", or points of view involving the existence of "other minds", etc. Even if there is not an external world, even if there are not other minds, it is very difficult to deny that we have points of view involving those things.

2.2   Defining Temporal Points of View

What is a TPoV? We can say that a temporal point of view is originated when different explicit non-conceptual contents of a point of view are identified, either non-conceptually or conceptually (or in a mixture of both sorts of identification), as changes of content in relation to distinct positions in an A-series.

The proposal is very simple, but it has important consequences. As we will see, our proposal can make sense of (1) the crucial difference between "histories" and "stories"; (2) the possibility in principle of a "variety of temporal perspectives" regulated by the normative requirement that in the end only one of them has to be the correct one; (3) the existence of "temporal experiences" with relative independence from "temporal concepts"; and (4) the notion of a "non-absolute but not merely subjective either fluent time", in contrast with a "merely subjective time". In addition, (5) our proposal would be capable of integrating in a single and unified way many of the ideas of McTaggart, Prior, Kant, Bergson, and Husserl.

We are going to define the notion of temporal points of view (TPoV). But we need the help of a conception of points of view (PoV) according to which any point of view can be seen as having the following canonical structure:

$PoV = <B, R, non\text{-}CC, CC, Cp>$, where

1. B is the bearer of the PoV (in personal PoV, a subject like us),
2. R is a set of relations connecting B with the explicit contents of the PoV,
3. non-CC and CC are the two kinds of contents that can be explicitly included in the PoV: non-CC is a set of non-conceptual contents and CC is a set of conceptual contents, and
4. Cp is a set of possession conditions for having the PoV.

Now, let us think of TPoV. It is neither necessary nor sufficient for the existence of a TPoV that there be more than one PoV, or that there be a change of PoV. In all these cases, we would have either a variety of PoV, or a PoV changing in time. However, strictly, we would not have a TPoV.

In order to have a TPoV, what we need is to identify, or recognise, certain "differences in content" as "changes of content". More precisely, we can define a TPoV in the following way:

<u>Temporal Points of View</u> (TPoV) are PoV with explicit contents EC*, either non-CC or CC, identifying certain differences in some explicit non-CC, let us call them EC, as changes in time, or permanencies in time, with respect to distinct positions in an A-series (past, present and future).

TPoV only focus on some explicit non-CC. That is, EC only contains non-CC. The other possible explicit contents of a PoV, its CC, do not change. There is no change when we have in perspective, for instance, that $2 + 2 = 4$. However, to have in perspective that we have in perspective $2 + 2 = 4$ is to have in perspective a non-CC, and hence to have in perspective a set of (actual and possible) changes. The conceptual world (concepts, propositions, sets, numbers, etc.) is "outside the perspective" of our TPoV. Only the world we experience is a changing world. However, this world includes us having in perspective all sorts of CC.

In the minimal case, we would have in temporal perspective two explicit non-CC contents, one of them being placed in the past and the other one in the present or in the future, or one of them being placed in the present and the other one being placed in the future. And our TPoV identifies a change. However, there may be TPoV with a much greater temporal complexity.

A TPoV also could take some non-CC as displaying a certain "permanence in time", i.e., as something continuing from the past to the present, or from the past to the future, or from the present to the future. We can deal with this issue very easily. We can consider that the identification of a permanence in time is dependent on the possibility of identifying changes. That way, to identify a permanence would be to identify possible but not actual changes.

The complexity of TPoV can give place to "histories" and to "stories". When intentional actions are involved, this important distinction can be defined as follows:

In the case of "<u>histories</u>", but not necessarily in the case of "<u>stories</u>", some of the non-CC contents EC, which are identified through some EC* as changes of content, also have to be contents, either explicit or implicit, of other PoV.

Both histories and stories are TPoV. However, in contrast with stories, histories need the existence of other PoV. A history is a TPoV about something that belongs to other PoV. The bearer of a history does not need to be the same as the bearer, or bearers, of those PoV. If it is the same bearer, then the history becomes a "biography".

In our definition, we have assumed McTaggart's idea that there is no TPoV without reference to a certain A-series. However, in principle, there could be more than only one A-series. According to the way we have introduced the notion of TPoV, there could be many pasts (not only many possible reconstructions of the past, but "many pasts"), many presents (not only many possible ways of living the present, but "many presents"), and many futures (not only many possibilities of imagining the future but, again, "many futures").

We have not required that TPoV logically entail only one unique present and one unique past, beyond the possibility of having an open variety of futures. In principle, the existence of a variety of TPoV in which the past, the present, and the future are not univocally identified is possible. We have to make room for these possibilities. Some supposed cognitive disorders consist precisely in having a number of TPoV offering more than one single "past", or more than one single "present", or more than one single "future".

These are serious possibilities. However, all of them are balanced by the normative restriction R regulating our attributions of temporal characteristics. According to R, nothing that change can have more than one different temporal characteristic in any of its temporal positions, and only the future can be had by the same things more than one time.

We said that restriction R is not relativised to any particular A-series. Now, we can say that it is not relativised to any particular TPoV either. Both things can be taken to be equivalent. Restriction R partially defines the way we see the "real time". It forces us to choose only one different temporal characteristic for each temporal position of a thing that suffers a change. We can have doubts about where to place in time the temporal positions of something that change. However, according to R, at the end, each temporal position of a thing that changes has to be placed either in the past, or in the present, or in the future; and only in the future it could to be placed more than one time. We attribute temporal characteristics trying to follow R. Only in that way can our temporal attributions be "correct" ones.

The normative restriction R regulates the constitution and dynamics of TPoV involving things that change.[29] In the long run, nothing that change can have at the same time more than one different temporal characteristic, and only the characteristic "to be in the future" can be had more than one time. If something has changed, then it has to be placed in the past. If something is changing, then it has to be placed in the present. If something will change, then it has to be placed in the future. And only things that will change can be placed more than one time in the future.[30]

The normative restriction R has a special relevance both when we consider our own "personal identity" through time and when we consider that temporal points of view also can be "intersubjective". We can define the last notion as follows:

Intersubjective Temporal Point of view (ITPoV) are TPoV shared by different subjects.

What is shared in ITPoV are certain temporal identifications. A number of different subjects identify in the same temporal ways some EC. Some differences in

---

[29]We have assumed that to identify permanencies in time is to identify "possible but not actual changes". Without the possibility of changes, we could not identify permanencies either.

[30]Perhaps something "changing" requires that "it has changed" a bit, and also that "it will change a bit". Being this true, the present will always need a small portion of past and a small portion of future. In other chapters of the book, it will be argued that this is just the case. More precisely, it will be argued that the present is part of a "now" that always includes a certain past and a certain future. In any case, when we attribute temporal positions in terms of A-series to a thing changing, we try to be maximally selective. We try to refer to the present in the narrowest way.

non-CC are taken as changes in time, or some non-CC are taken as permanencies in time. And these temporal identifications can be made either in a conceptual or in a non-conceptual way (or in a mixture of both).

ITPoV are crucial in our life. To share a TPoV is to "share a time". More precisely, it is to share a past, a present, and a future. And this shared time can exist, and have objectivity, even though there is no time except in relation to some PoV.

That way, a time with the structure of A-series, i.e., a fluent time, can be something "internal to some PoV", and have objectivity, without being something "internal to the subjects" having those PoV. Moreover, it can be internal to a number of PoV, possessed by different subjects, and have a "shared objectivity".

Other important consequences of our definition of TPoV are the following ones:

1. We have said that to have more than one PoV, or to have a change of PoV, does not entail having a TPoV. Changes in the bearer B of the PoV, or changes in the relations R connecting B with the explicit contents of the PoV, or changes in the possession conditions Cp, do not entail the existence of a TPoV either. To change the bearer B of a PoV, or to change the relations R (psychological attitudes in personal cases) towards the explicit contents of a PoV, or to change the possession conditions Cp of the PoV, are changes producing a different PoV. However, by themselves, those changes do not generate a TPoV.

2. In principle, the temporal identification can be made through explicit contents EC* which can be either non-CC or CC. This includes the possibility of identifying changes of explicit contents EC through the help of some explicit contents EC* which are non-CC. And this means that it would be possible to identify changes with respect to distinct positions in an A-series without possessing the concepts of "past", "present", and "future" in a fully developed sense. It would be possible to experience some contents as "past", or "present", or "future" events, or facts, or objects, etc., with relative independence from the full possession of these concepts.

   This would give a robust sense to the notion of "temporal experiences", and to the possibility of having those temporal experiences without the possession of "temporal concepts" in the sophisticated sense in which personal subjects, with high cognitive capacities, can have those concepts. That way, some non-human animals, pre-verbal children, etc., would be capable of having a TPoV.

3. We can give a very simple answer to the problem of whether temporal concepts (in particular, the concepts of "in the past", "in the present", and "in the future") are primarily applied to propositions, as Prior maintained, or whether they are primarily applied to events, or facts, or objects, etc. Prior's approach would be directly relevant when the explicit contents EC* involved in a TPoV are CC. And the other approaches would be directly relevant when the explicit contents EC* involved in a TPoV are non-CC.

4. According to the above way of understanding the internal structure of a PoV, it is possible to distinguish between what is "internal/external to a PoV" and what is "internal/external to the subject which is the bearer of that PoV". We have made use of that distinction in relation to ITPoV. This has a very special

relevance in relation to the notion of an "external time". Even leaving open the question of the absolute existence in reality of a time of kind A, in the sense of having an existence external to, or independent from, all PoV, we would have enough room for distinguishing between two sorts of temporal series of kind A:

(a) Temporal A-series internal to some TPoV, but external to the subjects which are the bearers of those PoV.
(b) Temporal A-series internal to some TPoV, and also internal to the subjects which are the bearers of those PoV.

Temporal A-series of the first sort could have enough "objectivity" (certainly a non-absolute objectivity, but enough objectivity) to make sense of a "not merely subjective" reality of time. Temporal A-series of the second sort would be obtained when the EC present in a TPoV involve only the subject which is the bearer of the TPoV.

The first sort of temporal A-series also would provide an important sense in which a fluent time can be real even though it is not real in the sense of having an "absolute objectivity". It can be real in the sense of existing inside some TPoV without being subjectively epiphenomenal, i.e., without being merely determined by the subjectivity of the bearers of the TPoV.[31]

As we are seeing, our distinction between those two sorts of temporal series of kind A can be of help in order to get a better understanding of the distinction between an "objective, but perhaps relative, i.e., non-absolute, fluent time", and a "completely subjective fluent time". We can find the need to introduce these two kinds of fluent time in many authors. In particular, we find it in Kant, Bergson and Husserl. The objective, but perhaps relative, non-absolute, time would be connected to temporal A-series of the first sort. The second kind of time would be connected to temporal A-series of the second sort.

5. At first look, there are close relations between TPoV and reflective PoV. In a certain sense, TPoV are reflective PoV. But, if reflective PoV entail the possession of "conceptual" capacities, then TPoV cannot be a kind of reflective PoV. In order to be precise, the thesis would have to be the following:

To have a TPoV entails adopting "something like" a reflective PoV in which (1) some differences in the explicit non-CC of a certain PoV are identified, perhaps only in a non-conceptual way, as changes in time, or some explicit non-CC are identified as permanencies in time, and (2) the bearer of that PoV is identified, perhaps only in a non-conceptual way, as being the same as the bearer of the TPoV.

---

[31]That way, we could maintain what D. Mellor called an A-theory of time, or what L. Baker calls a BA-theory of time, in opposition to a B-theory of time that only would admit the reality of temporal B-series. See [15] and Baker [2].

This has very important consequences. In order to have a TPoV, it is necessary to have certain explicit non-CC in perspective, identifying or recognising some differences as changes in time, or identifying in those non-CC some permanencies in time, with respect to distinct positions in an A-series. And it is necessary to identify the bearer of the TPoV as being the same as the bearer of the PoV having those explicit non-CC. Indeed, this entails a peculiar reflective move over the contents of our PoV. However, we have assumed that this reflective move can be made in non-conceptual ways. Hence, a TPoV would be "something like" a reflective PoV. But, strictly speaking, it would not be a reflective PoV. So, entities without conceptual capacities could harbour TPoV even though they cannot harbour reflective PoV.

In other words, it is possible to have reflective points of view which are not temporal. And it is possible to have TPoV without being able to have reflective PoV in the sense of requiring conceptual capacities. However, only subjects with a minimum of reflective non-conceptual capacities would be subjects capable of adopting TPoV.

6. The notion of an "intersubjective" PoV is different from the notion of a "collective" PoV. We have used the first notion in our discussion of ITPoV. There is something to say of the second one also. In intersubjective PoV, a number of different subjects share certain contents. Collective PoV have a collective subject as their bearer. The claim that there are collective subjects can have a more or less strong sense. In any case, many of the above points would apply to TPoV of a collective sort. We can talk about "individual TPoV", and also about "collective TPoV". We can talk about "individual histories" and "individual stories", and also about "collective histories" and "collective stories". We can talk about "individual temporal experiences", and perhaps also about "collective temporal experiences", etc.

2.3   From a Temporal Point of View

A TPoV is originated when different non-CC of a PoV are identified as a change in time, or when some non-CC are identified as a permanency in time, in relation to distinct positions in the past, the present, and the future. The contents so identified can belong to other PoV. And the identification can be made either in conceptual or in non-conceptual ways.

There are many important issues that can be unified and clarified by paying attention to that characterisation of TPoV, in combination with the distinction between understanding the existence and structure of time in reality (our previous problem A) and understanding the existence and structure of TPoV (our previous problem B). We will mention three of them, apparently disconnected. They have to do 1) with the role of science in order to understand TPoV, 2) with the lack of need to be engaged in the problem of understand time in reality (problem A) when we are interested in epistemological or logical questions involving temporal perspectives, and 3) with the so called "time travel paradoxes":

1. Disciplines like physics or neurology have little relevance with respect to the existence and nature of TPoV (problem B).

   We can say that physics, neurology, etc., only can be relevant with respect to the existence and nature of TPoV if they are relevant with respect to the existence and nature of PoV. However, it is not clear how they could facilitate an understanding of PoV. What can physics, neurology, etc., say about the existence and nature of things like "content"? Moreover, what can they say about the crucial distinction between the merely "subjective" aspects of a PoV and its "objective" aspects? It can be claimed that TPoV are sensitive to content and to the distinction subjective/objective in non-reducible ways.

   So, even though a natural discipline like physics is very important in order to understand the existence and structure of time in the physical world, and even though a natural discipline like neurology is very important to understand the existence and structure of time in the context of neurological processes, all of that is far from providing a complete, even clear, understanding of the existence and structure of TPoV.

2. We do not need to be engaged in discussions about the existence and structure of time in reality (problem A) when we are dealing with epistemological or logical questions involving temporal perspectives; for instance, when we are trying to combine temporal and epistemic components in order to logically analyse temporal discourse.

   The last projects would not try to understand time in reality (problem A), but only TPoV (problem B). It is plausible to argue that the problem of understanding what time is in reality has to be answered from the basis of all we know, and aim to know about reality. In contrast, to understand TPoV only has to do with some parts of reality: some peculiar sorts of PoV. Neither the epistemology of attributions of temporal features, nor the logic of time, needs to understand "previously" the existence and structure of time in reality.

3. The discussion of the ontological and epistemological aspects involved in "time travel paradoxes", in particular the ones derived from the possibility of travelling to the past, can be clarified paying attention to our characterisation of TPoV.

Let us consider the temporal paradox of myself going to the past in order to kill my grandfather before he knew my grandmother. It involves

1. A TPoV intending to identify some particular non-CC as past, present, or future. In particular, a TPoV intending to identify a certain future killing, planned by me now, as coming to occur "in the past".

2. A TPoV according to which a certain killing seems to affect the same person who, from other PoV, would count as "my grandfather".

The paradox only appears when 1 and 2 are interpreted in some peculiar ways. The paradox appears

- when 1 is interpreted as really involving the killing coming to occur "in the past", and
- when 2 is interpreted as really affecting "my grandfather".

However, 1 and 2 also can be interpreted in other different ways. For instance, the time travel can be interpreted not as travel "to the past" but as travel "to the future", to a certain very unexpected future. If we reinterpret 1 in that way, and we so modify our TPoV, then there is no problem with 2. In that unexpected future I can perfectly well kill my grandfather. He could be alive again and I could kill him.

We can also reinterpret 2. The person that seems to count as my grandfather can be taken to be only some kind of "twin-grandfather", i.e., someone close to being qualitatively identical to my grandfather, but in any case not numerically identical to him. Under that reinterpretation, there would not be any problem with 1. From that new TPoV, I could perfectly well travel to the past and kill that other person.

Perhaps we could not travel to the past. But, even if we could travel to the past, there would not be any paradox if we reinterpret 2 in that way. In general, if we adopt any of the two alternative TPoV indicated, the paradox generated by the supposed possibility of travelling to the past and killing my grandfather (or my father, or altering the past in any other problematic way) would disappear.

The last issue is connected with the other two. It is not clear at all how physics or neurology could decide how to interpret the two TPoV mentioned in the third issue. Epistemological and logical questions like the ones posed by time travel paradoxes have to be answered in their own terms.

3   The Dynamics of Points of View

Reflection about points of view shows that they are temporal entities. Points of view also change with time. Mainly, they may change according to changes in their explicit non-conceptual and conceptual contents, they may change according to how those explicit contents interact each other, and they may also change with changes in the relations that the bearer of the point of view maintains with those contents.

We can have reflective points of view about our points of view. And we can also have temporal points of view (TPoV), or temporal perspectives about them. Whereas the first ones have always a conceptual character, the second ones can be non-conceptual. Without conceptual capacities, we could not reflect about our own points of view. However, we could have the capacity of experiencing and feeling the tensed nature of our perspectives even though we would not have conceptual capacities.

Many of the dynamical peculiarities of points of view come with the implicit non-conceptual contents linked to the peculiar "attitudes" that the subjects are maintaining towards the explicit contents of the point of view. This is especially important in personal points of view.

The role of the implicit non-conceptual contents linked to the attitudes is manifold and complex. In relation to time, those implicit contents accomplish an essential function: they can counterbalance both the changes due to the explicit contents of the point of view, and the changes due to how these explicit contents are interacting each other.

Just as there may be compensations, and situations of equilibrium, among the explicit non-conceptual and conceptual contents of a point of view, there may be compensations, and situations of equilibrium, between all those explicit contents and the implicit contents of the point of view, the contents linked to the attitudes.

Here, a very relevant distinction has to be made between, on the one hand, changes "in" a point of view and, on the other hand, changes "of" point of view. There may be changes in the explicit contents of a point of view without any change of point of view, and there may be changes of point of view without any change in the explicit contents of the point of view.

In the first case, the changes in the point of view would compensate each other in such a way that they do not cause any change "of" point of view. In the second case, the change of point of view would be caused by changes in the attitudes articulating the point of view, with independence from the explicit contents included "in" it.

The second case is very important. There could be three main sorts of changes "of" point of view which would be crucially promoted by the implicit non-conceptual contents linked to the attitudes:

1. Changes in focus: Here, some of the explicit contents of the point of view become more salient than others, as a result of changes in the attitudes involved. Wittgenstein's discussion of "the duck-rabbit drawing" offers a classical example of that kind of change.[32]
2. Radical changes of perspective: The point of view becomes completely different even though there is no change in the explicit contents, either non-conceptual or conceptual, involved in it. Some cases of religious conversion are of that kind.[33] Also, some radical changes in political perspective could be included here. Wittgenstein is referring to that kind of change when in the *Tractatus* he says that, "the world of a happy man is a different world than the world of an unhappy man".
3. Structuring changes: Some changes in the implicit contents linked to the attitudes of the point of view originate changes in the internal structure of the explicit contents of the point of view. For instance, a change in the attitudes towards logics, as an effect of the improvement of logical skills, can give place to very different ways of organising our thoughts and discourses. Structuring changes can even change the explicit contents of a point of view. Here, to change the ways of "seeing the world" entails changes in "the world that is seen".

---

[32]Wittgenstein (1953).

[33]See James [11], and Unamuno [23].

The three kinds of changes are worthy of emphasis. They call our attention to some very significant dynamical phenomena that cannot be reduced to changes "in" the explicit, non-conceptual or conceptual, contents of the points of view.

Furthermore, the third sort of change could have an important explanatory power in relation to the "constitution" of the explicit contents of a point of view.

In that sense, a very suggestive hypothesis is that the implicit non-conceptual contents of a point of view put all kinds of pressures over the ways in which the point of view can have some peculiar sorts of non-conceptual and conceptual explicit contents. Through those pressures, the point of view becomes capable of having the particular sorts of non-conceptual and conceptual explicit contents it is able to have. In the last term, that process of, let us say, "modulation", or "tuning", would produce in a subject the various "types" of explicit contents that can be tokened in one way or another.

According to that hypothesis, the different types of explicit contents a point of view can have would be the result of a process of "modulation", or "tuning", of the implicit contents of the point of view. For subjects like us, the implicit contents of our points of view would be a "precondition" of their explicit contents.

The third sort of change would also be crucial in order to understand the formation of TPoV. The key feature of TPoV is to take some differences in non-CC as changes in time (and some possible but not actual differences as permanencies in time). How can we explain that transformation? Another suggestive hypothesis would be that it is a case of "structuring change" in the internal structure of a point of view.

That structuring change would be provoked by a peculiar kind of attitude toward the explicit non-CC of the point of view. We can call them "temporal attitudes". They are attitudes prone to identifying changes in time beyond mere actual differences in content, and prone to identifying permanencies in time beyond mere possible differences in content. Some subjects have the dispositions to have these "temporal attitudes", and other entities do not have them.[34]

Of course, we only can say that, i.e., we only can guess one such explanation, from a speculative stance. We adopt a reflective point of view about our points of view, and about other points of view. But, there is nothing necessarily wrong in that.[35]

---

[34]The generation of conceptual contents in "conceptual spaces" of qualitative dimensions, and the formation of these qualitative dimensions from identifications of similarities and differences among experiential contents, would offer a very interesting approach in order to understand the last two points. See Gärdenfors [6] and Hautamäki [7].

[35]We have suggested that the constitution of TPoV could be understood as the formation of a new qualitative dimension in a conceptual space. Some differences in non-CC are taken as temporal differences according to a past, a present, and a future. As any other qualitative dimension, that temporal dimension could be interpreted phenomenally (for instance, from the temporal values of a psychologically extended "now", including a certain past, present, and future) or scientifically (for instance, using the theoretical values of some metric applied to brain processes). The comparisons between the two interpretations would be comparisons between "two different conceptual spaces". See again Gärdenfors [6] and Hautamäki [7].

We have considered changes "in" the explicit contents of a point of view, non-conceptual and conceptual ones, provoked by the implicit non-conceptual contents linked to the attitudes (changes in focus). Also, we have considered changes "of" a point of view produced by those implicit contents without any change in the explicit contents of the point of view (radical changes of perspective). And we have considered changes "in the internal structure" of the explicit contents of a point of view produced by those implicit contents (structuring changes). In all these different kinds of changes, the interactions among different points of view can have a very relevant role. They can have direct effects on the attitudes involved in each point of view. In many cases, those interactions are the main source of changes both "in" a point of view and "of" point of view.

The significance of the normative restriction R is clear when we consider ITPoV. In the long run, no temporal position can have at the same time more than one different temporal characteristic, and only the characteristic "to be in the future" can be had more than one time. Nothing can be past and present; nothing can be past and future; nothing can be present and future; and only the future can be such that something can have more than one position in the future. In the case of our "personal identity" through time, R also has a very important role. We can never have more than one past; we can never have more than one present; and perhaps our future is open, or perhaps we simply do not know all the details.

Ontologically, all of that entails that, if TPoV exist in reality, and they are the source of an objective fluent time that is internal to them, but not merely internal to the subjects, then in the long run (in the very "long run" of the whole of reality displaying all its potentialities) there would have to be only one such fluent time!

References

1. Areces, C., & Ten Cate, B. (2006). Hybrid logics. In P. Blackburn, J. van Benthem & F. Wolter (Eds.), *Handbook of modal logics* (821–868). New York: Elsevier Press.
2. Baker, L. R. (2007). *The metaphysics of everyday life*. Cambridge: Cambridge University Press.
3. Bergson, H. (1911). *Creative evolution. 1998*. New York: Dover.
4. Bergson, H. (1896). *Matter and memory. 1994*. New York: Zone Books.
5. Dummett, M. (1960). A defence of McTaggart's proof of the unreality of time. In *Truth and other enigmas. 1978* (pp. 351–357). Harvard: Harvard University Press.
6. Gardenfors, P. (2000). *Conceptual spaces. The geometry of thought*. Cambridge: MIT Press.
7. Hautamäki, A. (1992). A conceptual semantics approach to semantic networks. *Computers & Mathematics with Applications, 23*(6–9), 517–525. [Also In F. Lehmann (Ed.), *Semantic networks in artificial intelligence* (pp. 517–525). Oxford: Pergamon Press].
8. Heidegger, M. (1962). *Being and time* (p. 1990). Albany: State University of New York Press.
9. Horwich, P. (1990). *Asymmetries in time*. Cambridge: MIT Press.
10. Husserl, E. (1928). *On the phenomenology of the consciousness of internal time. 1990*. Dordrecht: Kluwer.
11. James, W. (1902). *The varieties of religious experience. A study in human nature. 2012*. Oxford: Oxford World's Classics.

12. Kamp, J. A. W. (1968). *Tense logic and the theory of linear order*. Ph.D. thesis. Los Angeles: University of California.
13. McTaggart, J. (1908). The unreality of time. *Mind, 17*, 457–474.
14. McTaggart, J. (1927). *The nature of existence*. Cambridge: Cambridge University Press.
15. Mellor, D. H. (1998). *Real time II*. London: Routledge.
16. Merleau Ponty, M. (1945). *Phenomenology of perception* (p. 2012). New York: Routledge.
17. Nyiri, K. (2008). Hundred years after: How McTaggart became a thing of the past. *Proceedings of the 6th European Congress of Analytic Philosophy (August 21–26, 2008)*. Kraków: European Society for Analytical Philosophy.
18. Øhrstrøm, P., & Hasle, P. (1995). *Temporal logic: From ancient ideas to artificial intelligence*. Dordrecht: Kluwer Academic Publishers.
19. Prior, A. (1957). *Time and modality*. Oxford: Oxford University Press.
20. Prior, A. (1967). *Past, present and future*. Oxford: Oxford University Press.
21. Prior, A. (1968). *Papers on time and tense*. Oxford: Oxford University Press.
22. Russell, B. (1918). The relation of sense-data to physics. In *Mysticism and logic, and other essays* (pp. 113–140). London: George Allen & Unwin.
23. Unamuno, M. (1912). *Del Sentimiento trágico de la vida* [*The tragic sense of life. 1954.* New York: Dover].

Chapter 4
Fluent Time, Minds, and Points of View

Antonio Manuel Liz Gutiérrez

Abstract It is argued that the existence of a fluent time with a past, a present, and a future is linked to the existence of experiential points of view with non-conceptual contents. There cannot be a fluent time without points of view with non-conceptual contents, and there cannot be such points of view without a fluent time. The main components and consequences of these ideas are analysed. The non-conceptual contents of our experience are tensed entities capable of making true some tensed truths. A crucial distinction is assumed among being external to all points of view, being internal to some points of view, and being internal to the subjects having those points of view. Fluent time would be internal to some points of view without being internal to the subjects. That way, even if fluent time cannot be placed in the physical world, such as that physical world is conceptualised by our physical sciences, it could not be said to be merely subjective either. The notion of a fluent time, always internal to our experiential points of view, is also distinguished from the notion of having a temporal point of view. Temporal points of view are a peculiar kind of, in a certain sense reflective, points of view. It is by adopting some temporal points of view that we come both to identify a fluent time and to be able to postulate a physical time.

I want to present and defend a very conservative conception of time, a realist conception of a fluent time in which things can have a position either in the past, or in the present, or in the future. The real existence of such a time will be based on the objectivity displayed by some of the contents of our points of view: namely, their non-conceptual contents. Firstly, I will analyse the sorts of facts that are capable of being appropriate truth-makers of tensed truths. Then, I will argue that even though

A.M. Liz Gutiérrez (✉)
University of La Laguna, Tenerife, Canary Islands, Spain
e-mail: manuliz@ull.es

© Springer International Publishing Switzerland 2015
M. Vázquez Campos and A.M. Liz Gutiérrez (eds.), *Temporal Points of View*,
Studies in Applied Philosophy, Epistemology and Rational Ethics 23,
DOI 10.1007/978-3-319-19815-6_4

those facts could not be found in the objective world of science, in the last instance in the physical world, nor in a merely subjective domain internal to the subjects either, they can be facts external to the subjects but internal to their points of view.

To begin with, let us note a crucial distinction between

1. A is true iff B is the case

and

2. A being true is an actually existing fact in virtue of C being an actually existing fact.

In 1 and 2, A is a truth-bearer: perhaps a proposition, or a statement, or a belief, etc. For simplicity, we will generally speak of truth-bearers (TB). B is the truth-condition (TC) of A. The TC of a TB is the fact that would have to exist actually in the world iff the TB were true.

1 states a semantic (we can also say, Tarskian) fact: an equivalence between that A is true and that B is the case. This means that B in "B is the case" can perfectly well be a possible but not actually existent fact. There is nothing in 1 entailing existence. In particular, none of the two copulas in 1 entails existence.

2 is very different from 1. In 2, a connection is stated between two actually existing facts: A being true and C being the case. In 2, C is the actually existing fact in virtue of which there is another actually existing fact: A being true. We can say that C is a truth-maker (TM) of A.[1]

---

[1] In general, I will take "existence" and "actual existence" as equivalent. My use of the notion of truth makers does not follow any of the recent approaches to the subject. It is largely based on McTaggart's ideas about truth-making in the first chapters of his book *The Nature of Existence*, vol 1 (1921). We can define a truth maker as follows:

Something S is a truth-maker TM of the truth-bearer TB = the TB being true is an existing fact in virtue of the fact that S exists.

Existing facts are facts that actually exist or occur in the world, as opposed to facts that simply may exist or occur, mere possible facts, or facts that cannot occur, impossible facts. If S is a TM of a certain TB, then we have two different existing facts and a connection between them. The two existing facts are the existing fact that the TB is true and the existing fact that S exists. The connection is that the first fact occurs "in virtue of" the occurrence of the second fact. This entails a crucial claim: It can be an existing fact that a TB is true without the TC of that true TB being an existing fact. So, we could have truths (i.e., true TB) about possible but non actual things, or about fictional entities, or about impossible things, or about the past and the future, etc., even though their TC are not, or even cannot be, actually existing facts. The TC are entailed by the truth of the TB. However, this is only a "semantic fact". And semantic facts do not need to be "actually existing facts". The TC do not need to be actually existing facts.

The idea behind this approach is that we can conceive the world as a complex net of highly interconnected existing facts. Some of these facts consist in that certain TB are true, and some of these true TB are true in virtue of other existing facts. We can call them the TM of those true TB. In some cases, the TC of the true TB also are existing facts. But this is not necessarily so. The interconnections among the existing facts make it possible that some existing facts are TM of some true TB without the TC of those true TB being existing facts. The locution "in virtue of", in the

In order to illustrate the distinction between 1 and 2, let us see some divergences in their logical behaviour. From 2, and

3. C is an actually existing fact

we can infer

4. A being true is an actually existing fact.

Now, from 4, we can infer directly

5. A is true.

And, from 5 and 1, we can infer

6. B is the case.

But, from 6 alone, we cannot infer

7. B is an actually existing fact.

4 and 5 are very different. 4 entails 5, but 5 does not entail 4. Claim 5 only states a "semantic fact". The "is" in 5 entails nothing about actually existent facts.

6 and 7 are also very different. Now, 7 entails 6, but 6 does not entail 7. Again, claim 6 only states a "semantic fact". It is a semantic fact because it is obtained exclusively through a derivation from other semantic facts: 5 and 1. As it is presented in 6, B can be a possible but not actually existent fact. It can be so because the "is" in 6 entails nothing about actually existent facts. In contrast, 7 is a claim about the actual existence of a certain fact. And that existent fact is B.[2]

TC are semantic facts, TM are existing facts. TC and TM are very different things. And we can exploit that difference to argue that "tensed truths", i.e., true TB with TC that must have a position either in the past, or in the present, or in the future, especially true TB with TC conditions in the past or in the future, and so with TC not having the ontological status of actually existent facts, can nevertheless have TM that always are facts actually existing in the world, facts that are existent facts.[3]

But, which kinds of existent facts could be the TM of tensed truths? On the one hand, they do not seem to be existent physical facts. It is very difficult to find in the

---

(Footnote 1 continued)

above definition, refers to those interconnections. Many times, "in virtue of" is equivalent to "because of", and it can involve causal components. However, truth-making can have a highly heterogeneous ontology. Moreover, the most important aspect of a truth-making approach is not the ontological "nature" of TM, but their ontological "role".

[2]More precisely, there is a systematic ambiguity in expressions like "X is true" and "Y is the case". They can refer either to purely semantic facts, or to existing facts. This is one of the main sources of thinking erroneously that 1, together with some true TB, can have direct ontological implications about which existing facts are in the world. The point that will be important for us is that, "B is the case", in 6, can occur either in the past or in the future, not only in the present.

[3]Tensed truths with TC in the present always will have as TC existent facts. However, not all truths with existent facts as TC are tensed truths. Consider, for instance, "There is something".

physical world, such as it is objectively described by basic science, appropriate TM for a "fluent time". Moreover, as we will see, it is not easy to find physical TM for the truth of a TB saying that something is happening just "now". But, on the other hand, the relevant existing facts could not be purely epiphenomenal, subjective facts either. In that case, they would not have the required "objectivity" to play the role assigned to TM.

Aren't there other options? Yes, there are. I will argue that the relevant existing facts can be some facts "internal to points of view" without being facts "internal to the subjects having those points of view". That way, we could have an objectivity different from the objectivity of the physical facts described by science. And the fluent time linked to the TM of the corresponding tensed truths would not be a merely epiphenomenal, subjective time either.

Which facts "internal to points of view" can play that role? My hypothesis is that they are facts in which the point of view has certain "non-conceptual contents in perspective". Points of view can have contents of a conceptual kind and contents of a non-conceptual kind. The first ones are abstract entities: namely, concepts and propositions. The second ones constitute the world of objects having properties that we are able to experience.[4] We discover that world through the contents of our experience. It is not the world such as it is described by basic science, in the last instance by physics. But it is not a world only existing "inside us" either. And that world is a world in "continuous change".

Moreover, it can be argued that change is not something derived from objects and properties. Properly, we do not experience objects having some properties and ceasing to have other properties. We directly experience change, objects and properties being something derived from our experience. The non-conceptual contents of our points of view, those contents constituting the world we have in perspective, are always experienced as having an existence in the form of a, let us say, "gerund". For those contents, to exist is always an "exist-ing".

That the contents of our experience, some non-conceptual contents, have always an existence in the form of a gerund has important consequences with respect to the notion of a "now". We can say that the "now", the time in which those non-conceptual contents are displayed, never is like a "point" with no temporal extension. The "now" also has the form of a gerund. The now is always a "now-ing". Moreover, the "now" is always a gerund containing many other gerunds.

Are we simply adding some weird things, called "points of view", to the physical world? This would be to misinterpret our proposal. It is not a mere addition. Points of view are indispensable elements of reality. Epistemologically, a physical world

---

[4]We experience the world. And we also experience that we are experiencing the world, i.e., that we are having a world in perspective (some contents in perspective). That way, we become conscious of a peculiar sort of entity (a person, ourselves) having some peculiar properties (mental properties). Even when the contents in perspective are purely conceptual ones, and not the non-conceptual contents associated with having a world in perspective, to experience that we are having those conceptual contents in perspective would have a very important kind of explicit non-conceptual content.

external to all points of view, a physical world involving a physical space-time, is something postulated from some of our points of view. Ontologically, it is obvious that points of view exist. More precisely, to argue both that they exist and that they do not exist would show that they exist. The important problem with points of view is to understand them. And the most honest thing to say is that, in the last instance, we do not know how to put together the physical world and points of view.

This work has five sections. In Sect. 1 (*Temporal Truth-Makers*), we explain the notion of temporal truth-makers. In Sect. 2 (*Fluent Time and Points of View*), we argue that some of the contents of our points of view can play the role of temporal truth-makers for truths about the past, the present, and the future. In Sect. 3 (*McTaggart's Claims*) we compare our results with McTaggart's claims about the unreality of time, and we reject his arguments. In Sect. 4 (*Two Other Confusing Approaches*), we compare our results with two more recent approaches, representative of two very influential conceptions of time. We also consider that both approaches are wrong. Section 5 (*Tensed Truths and the Sea-Battle Problem*) offers some conclusions and an application of our results to the classical Aristotelian problem of the sea-battle.

1 Temporal Truth-Makers

McTaggart's problematic about A and B temporal series can be reformulated using the notion of temporal truth-makers determining the truth of some answers to when-questions. That way, we will obtain very directly some important results concerning the existence of temporal truth-makers associated to A-series. Such truth-makers would have to be objective even though they were to be placed outside the world of physics. In the next section, I will argue that the non-conceptual contents of points of view, or perspectives, can play such a role.

1.1 When-Questions

Temporal concepts are needed to give answers to when-questions such as

When did it happen?
When will we do it?
When do you want it?

Certainly, there are many when-questions in which the word "when" is not used. However, it is arguable that every when-question can always be rephrased as a question with an explicit "when".[5]

It is very difficult to imagine a language without when-questions. That language would not contain any sort of verbal tenses, nor any sort of temporal adverbs either. Even if representation, expression, communication, and other linguistic actions, were possible in one such a language, none of the things represented, expressed, communicated, etc., could be represented, expressed, communicated, etc., as a temporal entity, as something having a beginning, and an end, as something taking time, as something that is occurring just in the present, or in the past, or in the future. And this would also apply to the representation, expression, communication, etc., of those very acts of representation, expression, or communication.

In an intuitive way, we can define time, a fluent time, as what makes it possible to answer when-questions. More precisely, that fluent time has to be what makes it possible that some when-questions can have non-arbitrary correct answers.

1.2   A-Answers and B-Answers

We must distinguish between two different kinds of answers to a when-question. One kind of answer states that some things are, or happen to be, in the present, in the past, or in the future. The other one states that some things are, or happen to be, simultaneous to other things, or before other things, or after other things. These two kinds of answer correspond to McTaggart's series A and B. So, we can call them, respectively, A-answers and B-answers.

Each kind of answer would have a peculiar sort of TM. And the TM of A-answers are very different from those of B-answers. Whereas in order to account for the non-arbitrary correction of an A-answer, something capable of making real differences among the present, the past, and the future is needed, in order to account for the non-arbitrary correction of a B-answer, something capable of making real differences among things occurring simultaneously with other things, things occurring before other things, and things occurring after other things is needed.

So, we can distinguish between the temporal TM of A-answers and the temporal TM of B-answers. Let us call them, respectively, A-TM and B-TM.

---

[5]This is pretty clear with interrogative expressions like "At what time/date/instant/moment …?", or "At which time/date/instant/moment …?". I assume that the same happens with other more complicated or unusual expressions.

1.3 B-Truth Makers

So, we pose when-questions. And we want non-arbitrary correct answers. Temporal TM select those answers. There are two kinds of answers, and two kinds of temporal TM: A-TM and B-TM.

How to look for those temporal TM? It is not difficult to find B-TM in the objective world such as it is described by science. Here, there are objective facts supporting the claim that some things occur before, or after, or simultaneously with other things.

Simultaneity seems to pose some problems. However, it does not pose any serious problem concerning B-answers. Certainly, according to Special Relativity, simultaneity is relative to a reference frame. We can say that simultaneity becomes a relative notion. Moreover, this entails that "before" and "after" also become relative notions. However, there are objective facts accounting for the relativised relations of simultaneity. And there are objective facts accounting for the relativised notions of "before" and "after".

1.4 A-Truth Makers

It is not so easy to find A-TM. The objective world described by science, in the last instance by physics, does not permit distinguishing present things from past or future things, at least not in a non-indexical way. Fluent time, a time involving the notions of present, past, and future, seems to be beyond physics.

Let D be a description intending to apply to all and only those things that are in the "present". And let us suppose that n satisfies D. This is consistent with imagining an n' qualitatively identical with n, and so being able to satisfy D, but placed not in the present, but in the past or in the future. In such a situation, it seems that we would have to think that D is not a complete description, and that a complete description would have to specify that the item placed in the present is n, but not n'. However, there is no scientific (or at least, non-indexical scientific) way to construe such complete descriptions, excluding possibilities such as cyclic universes, episodic repetitions of some present things, parallel universes, etc. In all of these possibilities, there would always be more than one unique thing satisfying D, with the only difference that those other things would not be placed in the present.

In the objective world described by science, there is no room for "all and only those things that are in the present". In the last instance, there is no such class of things in the physical world. And without a place for things that are in the present, there seems to be no room either for past things or future things. There are no such classes either. Of course, we can define present things, and so past things and future things,[6] making indexical references to our actions and thoughts. But this would not count to making objective scientific descriptions.

Hence, we can find B-TM in the objectivity of science, but not A-TM. And this suggests the following question: Is subjectivity the adequate place of A-TM?

B-TM can be understood as the "physical time". Can we maintain that A-TM are merely subjective entities? Would those A-TM constitute what sometimes is called the "subjective time"? That being the case, past things, present things, and future things would have an adequate place in the subjectivity of our minds. There are, however, two serious problems with this approach.

The first one is that there is something really odd with the idea of "purely subjective" truth-makers, and therefore that there has to be also something very odd with the idea of purely subjective A-TM.

The second problem is that, in any case, the idea of a merely subjective time linked to A-TM completely different in kind from B-TM, does not make sense either.

1.5 Purely Subjective A-TM?

For convenience, let us assume that a truth-maker is the "minimal" state of affairs, or the minimal fact, in virtue of which a truth bearer has the truth value it has.[7] Temporal truth-makers would determine the truth of certain A and B answers to some when-questions. What can it mean that A-answers only have "subjective" truth-makers, i.e., that A-TM are purely subjective?

It cannot mean only that A-TM are in some way "dependent on" subjective facts. If that were the case, then that dependence could be supported by "objective" facts. Both things are perfectly compatible. Subjective facts would be subjective, of course. But, there could be objective facts about the sort of dependence that A-TM would have on those subjective facts. And those objective fact could provide the required (minimal) A-TM. At the end, the A-TM could be of the same kind as the B-TM.

Purely subjective A-TM would have to be subjective in a much more radical sense. In order to be purely subjective, the following requirement has to hold:

Q. The existence of a subjective A-TM cannot entail the existence of any objective fact.

---

[6]Present, past, and future are interrelated concepts. But the present has a certain primacy. Suppose we were to know all correct B-answers. Then, in order to place something in the past or in the future we would need to know how to place other things in the present. However, in order to place something in the present we would not need to know how to place other things in the past or in the future.

[7]As I said, I am not endorsing any particular theory of truth-makers. Here, "minimal" only means that the truth-maker has no proper part that is also a truth-maker of the same truth bearer.

However, this requirement poses a crucial problem. Let us consider the following statement

(a) A certain A-TM, let us call it M, exists for some A-answer T.

Requirement Q can be fulfilled only if (a) itself does not have any TM!

The claim (a) cannot itself have any, let us say, "objective" TM. If there were to be one such objective TM for (a), then there would be also one (at least one) objective fact entailed by the existence of M for T, and the requirement Q above introduced would not be satisfied.

There cannot be purely subjective TM. To suppose them leads to contradictions. If there were to be purely subjective TM for a certain TB, for instance, T, then there could not be any objective facts in virtue of which statements like (a) can have a truth value. And without any such TM for (a), T itself becomes unable to be an adequate TB. In other words, T is a TB only if (a) has a TM.

Without (a) having a TM, T becomes something with no cognitive value.

Another equivalent way to express the last idea is by saying that, in that case, any cognitive value, and any truth value, associated to T would be completely "arbitrary". T could have any truth value we want.

1.6 A Merely Epiphenomenal Time?

In virtue of some A-TM, certain A-answers to some when-questions have to be true. Can the fluent time associated with those A-TM be only a purely subjective, "merely epiphenomenal" time?

I think that the answer has to be negative. The fluent time originated by A-TM cannot be an epiphenomenal time. This result would follow from the very meaning of "epiphenomenal" in conjunction with the fact that A-TM are different in kind from B-TM. The argument is the following one:

1. Something is "epiphenomenal" only if it can be fully explainable, including its intended causal effects, through other phenomena which are not epiphenomenal. A rainbow in the sky, a bent rod in the water, a train coming toward us in a movie, are typical examples of epiphenomena. In all these cases, there are other phenomena capable of "explaining" the respective appearances, including all their intended causal effects.
2. The fluent time associated with A-TM cannot be reductively explainable through what is offered by B-TM.
3. There is no other way to explain reductively the fluent time associated with A-TM.
4. Hence, the fluent time associated with A-TM cannot be epiphenomenal.

To be epiphenomenal is not simply "not to exist". We are not epiphenomenalists about unicorns or about Santa Claus. To be epiphenomenal is to be fully explainable, in a reductive way, through other kinds of phenomena.

Let us compare ladders, rainbows, colours and time. To be a ladder is something epiphenomenal with respect to other things. In order to explain what a ladder is, we can say

A ladder is a structure consisting of two long sides crossed by parallel rungs, used to climb up and down.

Something like that may constitute a complete, or at least easily completable, reductive explanation of what a ladder is. If we understand the words used in the explanation, we understand what a ladder is. Assuming that things like "to have two long sides", "to have parallel rungs", "to climb up and down", etc., are not epiphenomenal, we explain what a ladder is by saying that. To be a ladder can be taken as something epiphenomenal in the sense that it can be fully explainable, and its causal effects can be fully explainable, through other phenomena which are not epiphenomenal.

Now, consider a rainbow in the sky. Assuming that we can explain what colours are in relation to some properties of the light and some properties of our perceptual systems, we could explain what rainbows are by saying

A rainbow is an arc of spectral colours, usually identified as red, orange, yellow, green, blue, indigo, and violet, that appears in the sky opposite to the sun as a result of the refractive dispersion of sunlight in drops of rain or mist.

Assuming that we can explain what colours are, rainbows are epiphenomenal. Again, under that assumption, we can fully explain what rainbows are, including all their supposed causal effects. Rainbows are epiphenomenal in the sense that they can be fully explainable through other phenomena which are not epiphenomenal.

Can we, in just the same way, explain what colours are? If we identify colours with certain combinations of properties of the light and properties of our perceptual systems, it is easy to do it. However, it is not so easy if we consider colours as a kind of "subjective experience". Simply, there is no complete explanation of what colours are in that phenomenal sense. Colours in that sense, as a kind of subjective experience, are not fully explainable, nor are their causal effects fully explainable either, through other phenomena which are not epiphenomenal. And this being so, colours cannot be something merely epiphenomenal.

The same happens with fluent time, with the time associated to A-TM. To the extent that the appearances of a fluent time exist, i.e., to the extent that we really have them, and they are not fully explainable through other phenomena which are not epiphenomenal, that fluent time cannot be a merely subjective epiphenomenon.

In the last section, we have claimed that A-TM could not be purely subjective without being completely "arbitrary". Now, we can claim that the fluent time associated to A-TM could not be a merely subjective time in the sense of being "epiphenomenal".

1.7   Objective A-TM Outside the Scope of Physics?

Let us summarise our main results:

(a) A-TM are very different in kind from B-TM. In particular, no combination of B-TM would constitute an A-TM.
(b) In the world objectively described by science, in the last instance in the physical world, there are only B-TM.
(c) If A-TM are purely subjective, then they become completely "arbitrary".
(d) A-TM cannot be merely "epiphenomenal" either. The temporal A-answers associated to those A-TM cannot describe only a merely subjective epiphenomenon.

How to maintain these four claims together? This is a very difficult problem. (c) invites us to place A-TM in the same world in which we can find B-TM. However, as (b) states, if that world is in the last instance the physical world described by science, then A-TM cannot be placed there. Even if B-TM have a place in that physical world, A-TM do not. More precisely, they cannot have such a place if they are as different in kind from B-TM, as (a) claims.

Another option is to consider that in some way A-TM are placed in our subjectivity. However, according to (c), they could not be purely subjective without being completely arbitrary. Moreover, according to (d), they could not be considered a merely subjective epiphenomenon, nor could the fluent time described by the respective A-answers be so considered.

2   Fluent Time and Points of View

If some A-answers have TM, and so can have a cognitive value, then there has to be something which can have an objectivity different from the objectivity of the physical world such as it is described by science. My hypothesis is that the non-conceptual contents of our point of view, or perspectives, can play such a role.

2.1   Points of View and Their Non-conceptual Contents

Points of view (hereafter PoV) can have contents of two kinds: conceptual (CC) and non-conceptual (non-CC). The second ones, non-CC, constitute all we can come to experience. They are the contents of our experience.

A very important claim is that our experiences are always experiences of a changing world. We do not experience objects and properties separately. Strictly speaking, we do not experience objects "that have properties", nor properties "that are had by some objects". And we do not experience objects "that change" either. We experience objects having properties, properties being had by some objects, and objects changing their properties. This would be the correct way to describe our experience.

There is a very simple sort of consideration that can support that crucial claim. Think of the apparent movement of a light through a lineal sequence of bulbs. The bulbs go on and off creating a visual illusion of movement. What is the "content" of our perceptive experience? Some bulbs going on and off? Or a light moving from one position to other in the lineal sequence of bulbs?

Suppose that we opt for the first answer. Now, imagine that the bulbs themselves are only apparently "the same bulbs". There are many bulbs occupying successively the respective positions. And they create the visual illusion of "a single bulb" going on and off. We can repeat our question. What is, in this new case, the "content" of our perceptive experience? A number of bulbs, on each position of the sequence, creating the visual illusion of being single bulbs, which in turn are creating a visual illusion of movement? Or, a light moving from one position to other in the lineal sequence of bulbs?

If, faced with these last questions, we continued opting for the first sort of answer, the situation could be repeated again and again. In the end, the non-CC of our experiences would become something beyond any familiar specification. Moreover, indeed, we would have "no way to know" the contents of our experiences!

2.2 Objectivity in Non-CC

The non-CC of our PoV are the contents of our experiences. Can we "share" our experiences? In a certain sense, we obviously can, and very often we do it. To the extent that we distinguish the subject having a PoV from the contents of the subject's PoV, there is nothing in the notion of PoV excluding the possibiliy that different PoV's can have "the same contents" in perspective, both non-CC or CC.

There are "intersubjective PoV", both conceptual and non-conceptual. Different subjects can have the same experiential non-CC under their respective PoV, in the same sense in which different subjects can have the same landscape in perspective. The subjects are different. And, consequently, the relations they maintain with the contents of their respective PoV can be different. But the non-CC can be the same. In fact, when we say that all of us are living "in the same world", we are pre-supposing that the non-CC of our PoV can be the same, or approximately the same.

To say that all of us are living "in the same world" is not to say (in any case, it is not only to say) something like

All of us are made of atoms and molecules of the same kinds.

It is to say

All of us are facing the same chairs and tables, the same trees, the same sky, etc., and the same atoms and molecules.

The non-CC of our PoV are "objective" in that sense. And that sense is not only an epistemic sense. That a variety of PoV can "share" the same non-CC, and also the same CC, and thus that a variety of subjects can have similar PoV, or that they can compare their PoV in relation to their contents, etc., are not only epistemic facts. They are ontological facts about the nature of PoV. They are very important ontological facts about what PoV are.

All of that has relevant consequences in relation to the existence of a fluent time. The fluent time associated with the non-CC of our experience also could be "objective" in that sense. Also, there could exist a "shared fluent time". It could be as shared as chairs, tables and trees can be. And as shared as the sky.

2.3   Are Points of View Only a Weird Ontological Addition?

We are looking for some "objectivity" different from the objectivity of the physical world such as it is described by science. And my hypothesis is that we can find such objectivity in some of the contents of PoV, more concretely in their non-CC. However, there is something wrong in thinking of PoV as something that perhaps would have to be simply "added" to the objective world described by science, in the last instance by physics.

PoV are not a "possible addition" that has to be carefully assessed. The status of PoV is very different. PoV are unavoidable. Both to assume and to reject anything entails adopting a PoV. The very notion of assessing something entails adopting a certain PoV. We can say that PoV are what make experiencing, knowing, and acting possible. In particular, science could not exist if PoV did not exist.

Through science, we construe an objective image of the world. Also, we construe an objective image of ourselves. And in so doing, many things can be rejected as not having a real existence. But in no way can we reject the real existence of PoV.

2.4   Internal/External to a Point of View, Internal/External to a Subject

There is a crucial distinction between

the subjects, and
2. their PoV

and a correlative distinction between

3. what can be "internal to the subjects", and
4. what can be "internal to their PoV".

To have certain contents in perspective, to have certain non-CC or CC contents, can be something "internal to the PoV of a subject" without being something "internal to the subject" individualistically considered.[8]

To take seriously the notion of PoV leads to a rejection of the claim that PoV are something merely internal to the subjects. PoV redefine the "spaces of possibilities" relevant for a subject. Many PoV open "spaces of possibilities" that would not be available if the subject were not to take those PoV, or if the subject were to abandon them. In that sense, PoV are not merely internal to the subjects. They have "objectivity".

Now, we can apply the above distinction to A-answers, to their respective A-TM, and to the peculiar kind of fluent time linked to those A-TM. In principle, there could be temporal TM in the following fields:

1. A field external to all the subject's PoV.
2. A field internal to some of the subject's PoV, but external to the subject.
3. A field internal to some of the subject's PoV and also internal to the subject.

My claim is that in 2, and only in 2, we can find adequate A-TM capable of supporting A-answers to when-questions. In 1, we can only find B-TM, but not A-TM. In 3, we could only find arbitrary truth-makers, and so very questionable A-TM.

Indeed, 2 offers a very important sort of fluent time. It is a time different from the time present in the physical world, and it is different too from what can be found in our subjectivity.

Two further remarks are worthy of attention. Firstly, we can call the time associated with A-TM "internal time", and we can contrast it with the "external time" associated with B-TM. However, even in that case, the internal time could not be a purely subjective time. And it could not be a merely subjective epiphenomenal time either.

Secondly, even though A-TM are very different in kind from B-TM, the existence of A-TM entails the existence of B-TM. So, in field 2 we could find both A-TM and B-TM. And it is very suggestive to suppose that it is from the basis of these B-TM that we construe the scientific image of the world, and of ourselves.[9] Things are not so in field 1. The existence of B-TM does not entail the existence of A-TM. And this would be the reason why it is not possible to understand A-TM exclusively from the "scientific image" of the world, and of ourselves.

---

[8]We can say that PoV are "modal ways of being in relation" to the world and to ourselves. The highly relational and strong modal character of PoV, together with their heterogeneous nature, make it very difficult to think that they can be reducible to subjectivity, or to information, or to physics.

[9]There is in Russell [14] a very elaborated example of that sort of construction. It is explained in other chapters of the book.

2.5   *The Non-conceptual Contents of a "Now"*

Non-CC can be internal to PoV without being internal to the subjects having those PoV. Non-CC also are tensed entities. They have a past, a present and a future. Non-CC can be so because the "now" in which we experience any non-CC is not like a "point" without any extension, but something with the form of a gerund. The "now" we experience always is a temporally extended "now-ing". It includes a certain past, present and future.

The "now" we experience is always a now in which we experience changes in a non-reducible way. As we said, we do not experience objects "that have some properties", or properties that "are had by some objects". We experience "objects having properties". These objects having properties constitute time in the "now" of our experience. And that time is a fluent time.

The pivotal question is:

Does it make sense to say that we are experiencing something "now" as past or as future?

My answer is affirmative. It makes sense because to experience any non-CC is to experience it as "something changing". And to experience something changing is to have an experience of something being past, present, and future. We are experiencing many things "now" being past, present, and future.[10]

Is, then, the present a part of the "now"? Yes, it is. The present is a part of the experienced "now". And some very peculiar little pieces of past and future also are parts of the "now". Properly, there is no other way of having an "experience of change", instead of having simply a set of, let us say, "pointillist" experiences.[11]

We are introducing a very important distinction between, on the one hand, the "now" in which we have an experience and, on the other hand, the "present" that results from contrasting some things actually existent with other things that are past or future. That distinction between the "now" and the "present" is crucial. The source of the two notions is very different. Whereas the "now" is something we experience, the "present" comes from a contrast with the past and the future.

We could say that, by itself, the "now" is not tensed. We could say that it is an "absolute happening", a happening not reducible to anything else. But this would need much more elaboration. In any case, the "now" is the source of a fluent time. What is clearly tensed is the "now" under the perspective of a temporal point of view. The "now" is tensed when we see it from the (in some peculiar sense

---

[10]This is a fundamental claim in order to find in the non-CC of our PoV some adequate A-TM. My final aim is to suggest that the appropriate A-TM can be found in that sort of "now".

[11]This suggests a relevant difference between the non-CC and the CC of our PoV. When we have in perspective, for instance, that $2 + 2 = 4$, we are not faced with any change. Of course, as we noted, to have in perspective that we are having in perspective that $2 + 2 = 4$ is to have in perspective a non-CC, and hence to have in perspective a set of changes.

reflective) perspective of a temporal PoV. From what is contained in the "now", a temporal PoV generates, and extrapolates, a past, a present, and a future.

The "now", what is contained in the "now", is what makes possible the existence of a fluent time in which we come to identify some "things" as going from the future to the present, and then to the past. Moreover, this is what makes possible any extrapolation to a remote past and to a distant future.

From a reflective perspective, we can also speak of the "now" itself, instead of speaking of "the things or events that are now", and say that we experience the "now" as moving from the past to the present, and then to the future. But that is only a loose way of speaking.

The notion of a "now" is really peculiar. It is "orthogonal" both in relation to A-series and in relation to B-series. On the one hand, the "now" can be contrasted with what is "before", and with what is "after". But "to occur now" is not the same as "to be simultaneous to". Whereas "to be simultaneous to" is a dyadic relation, "to occur now" is not a relation. Moreover, "to be simultaneous to" can be placed in the past and in the future, and not only in the present.

On the other hand, "to occur now" is not the same as "to be present" in contrast with "to be past" and "to be future". Some little pieces, or fragments, of past and future are occurring "now". The argument again is that, if the now consists in experiencing changes, then some little fragments of past and future have to occur just "now".

So, I have "now" some experiences of the past and the future. And I have also some "present" experiences. The "present" is, we can say, the fully existing part of a "now", that part which is neither past nor future. However, a "now" also has other existing parts: that part that is ending being present, and that part that is beginning to be present. The first part is the little fragment of past we are experiencing "now", and the second part is the little fragment of future we are experiencing "now".

Memory and imagination can be thought of as part of the mechanisms through which we are capable of experiencing a "now". They make possible our experiencing a "now". However, properly, they are not the experiencing of a "now". There is a big difference between the mechanisms and what they make possible. The past in a "now" cannot be only a remembered past. And the future in a "now" cannot be only an expected future either. If we experience change, then we have to experience not only a remembered past and not only an expected future, but a "real past" and a "real future".

The present is not something like this (it is the definition given in my dictionary):

The period of time that is happening now, in the moment of the speech.

This is a sort of definition many philosophers have used in their philosophical theories. However, the present is not like that. Let us put aside the confusions that may be involved in an expression such as "period of time that is happening now". The point I want to note is that the present cannot be a "happening". If the present were something that is "happening", it would entail a change. And it would have to contain a certain past (perhaps a very small piece of past) and a certain future (perhaps a very small piece of future). But, the present is not a change. Strictly, the present cannot change. And it does not contain any past, or any future. What is a happening, what clearly entails a change, and what has a certain past and a certain future, is the "now".[12]

The issue is not only terminological. There is a real distinction to be made, and a very important one. The "now" and the "present" involve quite different temporal perspectives. Consider someone waiting for something with anxiety, and saying either

(a) "now ..., now ..., now...,"; or
(b) "in the present ..., in the present ..., in the present ...,"

Only (a) makes sense! The reason is that only the "now", but not the "present", is a happening. Only the "now" entails a change.

What is a happening, and a very special kind of happening (perhaps an "absolute" happening), is the "now". What is a change (perhaps an "absolute" change) is the "now". The "now" is constitutively a happening containing many other happenings: namely, its non-CC. The "now" is a change because it is always a "now-ing", full of other "...-ings", constituted by many other "...-ings".

We can claim that the primary source of fluent time is the (absolute) happening of a "now". And the "present", whatever "present", is always part of a "now". Primarily, the "present" is that part of the "now" contrasted, from a certain temporal point of view, with the close past and with the close future included in that very "now". This is the source from with any other past and any other future is extrapolated.

The "now" and the "present" involve very different temporal perspectives. Also, the semantics of "occurring now" is very different from the semantics of "is present":

"Occurring now", on each occasion of use, refers (primarily) to what is experienced as a changing existing fact by the subject saying it.[13]
"Is present", on each occasion of use, refers (primarily) to what is neither past nor future in what is experienced as a changing existing fact by the subject saying it.

As we said, experiencing something as an existing fact is always experiencing it as a "changing thing". Properly, only beings capable of experiencing changes are beings capable of having experience. It is because we experience changes that it makes sense to say that the present becomes past, and that the future becomes present. But, again, this is only a loose way of speaking.

---

[12]And, of course, at this point it would be viciously circular to define the "now" as the period of time that is happening now, in the moment of the speech.

[13]In that token-indexical referential rule, a persisting thing (a thing that continues to exist) would count as a part of a changing thing. The changing thing would be the whole constituted by that persisting thing and its environmental, or our psychological, conditions. Without something changing, there cannot be persisting things either. Of course, this would be the primary sense of "persisting thing". In a non-primary sense, it may be enough that some change is seen as possible.

However, even if it can make a certain sense to say that the present becomes past, and that the future becomes present, it makes no sense to say that a now becomes past, or that a future becomes now. The "now" cannot become because it is a becoming.

Another interesting difference between the "now" and the "present" is this. There can be a first "now", but not a first "present". We can experience a "now" for the first time, but we cannot experience a "present" for the first time. To experience a "present" always entails to experiencing a certain past that was present, and also to experience a certain future that will be present. This is to have experiences of a fluent time.

We can say that the world we encounter in our experience had a first "now", but we cannot say that the world we encounter in our experience had a first "present". For every intended present, there is always a certain past that was present.

In some cases, the fluent time itself is experienced, even occasionally conceptualised, as an existing fact by the subject who is having in perspective the relevant non-CC. In these cases, "the flux of time itself" would be an existing fact. But this would be so only for beings capable of having a peculiar kind of "reflective" PoV about the contents of their experiences.

2.6   Non-CC as A-TM

How can the non-CC of our experience have a truth-making role in relation to some A-answers? How can those non-CC be the A-TM of tensed truths?

Let us begin by introducing a meaningful distinction between two kinds of TM:

A <u>direct TM</u> for a statement S is a TM that includes the TC of S.
An <u>indirect TM</u> for a statement S is a TM that does not includes the TC of S.

This distinction also applies to A-TM. I want to argue that the non-CC of our experience can be direct A-TM for some tensed TB. Which ones? Here is the answer:

The non-CC of our experience can be direct A-TM of those tensed TB stating something we are experiencing "now" either as past, or as present, or as future.

"I am finishing my work", "Look! The car is approaching us", "I am going to cry", can be examples of tensed truths of that kind. We can have experience of their respective TC, and those TC are their A-TM, or at least part of their A-TM.

However, this is so only with respect to some tensed TB. With respect to other tensed TB, things have to be done much more "indirectly". Consider, for instance,

Tomorrow, there will be a sea-battle.

The TM we can find in the non-CC of our experience could not be direct A-TM of the truth of tensed TB like those. At least for subjects like us, for subjects capable of having the sort of experiences we have, i.e., the sort of experiential "now" we have, these are not truths supported by any direct A-TM.[14] The A-TM of these statements do not include their TC. These truths can have only indirect A-TM. Moreover, in many cases, these A-TM will be some non-CC of our experience in combination with the TM of some of our true conceptual contents about the world.[15]

Let us consider the above statement about a certain sea-battle. Its TC is placed in the future. It is placed tomorrow. So, the TC of the statement is not an existing fact. We can say that from a temporal perspective that includes the non-CC of our imagination and the CC of our forecasting. However, the A-TM of that statement can be an existing fact. This A-TM can be a combination of some of the non-CC we are experiencing "now" and the TM of some of the conceptual truths present in my conceptual points of view (for instance, about the unavoidable character of the sea-battle in the circumstances in question).

We can compare the above statement with other ones having directly some non-CC of our experience as their TM. For instance,

The sea-battle is starting.
The sea-battle has just finished.

We could have non-CC in our experience, in the "now" of some of our experiences, capable of being direct A-TM of these statements. The TM of these statement can include their TC.

To experience a "now" is to have in perspective a changing world full of a huge variety of events. Some differences are taken as changes. This is the key move. We can say that to experience a "now" entails embracing a temporal point of view. Only entities able to embrace temporal points of view can experience a "now".

2.7   Temporal Points of View

We have offered a way of understanding A-TM capable of supporting the existence of a fluent time, which can be internal to our PoV but external to the subjects having those PoV. The existence of a fluent time depends on the existence of certain PoV with some peculiar kind of explicit non-CC. There is a fluent time when the non-CC of our experience are seen from a temporal perspective.

We can define a <u>temporal PoV</u> (in short, TPoV) in the following way:

---

[14]Could God's "experience" contain direct A-TM for any tensed TB? I will leave open this question. In any case, it is not necessary that the A-TM of tensed TB, either direct or indirect, are the same forever. They are existent facts. Hence, they can change.

[15]Those conceptual contents can involve, in various ways, other non-CC. However, these non-CC would not have here the role of being A-TM, or parts of the A-TM, of the relevant tensed truths.

A TPoV is a PoV in which some differences in non-CC are identified, or recognised, either conceptually or not conceptually, as changes in time with respect to distinct positions in an A-series.

In TPoV, some differences in the contents of experience are seen as temporal differences, i.e., they are seen as a change. Furthermore, as we said, it is only from a background of changes, actual or possible, in the contents of experience, that we come to identify, or recognise, "permanencies" in time, "persisting things".[16]

Only subjects able to have experiences and with the reflective capacities to have TPoV are capable of detecting and knowing the internal time of their PoV, and also the internal time of the PoV of other subjects.[17]

For the existence of a TPoV it is neither necessary nor sufficient that there be different PoV, or that there be a change of PoV either. In both cases, we would have either a variety of PoV, or a variety of PoV changing in time. However, strictly, we would not have a TPoV. In order to have a TPoV, what is required is to take certain "differences in content" (in non-CC) as "changes of content".

In order to describe a TPoV, we need to make reference to certain A-answers with their respective A-TM. We need some conceptual resources. However, TPoV can exist without those descriptions and resources. They can exist in a "non-conceptual", or perhaps "pre-conceptual", level. The relevant identifications and recognitions can be made in non-conceptual ways.

Another very important feature is that TPoV can generate more than one "temporal perspective". This means that there could be many possible pasts (not only a number of different reconstructions of the past), many possible presents (not only a number of different ways to live the present) and many possible futures (not only a number of different imagined futures). So, in principle, in correspondence with different TPoV we could have a "plurality of fluent times".

The possibility of such "temporal pluralism" is entailed by our approach. However, it can be argued that there are normative principles regulating the adoption of TPoV, and also regulating the constitution of intersubjective TPoV, and that those normative principles are able to reduce that temporal pluralism. It can be argued that the dynamics of TPoV, and the dynamics of the interrelations among individual TPoV and intersubjective TPoV, are able to induce a progressive convergence.

---

[16]The two main ways of understanding "persistence in time", Perdurantism and Endurantism, would reflect more complexities in the relevant TPoV. For Perdurantism, persistent things have temporal parts in different positions in the past, the present, and the future throughout their existence (a usual way of explaining this is saying that they are like "spatio-temporal worms"). For Endurantism, persistent things are wholly present at every position in the past, the present, and the future throughout their existence (they are like some kind of "substrata of change"). We will not discus these proposals here. See the seminal work of Lewis [6]. More recently, see Sider [15]; and McKinnon [8].

[17]We have to note again that to experience (reflectively) that some CC are in perspective entails having some non-CC in perspective.

That way of speaking may seem odd: a plurality of fluent times being progressively reduced thanks to some normative constraints over the adoption of TPoV and over the constitution of intersubjective TPoV! However, it has to be emphasised that it only will seem odd if we continue seeing the problem of understanding a fluent time as a problem about something existing outside "all points of view". From this perspective, the above way of speaking would be really odd. But, it is not our perspective.[18]

That temporal pluralism and temporal convergence would not be a process taking place outside "all points of view". However, it would not be either a mere subjective process. What we are suggesting is not a mere epistemological reduction. In that case, the problem of temporal pluralism only would be a psychological problem, and it would not have to provoke any reaction of "oddity". But, it is not a psychological problem!

Temporal pluralism and temporal convergence of a fluent time would be ontological phenomena taking place inside our PoV. They would take place in that field of "objectivity" that is internal to some PoV but external to the subjects having those PoV.

We can say that TPoV are a peculiar kind of reflective PoV. They entail some sort of reflective movement. However, we have to insist, TPoV do not require highly developed conceptual capacities. The relevant identification or recognition can be made through non-conceptual recourses. This means that subjects without conceptual capacities can be able to have in perspective a fluent time. But, it also means that only subjects with a minimum of reflective non-conceptual capacities would be subjects capable of having in perspective such a time.

2.8 Going Upstream

We begin by focussing on when-questions. And we distinguished between A-answers and B-answers. The A-TM of A-answers were very special. They did not have a clear place in the world objectively described by science, in the last instance the physical world. And they did not have a clear place in our subjectivity either. They could not be purely subjective without being completely arbitrary. Moreover, they could not be mere subjective epiphenomena. We found something capable of playing the role of A-TM in the non-CC of our PoV. And we have argued that time exists, a fluent internal time associated with A-TM, because there are TPoV.

That fluent time is internal to some PoV qualified as TPoV, but it is external to the subjects having those TPoV. It is not the time of the physical world. But it is not

---

[18]That perspective would be an "absolute" perspective. With respect to this notion, and other related notions, see other chapters of the book.

a time merely subjective and epiphenomenal either. It has an "objective reality" inside our TPoV.

It is important to note that I am defending what D.H. Mellor called an A-theory of time.[19] "To be past", "to be present", and "to be future" are taken to be real properties.[20] Mellor argues that there is in reality no such thing as being past, present, or future. There is in reality no such thing as a "fluent time". The only "real time" that exists is the one constituted by McTaggart's B-series. According to Mellor, to say this is to maintain a B-theory of time. He argues that a B-theory is the correct ontology of time. Also, that a B-theory is semantically enough in order to understand how statements about the past, the present, or the future can have truth values. What makes "e is past" true at any time t, he claims, is the fact that e is earlier than t; what makes "e is present" true at any time t is the fact that e is located at t; and what makes "e is future" true at any time t is the fact that e is located after t.

We have argued that this strategy cannot identify a particular "now". This can only be done from some TPoV. "To be past", "to be present", and "to be future" are real properties. Their source is a "now". They exist inside our TPoV without being mere subjective epiphenomena. They have an objectivity internal to our TPoV. And our TPoV are neither eliminable nor reducible. They are a part of the world.[21]

It is also important to note that the kind of objectivity we have found inside PoV, in particular the kind of objectivity for a fluent time we have found inside TPoV, has similarities with the Kantian approach to time. For Kant, time is not "noumenal", it does not belong to "things-in-themselves", but it has objectivity. Space and time come from how subjects like us are capable of having a world in

---

[19]See Mellor [11]. I am adopting an approach opposite to the one defended by Mellor, but I consider that this book offers the best discussion of the many problems involving time.

[20]More precisely, what I want to defend is what Baker [1] calls a "BA theory of time". Let us quote her (pp. 149–50): "According to the BA theory, time has two irreducible aspects: one that depends on there being self-conscious entities (the aspect of the A-series, the ongoing now) and one that does not depend on self-conscious entities (the aspect of the B-series, simultaneity and succession). [...] it is part of the nature of time to be ordered by "earlier than", etc. (the B-series); but it is also part of the nature of time that it is experienced by self-conscious beings as ordered by past, present, and future (the multiple A-series). Everything that a self-conscious being is aware of—what someone else is saying, natural events, one's own thoughts, one's rememberings, what have you— is always experienced as being present. In the absence of self-conscious beings, we might say that the A-series is dormant (or merely potential, or not manifest)". Two points are worthy of attention. The first one is that, so understood, the A-series would be "multiple", even though they could converge too. The second one is that because the ontological "places" of those A-series are the multiple perspectives of conscious beings, there would not be any direct problem about simultaneity coming from the Special Theory of Relativity.

[21]According to Mellor, even though there is no "fluent time" in reality, we think and act as if there were. And we can say with truth that something is past, present, or future. Mellor's position would be in clear contrast with the following passage of Baker [2]: "A metaphysical theory should help us understand reality and our experience of it. It is difficult to see how understanding is served by the suggestion, for example, that it is never the case that, ontologically speaking, there is exactly one cat in the room. It is even more mysterious to add that we shouldn't worry about this since we still may truly *say* that there is exactly one cat in the room". More about this approach in Baker [1].

perspective (the same can be said of the notions of causality, object, etc.). There are, however, crucial differences between our approach and Kant's approach. We have introduced our main concepts through a reflective move about our PoV. But we have tried to avoid any kind of "transcendental mood". Our reflection has tried to transcend any particular PoV in a constructive way, without adopting transcendentalism. In contrast, Kant's approach is explicitly transcendentalist. Moreover, with respect to space and time, what Kant does is to "transcendentalize" the Newtonian conception of an absolute space and an absolute time.

We will finish by contrasting our approach with the classical approach of McTaggart. Then, we will contrast it with two other very influential approaches. One of them has been recently defended by Andy Clark, the other one is defended by Benjamin Libet.

3 McTaggart's Claims

Let us compare our results with McTaggart's claims against the reality of time. According to McTaggart, time is essentially constituted by A-series. But A-series cannot be real. They are mere appearances. We have used the notion of PoV to propose a way to understand the fluent time associated to A-series. And we have argued that it is possible to find adequate temporal truth-makers for the A-answers of when-questions in the non-CC of our PoV. In this section, we will see how our approach can offer important ways to resist McTaggart's arguments against the reality of time.

3.1 Three Arguments and the Pursuit of an Explanation

McTaggart[22] offers three main arguments against the reality of time. Two of them are positive and one is negative. According to one of the two positive arguments, attributions of different positions in an A-series always entail a contradiction. According to the other positive argument, the only way to try to escape from these contradictions would lead to vicious infinite regresses or circularities. The negative argument emphasises the need of taking some references which are external to the A-series, in order to avoid arbitrary attributions of temporal positions. The three arguments can be resisted in our approach.

There is something more in McTaggart's approach to time. Temporal A-series are not real, but they appear to be real. And this needs an explanation. McTaggart complements his three arguments against the reality of time with an explanation of

---

[22]McTaggart [9, 10].

the existence of temporal appearances. As we will see, McTaggart's explanation also is wrong.[23]

3.2   First Argument: Contradictions in the Attribution of Different Positions in an A-Series

The first positive argument of McTaggart can be presented as follows:

1. Being past, being present, and being future are incompatible characteristics. If something is past, then it is not present and it is not future; if it is present, then it is not past and it is not future; if something is future, then it is not past and it is not present.
2. But, everything changing has the three characteristics, or at least two of them. That is, every event has to be present and past; or future and present; or, future, present, and past.
3. Therefore, A-series always lead to a contradiction and their reality must be rejected.

1 can be accepted. 3 follows from 1 and 2. But, why accept 2? Why do all the three characteristics, or at least two of them if we consider the first events or the last events of an A-series, have to belong to each event? McTaggart claims that it has to be so in a "non-temporal", "absolute", "predicative" sense.

To be in the past, to be in the present, and to be in the future are incompatible features. Hence, if something changes in time, then in a non-temporal, absolute, predicative sense there would be something having at least two of those incompatible features. The intended conclusion is that there is no change. And, therefore, that A-series are not real because their reality would be contradictory.

According to McTaggart, there is no room for something "real but non existent". He argues that possibilities, including here temporal possibilities, cannot be real because any truth about them is reducible to differences in what in fact exist in a timeless sense. So, truths about the past, the present, and the future can transmit only these timeless differences. But, taken as such timeless differences, to be in the past, to be in the present and to be in the future are incompatible characteristics. Hence, nothing can change. And time cannot be real.

How to confront this argument? There is in our approach a distinction that has direct effects over McTaggart's claims. It is the distinction between the TC of A-answers and the A-TM of some of these A-answers.

On the one hand, the descriptions of the TC of A-answers can be tenseless, and they can be absolute. Even truths about being in the past, being in the present, and being in the future can have such a "tenseless" and "absolute" semantical sense.

---

[23]A very interesting overview about the history of the commentaries and criticism of McTaggart's arguments, and also of the complexities of his philosophy, can be found in Nyiri [12].

But, on the other hand, the descriptions of the A-TM of those A-answers have to be "tensed" and "relative". And the facts in virtue of which some A-answers are true, the facts in virtue of which some TB about the past, the present, and the future are true TB, have to be "temporal facts" having a certain position in A-series constituted by the non-CC of our experiences.

That something is in the past is semantically incompatible with being in the present and with being in the future. That something is in the present is semantically incompatible with being in the past and with being in the future. That something is in the future is semantically incompatible with being in the past and with being in the present. Truths about being in the past, being in the present, and being in the future have to maintain these semantical relations of incompatibility. And the same holds for the truth-conditions that the respective TB would seem to have "essentially". We can continue maintaining that semantics. However, things are very different with respect to the TM of those TB.

A crucial point is that, in contrast with TC, the TM do not need to be "essential" for those TB. So, it is possible that many TB about the past, the present, and the future would not have any TM. And when they have TM, some of these TM could have a tensed existence excluding the tensed existence of other TM.

This is our way to confront McTaggart's first argument. The relevant TB are about temporal possibilities. They are "possible temporal truths". And they may have a tenseless and absolute semantical sense. But the TM of those TB are tensed and relative. They are temporally relative facts. And that way, the true TB also can be "ontologically tensed".

TM are not TC. Obviously, TM always maintain some relevant relations with TC. However, they are very different entities. And they have a very different status and role in relation to A-series. The A-TM of an A-answer, if it exists, is a fact. It is a tensed fact having a particular location in an A-series.

We will make another important remark about the distinction between TC and TM. There are TB saying something about A-TM. These TB will have essentially some TC, and some of them would have TM in virtue of which they are true TB about some A-TM. Our results also will hold with respect to these new TB. Their TC could be timeless and absolute. However, the TM in virtue of which some of these TB are true TB (true TB about some A-TM) could be tensed and relative.

3.3 Second Argument: Infinite Regresses and Circularities

It is very important to distinguish between the positive argument above analysed and the positive argument we are going to face now. McTaggart's second positive argument is about a way to avoid the supposed contradictions pointed out in the first argument.

If there is no contradiction, as we have argued, then there is not any need to go out of it. However, let us examine this second argument by itself.

McTaggart considers the most obvious way of trying to escape from the contradiction of the first positive argument. The characteristics to be in the past, to be in the present, and to be in the future only are incompatible when they are taken in a timeless, or absolute, sense. We can say that they are incompatible only when they are taken "simultaneously". But, there would not be any contradiction in the fact that those characteristics can be had "successively": something was future, is present, and will be past.

McTaggart rejects this strategy arguing that we cannot make sense of that "successive" character except in a circular or regressive way. The moment at which something was future is a moment having the characteristic of being in the past, the moment at which something is in the present is a moment having the characteristic of being in the present, and the moment at which something will be past is a moment having the characteristic of being in the future. And when those moments are taken in a timeless, absolute, sense, they become again incompatible.

In short, we can say, the objection is that it is not possible to "distinguish" in a non-regressive or circular way the past, the present, and the future. This is not the usual way to describe the situation. But it is at the bottom of it. We have a problem of "distinction", or "discrimination", among the past, the present, and the future.

But, is this an objection? In fact, in the case of conceptual identifications, we never can conceptually identify things that are F except by applying the concept of being F, or another concept having, in the relevant circumstances, the same conceptual content as the concept of being F. To do this cannot be wrong. What here is different from that?

The situation is the following. There are PoV with non-CC among which there may be A-TM for some A-answers. Those non-CC are constituted by things changing, or by things having a certain permanence or persistence in time. As we said, the adequate verbal form for describing the non-CC of our experience would be the gerund. Our "now", a "now" full of objects having properties, always is a "now-ing". Our TPoV identify some of those changes and permanencies, either in non-conceptual or in conceptual ways, trying to eliminate the contradictions we come to discover, or the contradictions we could come to discover, or even the contradictions we simply could imagine. As a matter of fact, and as a normative issue, some A-TM exclude others. And that way, there may be an elimination of the A-TM in virtue of which semantically incompatible TB about being in the past, being in the present, and being in the future could become true TB.

It is not important here whether those eliminations are made by our TPoV of a more conceptual kind, or whether they are the result of the adoption of TPoV of a non-conceptual kind, which our conceptual TPoV then take note of. The important thing is that our TPoV "readjust" A-series when there is any such problem. In general, we can guess, the "readjustment" is non-conceptual and automatic. This is how our TPoV seem to work.

If TPoV exist, if they are non-eliminable and non-reducible parts of the world we live in, we can assume all of that as "brute facts". Simply, reality is that way. Is there anything wrong in that assumption? I think that nothing is wrong here either.

3.4   Third Argument: External References

The negative argument of McTaggart against the reality of A-series has not received much attention. But it is a very important argument. McTaggart argues that, in order to avoid arbitrary results, we would need to adopt some relevant references which are external to the attributions of temporal positions. But, at the same time, he claims that it is very difficult to imagine what those references could be.

According to McTaggart, to be in some relevant sense external to the attributions of positions in an A-series would be the way to avoid arbitrariness in such attributions. But, we have in our approach enough resources for satisfying that condition. We have proposed to obtain the relevant A-TM from the non-CC of our PoV. And these non-CC are

1. internal to our PoV, but external to the subjects having those PoV;
2. they can be contents of other reflective PoV, which in turn could take "rational control" over them; and
3. they can also be contents of the TPoV of other subjects (so, they can be contents of intersubjective TPoV), which can have other reflective PoV over them.

That way, the A-TM can have an external and non-arbitrary character. In particular, according to 1, they would be external to the subjects and, so, not merely subjective in an arbitrary way. Moreover, according to 2, there may exist an important kind of "self-reflective rationality" about A-answers. And according to 3, there may also exist an "intersubjective fluent time", a "shared fluent time", and an important kind of "intersubjective reflective rationality" about it. We adjust reflectively not only A-series internal to our PoV, but also A-series internal to the PoV of other subjects.

This means that fluent time is essentially involved in crucial processes of "modulation" and "tuning", many times with an interpersonal character.

3.5   The Explanation of Temporal Appearances

McTaggart's efforts to explain the existence of temporal experiences of a fluent time has not received much attention either. That there are "temporal appearances" of a fluent time is not in question.[24] But, why there are such temporal appearances? Are temporal appearances a mere epiphenomenon explainable by other deeper phenomena?

---

[24]This is a very important point. To question the reality of a fluent time is not to question the existence of "temporal appearances of a fluent time". That we have those appearances is not in question. The problem is to understand those appearances, in particular to assess whether they can be merely epiphenomenal appearances, or something else.

The first thing to say is that this is a typical Platonic problem.[25] The second thing to say is that McTaggart has no success in explaining temporal appearances.

McTaggart's main proposal for such an explanation can be found in volume II of *The nature of existence*. It is based on the notion of C-series. One way to understand these C-series is through relations of inclusion. Positions in a C-series are achieved through non-temporal relations of inclusion. Some things are included in other things, or the second ones include the first ones. C-series offer irreflexive, asymmetric and transitive relations of inclusion. In that respect, there are important similarities with some aspects of temporal B-series, and also with some aspects of temporal A-series. According to McTaggart, reality is structured in a completely non-temporal, absolute, sense by C-series. And C-series, McTaggart maintains, make room for considering temporal appearances as a "*bene fundatum* phenomenon".[26]

There are non-temporal relations of inclusion between real things according to a number of C-series. And they give support to temporal appearances. We experience, or we have epistemic access to some parts of those C-series, and we erroneously take them as constituting B-series and A-series.

According to McTaggart, C-series determine temporal appearances. So temporal appearances would be "epiphenomenal" in the proper sense of the term. But, this entails a big problem. The problem is that whereas temporal appearances change, and involve a "moving now", a "now" that is a "happening", the C-series do not change. They are absolute, timeless. So, how could the C-series determine temporal appearances? How could the C-series determine that we are experiencing "some parts" of a C-series, and not other ones perhaps qualitatively identical? How could the C-series "distinguish" among a past, a present, and a future? Moreover, how

---

[25]For Plato, change and time belong to "the World of Appearances". However, it is not easy to characterise them as "mere epiphenomena" determined by the reality of the Ideal Forms. The World of Forms alone cannot explain why there are appearances. See the difficulties Plato faces in his *Timaeus*. This is McTaggart's problem. And it continues to be our problem.

[26]The expression comes from Leibniz's *Monadology*. According to Leibniz, the ordinary world, space, time, and causality do not exist in reality. But they are "*bene fundata* phenomena" in the sense that they are grounded on other things that in fact exist. A repeated example of "*bene fundatum* phenomenon" is the rainbow in the sky. Rainbows are epiphenomenal because they are fully explainable by other facts which are not epiphenomenal. In McTaggart [10, Ch LI, # 613], we can read:

> There is, on our theory, no time-series, for nothing is in time. There is no series of events, but a timeless series of misperceptions which perceive a series of timeless existents as being in time. [...] although time is not real, the appearance of time is a phenomenon *bene fundatum*, since the order of the apparent events in time is the same as the order in the inclusion series of the realities which appear as events in time. If, for example, the apparent event of my crime appears, in the time-series, between the apparent events of my temptation and my remorse, then the stage in the inclusion series which appears as the event of my crime will be really, in the inclusion series, between the stages which appear as the events of my temptation and my remorse.

could the C-series determine that "we are taking them erroneously" as a certain A-series?[27]

McTaggart's claim is that "the appearance of time is a phenomenon *bene fundatum*, since the order of the apparent events in time is the same as the order in the inclusion series of the realities which appear as events in time."[28] However, this is not enough if we require that the epiphenomenal appearance, and its supposed effects, are fully determined by something else which is not epiphenomenal.

Why it is not enough? The reason is quite clear. It is not enough because there may always be two, or more, very different temporal A-series in which the order of events is the same as the order in the inclusion C-series of the realities which appear as events in time. Let us consider a simple example:

1. In one A-series, something happens long after now.
2. In other A-series, the same thing happens just now.
3. In a third A-series, the same thing happens long before now.

We can perfectly suppose that the three A-series preserve the same B-order of events, and that this order corresponds to an inclusion order according to a certain C-series. However, to know only that C-series is not enough to decide "which A-series" is the case. In order to decide that, we need to place the "now" in the C-series. Unless we place the "now" temporally, we could not decide which A-series is the case. Is that to be a *"bene fundatum* phenomenon"?

The real problem is that "to occur now" is a non dispensable aspect of A-series.[29] And that "to occur now" is not the same as "to be simultaneous with certain things", or "to be in the same present as certain things". For them to be the same, it is necessary that some things are designated as occurring just "now". Hence, unless C-series can determine that some things are occurring just "now", C-series could not support temporal appearances in the intended sense. But C-series

---

[27]Temporal appearances are appearances of change. But, they are also changing appearances. Moreover, it is arguable that the fact that they are appearances of change entails that they are changing appearances. This feature is very important. In the first part of the text quoted in note 26, McTaggart refers to a "timeless series of misperceptions". This is deeply misleading. A "misperception" is a tensed entity. McTaggart strategy for arguing that temporal appearances are a *bene fundatum* phenomenon also depends on that point. We will not discuss it. In any case, "to perceive a series of timeless existents as being in time" only is a "timeless series of misperceptions" if we do not take it just as being an "appearance". If we take it as being an appearance, then "to perceive a series of timeless existents as being in time" has to be, itself, an A-series of misperceptions. Moreover, as we have been arguing, this feature is what constitutes our temporal experience, i.e., our adopting a TPoV. Our experience of time is a temporal experience!

[28]Again, McTaggart [10, Ch LI, # 613].

[29]Here, "now" and "in the present" could be interchangeable. The argument I am offering against the explanatory role of C-series can be articulated both in terms of a "now" and in terms of a "in the present". In ordinary language, sometimes we use "now" in the sense of "in the present". According to what has been argued in other sections, that ordinary use can be analysed as a case of metonymy: a whole (the "now") is used to refer to one of its parts (the "present").

cannot do that. And therefore, temporal appearances cannot be considered a *"bene fundatum* phenomenon".

An alternative way to put the above argument would be by saying that we cannot find in C-series any A-TM (any TM for true A-answers to certain when-questions). We cannot find in C-series any TM for something like "X is occurring just now", for any X.

We can conclude that temporal appearances are not a *"bene fundatum* phenomenon". They are not a *"bene fundatum* phenomenon" in the sense of determination, and no other sense would be enough.[30] Our experiences of a fluent time cannot be determined by any C-series. Temporal appearances of a fluent time, i.e., a time involving past, present, and future facts or events, are real and they cannot be explained by C-series.

4   Two Other Confusing Approaches

Now, let us compare our results with two more recent approaches, both of them representative of two very common, but quite different, conceptions of time. We will argue that these approaches also are wrong.

4.1   Andy Clark: Time and Mind

The theoretic context of Andy Clark's approach to the relations between time and mind[31] is the recent crisis of classical computationalism provoked by connectionism, neuroscience, and bio-ecological perspectives.

Clark makes a sharp contrast between a computational understanding of the mind and a dynamical understanding. And he argues resolutely for the second one.

The first one, directly associated with classical computationalism in the "golden years" of cognitive sciences, ignores the "real timing" in which mental phenomena occur. It only focuses on "sequences of discrete states computationally relevant".

The second one, the dynamical understanding, considers the mind as something essentially temporal. Here, "real timing" is crucial. According to Clark, the essence of the mind consists in complex causal interrelations of continuous states in real time. The mind cannot be explained using computational tools. We need the conceptual tools of "Dynamical Systems Theory".

Clark's approach is based on the following three observations:

---

[30]We could express the same idea not in terms of "determination", but in terms of the notion of "supervenience". However, this would introduce unnecessary technicalities.

[31]I will focus on Clark [3]. See also Gelder [17]; Port and Gelder [17]; Port and Gelder (eds.) [18]; Thelen and Smith [16]; Kelso [5]; and Varela et al. [19].

1. Mental processes take place in real time.
2. Moreover, the timing really matters.
3. But, computational models ignore real timing in favour of the artificial notion of sequences of discrete events.

According to Clark, mental processes are the result of a "continuous reciprocal causation". The expression is coined by him, and it gives a good idea of what it is being claimed. A "continuous reciprocal causation" occurs when some system S is both continuously affecting, and simultaneously being affected by, activity in some other system O.

Some clear examples would be those of a couple of dancers, or a group of jazz musicians playing together, or the predator-prey dynamics, etc. Clark's main thesis is that the multiple activities of our brains generating mental states fit that model.

Clark's theses are nowadays very influential, in particular in the fields of cognitive sciences, neurosciences and philosophy of mind. However, I think, they are wrong. Or at least, they need a serious reflection.

The basic idea underlying Clark's theses is that the existence and nature of the mind is essentially linked to the interrelations and complexities of our brain, considered as a physical dynamical system. In other words, and much more precisely, that neither the mind, nor a fluent time with a past, a present and a future, can be simply "virtual machines".

That approach has two main problems. The first problem is that the mathematical Theory of Dynamical Systems is just the sort of theory that, being in the background of Classical Mechanics, offers no room for making a clear difference between the past, the present and the future. In a nutshell, it offers no room for a "fluent time". In particular, as we have noted, it does not permit identifying "the present", or more generally the "now", in non-indexical ways. The only kind of time that we can find in that theory is a time defined through B-series. All the enrichments about control systems, feedback processes, attractors, bifurcations, chaos, etc., do not entail any change in that diagnosis.

The second problem is that, in the last instance, the computational understanding and the dynamical understanding can become equivalent descriptions. Moreover, perhaps the dynamical understanding can lead to the need of postulating irreducible computational descriptions. And both of these two possibilities suggest that other more accurate computational descriptions could do the work that Clark wants to assign only to dynamical descriptions.

So, in the end, the question is not one of choosing between a computational understanding of the mind and a dynamical understanding, or one of choosing between a computational time and a dynamical time. The crucial problem concerns our ways of understanding "fluent time", a time constituted by A-series of events, or facts, or objects, etc., placed in the past, in the present, or in the future. And this problem is not solved by opting for a dynamical perspective instead of opting for a computational one.

In other words, even if we were to assume Clark's approach, the deep philosophical problems about time would remain "untouched".

I would agree that neither the mind, nor a fluent time with a past, a present, and a future, can be simply "virtual machines". But I cannot but reject that they are no more that very complex "physical dynamical systems" like, for instance, the solar system.

4.2   Benjamin Libet: Mind Time

Benjamin Libet is a neurologist, very well known for his various experiments about how the brain produces conscious awareness. His first researches date back to 1950, and the most important ones are compiled in a recent book.[32]

The key result is that the brain needs approximately 500 ms to "elicit" awareness of a sensation. Conscious reports of having a sensation need electrical stimulations in the somatosensory cortex with a minimal duration of 500 ms. There is always that delay between sensory stimuli and sensory conscious experience.

This seems to have important consequences for "free will". It is said that voluntary acts of free will have to begin neurologically at least 500 ms "before" the subjects being conscious of their decision to act in that way. In other words, free will would not initiate the voluntary act. The act seems to be initiated instead by unconscious processes of the subject's brain.

But, let us focus on Libet's conception of the relations between time and mind, and about the nature of the "fluent time" we consciously experience. Mostly in an implicit way, that conception is even more influential than Clark's proposal. What Libet claims is that time, a fluent time with a past, a present, and a future, is simply a "subjective construction of our brains". Moreover, the mind itself is a "construction of the brain".[33]

We can make a very direct comparison between the approaches of Clark and Libet. Clark tries to find a place for mind and time in the objective world of science, in the last instance in the "objective" world of physics. In contrast, Libet tries to find such a place in a purely "subjective" world.[34] However, Libet's theses are not less wrong than Clark's ones.

---

[32]The book is Libet [7].

[33]As a matter of fact, Libet's conception of mind and time as "constructions of the brain" belongs implicitly to the materialistic ideology of many people. I thing that that conception is literally false; and that metaphorically it is highly confusing. For an explicit presentation and defence of that conception, see Wegner [20].

[34]In a recent review of Libet' [7] book, Tim Crane notes "After a lecture in Göteborg by the neuroscientist Benjamin Libet in 1993, the Göteborg-Post carried the headline, 'Now it has been proven: we are all somewhat behind'. The paper was referring to Libet's celebrated discovery that the neural precursors of some voluntary actions occur before the conscious awareness of the decision to act." This is just the influential conception about consciousness, and about fluent time, that underlies Libet's results, and that I want to reject: namely, that "we are all somewhat behind".

There is a crucial problem in Libet's approach. The combination of elements we find in the above expression "subjective construction of our brain" has no clear meaning. It is supposed that this expression makes sense, and that it is highly informative (moreover, that it is "scientifically" informative). But it is very difficult to see what that sense can be.

If the brain "construes" something, then it would have to make sense to ask when-questions such as the following ones:

When does that construction takes place?
How much time does the construction take?
Is it happening now?

These are when-questions that make sense. And their true A-answers would have to have some A-TM. But these A-TM cannot be themselves, in turn, a mere "construction of our brains"!

If that were the case, then the brain would construe the "time" in which it is "making the construction of the time". The brain would construe the "now" in which it is "constructing the now". But this cannot have any meaning at all! The situation is circular or regressive, and this time in a vicious ontological sense.

Occasionally, Libet's experiments are described as establishing that we are "living in the past". This would be so, it is said, because the causes of our actions in our brains occur earlier (500 ms earlier) than our conscious awareness of deciding to act. This illustrates how confusing Libet's conception of time is. "Living in the past" requires an objective temporal position in an A-series. But, that temporal position ("in the past") would not be possible if all A-series were merely epiphenomenal.

We can say that what the brain "construes" cannot be only a mere subjectivity. It has to have some sort of ontological objectivity. In other words, it cannot be a merely subjective epiphenomenon. The relevant A-TM (the A-truth makers of some true A-answers for certain when-questions like the ones above presented) have to be "ontologically objective". Moreover, as we have argued, they have to have an ontological objectivity different from the ontological objectivity attributed to the physical world described by science.

This has a considerable effect on Libet's problem about "free will". Consider again a claim such as the following

"A voluntary act will begin neurologically at least 500 ms before the subject being conscious of her decision to act".

In order to attribute these temporal positions, we would need to determine when the conscious decision to act occurs, and we would need to compare it with the occurrence of the neurological beginning of the act. But this entails two important things:

1. The conscious decision to act, closely linked to bring about some causal effects in the world (the pushing of a button, for instance), cannot simply be another neurological state, nor a causally inert effect of some neurological state either.

2. The conscious decision occurs as a non-CC in the PoV of the subject.

1 means that the conscious decision to act cannot be only a mere "subjective epiphenomenon". It has causal effects on the world. Furthermore, assuming that the non-CC of a PoV always have the temporal form of a gerund, 2 means that the "now" in which that conscious decision is made is not like a "point" without extension. It has to have, so to speak, a certain temporal extension. It has to have the temporal extension of the changes experienced in the PoV. Now,[35] we can state a crucial claim:

That temporal extension, "projected reflectively" over the dimensions deployed in our neurological descriptions, could perfectly well involve the 500 ms the brain needs to initiate the voluntary act.

We are saying that the temporal extension of a "now" can involve the 500 ms the brain needs to initiate the act. Indeed, this is a very odd way of speaking. But it would offer a very easy way to cope with Libet's problem. Moreover, as we have seen, Libet's problem, itself, also is formulated in a very odd way of speaking.

"Projected reflectively" is the key word. It would be a conceptual projection made from a reflective PoV. We try to make sense of the available data in the most plausible ways. If the conscious decision is not like a "point", if it has a certain temporal extension, if it involves to experience a change, to experience a past, a present, and a future, then our taking the decision of acting can involve those 500 ms of physical time.[36]

5   Tensed Truths and the Sea-Battle Problem

Tensed truths are true TB about either the past, or the present, or the future. Sometimes, the past and the future are so close to the present that they belong to the same "now". In that case, their TC are actually existing facts we can experience, and the TM of those tensed truths can include directly these TC. However, in general, tensed truths about the past and about the future go not only beyond the present, but also beyond the "now". This means that their TC are not existing facts, and that their TM cannot include those TC. However, in a much more indirect way,

---

[35]You can take this "now" as an example. In it, you can experience in the first person the sort of temporal extension and the sort of changes that can be involved in any "now".

[36]This would be to try a certain "stereoscopic understanding", in Sellars sense, of something coming from the competing "manifest" and "scientific" images, with the aim of finally reaching a "synoptic vision" of persons and their place in nature. See the chapter 1 of Rosenberg [13]. That strategy is very different from Dennett's [4] "heterophenomenology". According to Dennett, the objects of heterophenomenology are simply theorist's fictions. Sellars was much more realist than Dennett, and I am following Sellars in that respect. Even assuming that the objects of consciousness are theorist's fictions, they are not "simply" that. In many cases, we cannot but be strongly realist with respect to them.

they can also have TM. And their TM, or at least a part of them, also could be existing facts belonging to the "now". This is the important point. Their TM can be facts involving the non-CC of our experiences (the world we have in perspective when we have experiences). The non-CC of our experiences are always tensed entities. And because of that, be it in a direct or indirect way, they can play the role of appropriate TM for all kinds of tensed truths.

In what follows, we will better explain these ideas. And we will apply them to the Aristotelian problem of "the sea-battle".

5.1   Some Varieties of Tensed Truths

I have argued that there are some tensed truths having the following two features:

1. Their TC are not existing facts (i.e., the TC are placed either in a past or in a future beyond what can be contained in the perspective of a "now"), and
2. Their TM are existing facts (i.e., they are facts that actually exist in the world and to which we can have access from the perspective of a certain "now").

TB about the past, the present, or the future are, themselves, existing facts. And some of them are true TB. When this happens, it is in virtue of other existing facts, their TM. Sometimes, these TM involve the TC of those TB. Other times, they do not. In that case, we can have tensed truths having both 1 and 2.

The TM of tensed truths have to be quite peculiar. They have to be "fluent entities" placed in an A-series. B-series would not be enough. As we have said, it is not clear that we can find those fluent entities in the physical world such as it is described by basic science. But, we can find them inside the non-CC of our PoV. They are the contents of our experience. Over these non-CC, we continuously embrace some TPoV. And from that temporal perspective, those non-CC become existing facts in continuous change. This is what makes genuinely "tensed" the relevant tensed truths about them.

But those tensed truths are only a particular case of tensed truths. There are "many kinds" of tensed truths. And we can adopt different criteria to make relevant distinctions. Firstly, we have to distinguish among:

1. Tensed truths which are about a past and a future included in what is occurring "now". These truths will have TM among the non-CC of our experience. In virtue of these TM, some of these tensed TB will be true. And these TM could include the TC of the tensed truths in question.
2. Tensed truths which are about a past or a future that our TPoV place very close to what is ocurring "now". We will generally find appropriate TM for them inside the non-CC of our PoV, but perhaps in combination with the TM of other more conceptual contents (about the world).
3. Tensed truths which are about a remote past or a distant future that our TPoV place far away from what is occurring "now". Generally, we will not find

appropriate TM for them inside the non-CC of our PoV. These tensed truths have to be true in virtue exclusively of the TM of some conceptual contents (about the world).

Secondly, we have to distinguish between tensed truths "in a proper sense" and "more or less plausible truths", "more or less probable truths", "truths in a certain degree", etc., that can be interpreted in a temporal, tensed sense. Statements as, for instance, "It is probably that it rains" can be interpreted as "It will rain". However, it could be also interpreted in an epistemic way, as saying something about our state of knowledge and ignorance, or in other more ontological ways not involving references to time (to a fluent time). The same can be said of the other notions. These would not constitute tensed truths "in a proper sense".

The cases 1, 2, and 3, above introduced, are tensed truths in a proper sense. The TM of these other varieties of tensed truths would be different from the TM of the tensed truths in that proper sense. In general, the first ones will be more "abundant" (there will be many more TM for those truths) than the second ones.

Thirdly, we have to distinguish between, on the one hand, tensed truths and, on the other hand, TB about the past, the present, or the future. Many of those TB would not be tensed truths. They would be only "possible" tensed truths. We can consider that possible tensed truth is a kind of tensed truth, in the same sense in which we can consider that possible truth is a kind of truth. But, the TM of those possible tensed truths also would be different from the TM of the corresponding tensed truths. In general, the TM of possible tensed truths will be also more "abundant" than the TM of the corresponding tensed truth.

There are many kinds of tensed truths. And even if they may share the same general semantics, they do not have the same "ontology".

5.2 The Problem of the Sea-Battle

Let us finish by applying our approach to the classic problem of the sea-battle.[37] Let us consider the following claim:

---

[37]Many times, it is also called "the problem of future contingents": statements about future contingent facts. The first discussion can be found in Aristotle (Chapter 9 of his *De Interpretatione*), using a sea-battle as example. Further discussions are in Diodorus Cronus, Leibniz, Lukasiewicz, and Prior. It has received close attention in the last decades. The problem is to understand the structure of temporal propositions (statements, thoughts, etc.). Generally, the example is formulated in a negative way. Suppose that a certain sea-battle will not take place tomorrow. It seems that, if this proposition is true, then it has to have been also true in the past, for instance yesterday. But all past truths can be considered now necessary truths. Therefore, it is necessary that the sea-battle in question will not take place tomorrow. This seems to enter into conflict with our "free will", and Aristotle's solution was to reject bivalence in these cases. To say today that a certain sea-battle will not take place tomorrow is neither true nor false. These propositions are neither necessary nor impossible. They are contingent ones. Diodorus Cronos argued that they are either necessary or contingent. And Leibniz's approach about God and

1. Tomorrow, there will be a sea-battle

Claim 1 can be the A-answer to a when-question. Suppose that 1 is true. Moreover, suppose that the fact that 1 is true is an existing fact in virtue of other existing facts. These will be its TM. Suppose that there is at least one such TM.

So, 1 is true. And the fact that 1 is true has a TM. Does this entail that 1 is an existing fact (not that the fact that 1 is true is an existing fact, but that 1 is an existing fact)? Is 1 a fact actually existing in the world?

The answer is negative. That 1 is true only entails (via Tarski's equivalence)

There will be a sea-battle tomorrow.

But, that 1 is true does not entail

It is an existing fact, a fact actually existing in the world, that there will be a sea-battle tomorrow.

And the existence of a TM for the truth of 1 only entails

It is an existing fact, a fact existing in the world, that it is true that there will be a sea-battle tomorrow.

But, it does not entail either

It is an existing fact, a fact actually existing in the world, that there will be a sea-battle tomorrow.

Alternatively, we can say that expressions such as "it is an existing fact", or "it is a fact actually existing in the world", do not permit the substitution of equivalent expressions under their scope. From

It is an existing fact (A is true)

and

A is true iff p,

we cannot infer

It is an existing fact (p).

Exactly the same happens with "It occurs now", and with "In the present". For instance, from

Now (A is true)

and

(Footnote 37 continued)

"individual essences" places him very close to fatalism. Everything is necessary. So, in those propositions, there would not be contingence, but ignorance. Lukasiewicz's multivalued logics and Prior's temporal logics show a great variety of ways to give answers to the sea-battle problem.

A is true iff p,

we cannot infer

Now (p).

We have arrived at a very important result. We can assume the truth of 1, i.e., that it is true that there will be a sea-battle tomorrow, as an existing fact, a fact occurring "now" in the world, without assuming that it is "now" an existing fact, a fact actually existing in the world, that there will be a sea-battle tomorrow!

We can obtain a similar result with respect to the "present", that part of the "now" that is contrasted with the past and with the future. We can assume the truth of 1, i.e., that it is true that there will be a sea-battle tomorrow, as a fact existing "in the present", without assuming that it is a fact existing "in the present" that there will be a sea-battle tomorrow.

Problems about "Fatalism" (roughly, if p then necessarily p), and also many arguments for "Presentism" (roughly, only the present exists), are very often formulated from the supposed need to assume that if there are actually existing facts, facts existing "now", or facts existing "in the present", such as

It is true that there will be a sea-battle tomorrow,

then there would have to be actually existing facts such as

There will be a sea-battle tomorrow.

However, there is no such need. Or so we have argued.

We can assume the truth of 1 as being an existing fact in virtue of other existing facts, namely its TM, without assuming that it is "now", still less "in the present", an existing fact, a fact actually existing in the world, that there will be a sea-battle tomorrow!

Which TM can 1 have? This question leads to another very important result of our approach. We have maintained that, in many cases, we can find appropriate TM for some A-answers to certain when-questions in the non-CC of our PoV. Also, we have maintained that these non-CC are always "fluent entities" placed in a temporally extended "now-ing". With this in mind, consider the following claims

2. A sea-battle is just beginning.
3. Immediately, a sea-battle will begin.
4. Within a year, a sea-battle will take place.
5. At some very distant future time, a sea-battle will take place.

Thanks to the adverb "tomorrow", 1 would occupy an intermediate temporal position between 3 and 4. From 2 to 5 (with 1 between 3 and 4), it will be more and more difficult to find appropriate TM for their truth in the non-CC of our PoV. The TC of those true claims only would be clearly existent facts, facts actually exist in the world, in the case of 2. And only the TM of 2 could include, or even be identical with, its TC. It is a serious error, and the source of many other errors, to consider that because 1–5 can have the same "semantic", they must have the same "ontology".

Also, we could say that there are important differences between the TM that can be found for TB about the past (or about probable past, or about plausible past, etc.) and the TM that can be found for TB about the future (or about probable future, or about plausible future, etc.). As a matter of fact, the first TM are much more abundant than the second ones. This would introduce a very important difference between the past and the future. It is a difference based on "truth-making". So, it is an ontological difference, not an epistemic one.

It is not a difference in predictive possibilities, or a difference based on considerations of knowledge and ignorance, etc. Between the past and the future there are many ontological differences. And one of them has to do with "truth-making". That there are many more TM for truths about the past than TM for truths about the future is an important fact concerning the structure of the world.

Simply, as a factual matter about the world we have in perspective in our experience, there are many more TM for the past than for future. The world we live in is made that way. And this offers a very natural and suggestive sense in which the future can be said to be "open" and the past can be said to be "closed", or in any case much more "closed" than the future.

What actually exists is the world we are experiencing. And it is a world of interconnected existent events (facts, objects, etc.). Truths (i.e., true TB) have actual existence in virtue of other existent events (facts, objects, etc.): their TM. It may occur that these TM do entail the actual existence of the respective TC. Many times, TC have no actual existence. Tensed truths are only a special case of this.

Acknowledgments This work has been granted by Spanish Government, "Ministerio de Economía y Competividad", Research Projects FFI2008-01205 (*Points of View. A Philosophical Investigation*), FFI2011-24549 (*Points of View and Temporal Structures*), and FFI2014-57409-R (*Points of View, Dispositons, and Time. Perspectives in a World of Dispositions*).

References

1. Baker, L. R. (2007). *The metaphysics of everyday life*. Cambridge: Cambridge University Press.
2. Baker, L. R. (2008). A metaphysics of ordinary things and why we need it. *Philosophy, 83*, 5–24.
3. Clark, A. (1998). Time and mind. *Journal of Philosophy XCV*(7), 354–376.
4. Dennett, D. (1991). *Consciousness explained*. Boston: Litle Brown.
5. Kelso, S. (1995). *Dynamic patterns*. Cambridge: MIT Press.
6. Lewis, D. (1986). *On the plurality of worlds*. Oxford: Blackwell.
7. Libet, B. (2004). *Mind time: The temporal factor in consciousness*. Cambridge: Harvard University Press.
8. McKinnon, N. (2002). The endurance/perdurance distinction. *The Australasian Journal of Philosophy, 80*(3), 288–306.
9. McTaggart, J. (1908). The unreality of time. *Mind, 17*, 457–474.
10. McTaggart, J. (1921–1927). *The nature of existence* (vol. I, 1921; vol. II, 1927). Cambridge: Cambridge University Press.
11. Mellor, D. H. (1998). *Real time II*. London: Routledge.

12. Nyiri, K. (2008). Hundred years after: How McTaggart became a thing of the past. In *Proceedings of the 6th European Congress of Analytic Philosophy (August 21–26, 2008)*. Kraków: European Society for Analytical Philosophy.
13. Rosenberg, J. (2007). *Wilfrid sellars: Fusing the images*. Oxford: Oxford University Press.
14. Russell, B. (1918). The relation of sense data to physics. In *Mysticism and logic, and other essays* (pp. 113–140). London: George Allen & Unwin.
15. Sider, T. (2001). *Four-dimensionalism*. Oxford: Oxford University Press.
16. Thelen, E., & Smith, L. (1944). *A dynamic systems approach to the development of cognition and action*. Cambridge: MIT Press.
17. van Gelder, T. (1995). What might cognition be, if not computation? *Journal of Philosophy, XCII*(7), 345–381.
18. van Gelder, T., & R. Port. (1995). It's about time: An overview of the dynamical approach to cognition. In R. E. Port & T. van Gelder (Eds.) *Mind as motion: Dynamics, behavior, and cognition* (pp. 1–44). Cambridge: MIT Press.
19. Varela, F., Thompson, E., & Rosch, E. (1991). *The embodied mind*. Cambridge: MIT Press.
20. Wegner, D. (2002). *The illusion of conscious will*. Cambridge: MIT Press.

Chapter 5
Branching Time Structures and Points of View

Margarita Vázquez Campos

Abstract In this paper I analyze the temporal structures that are appropriate to study the notion of point of view. When we analyze the points of view and their structure, it seems clear that we must take into account the time t in which a point of view is attributed to a subject. A two-dimensional temporal logic which combines a modal dimension for possibilities and a temporal one for the flow of time, offers a clear view of the temporary location of a point of view. In this logic, we have histories, thanks to the temporal dimension, and evaluation is in two indices, time and history. These stories can be seen as different scenarios, providing a clear advantage when applying to the analysis of the notion of point of view. The conclusion is that, in order to give a proper approach for the notion of point of view, all these aspects should be combined.

1 Introduction

In Chaps. 2–4, we have analyzed the notion of point of view and we have seen its objective, subjective and temporal aspects.[1] Now, it is time to show in more detail some of these temporal aspects, that is, the temporal structures that are more relevant when analyzing the dynamics of points of view.

One idea we have emphasized when talking about temporal points of view[2] is that our points of view are full of indexical ingredients. These indexical ingredients

---

[1]More about this topic in Liz [13, 14], Liz & Vázquez [15, 16].
[2]Section 4.2.

M. Vázquez Campos (✉)
University of La Laguna, Tenerife, Canary Islands, Spain
e-mail: mvazquez@ull.es

© Springer International Publishing Switzerland 2015
M. Vázquez Campos and A.M. Liz Gutiérrez (eds.), *Temporal Points of View*,
Studies in Applied Philosophy, Epistemology and Rational Ethics 23,
DOI 10.1007/978-3-319-19815-6_5

can involve an emplacement in space. Other times, they can involve a relative position concerning some properties and relations. But, many times, they can involve an emplacement in time. This last kind of indexical is the one this chapter is interested in.

This is a central point for us because emplacement in time is especially important for subjects which are "persons". We know that a person can become massively confused about his/her position in space and about his/her relative position regarding many of the properties and relations instantiated with the environment. But, a person cannot become massively confused about his/her position in time. Being a person "like us" entails having a temporal perspective with a minimum of correction. We have said that this correct temporal point of view could only be internal to our points of view. That is, it could exist without any external or objective support. In any case, it has to exist.

We have seen all the arguments of McTaggart against the reality of time and that, in contrast, Arthur Prior took very seriously temporal appearances. In any case, as we have said, it is important to distinguish between the problem of understanding the existence and structure of time and the problem of understanding the existence and structure of temporal points of view. Here, we are trying to understand the existence and structure of temporal points of view (TPoV).

We have TPoV because our thoughts and languages are tensed. In our definition,[3] TPoV are points of view (PoV) in which some differences in content are identified as changes in time.

The problem, then, is how to understand time in order to represent those changes in time. Coming back to McTaggart,[4] he is the classic reference of two different ways of understanding time, A-series and B-series:

- A-series would place the events in the past, present or future.
- B-series would place them according to a before-after relation.

McTaggart argued that the A-series are essential for the existence of time. But he also argued that, for this same reason, time is unreal. A-series are, according to McTaggart, deeply contradictory, or, at least, paradoxical. McTaggart's approaches have shaped subsequent discussions till nowadays, even in this book. So it is very usual to distinguish between conceptions, or theories, of time of type A and type B.

And, also, the idea is alive that a time series of type A, although it is essential to have proper time that "flows" (not merely spatialized time), confronts us with conceptual problems, if we understand it as something more than mere internal time.

---

[3]Chapter 4.
[4]McTaggart [17].

2 Prior and Temporal Structures

Using McTaggart's terminology, in a first approach we could think that TPoV could be related to temporal structures of A-series. In fact, classical temporal logic approaches, such as Arthur Prior ones, are usually understood as linked to structures type A-series. These temporal structures, linked to the A-series, define a kind of "internal time" highly directed, structured and constitutive for the points of view involved.

Scholastic philosophy was deeply influential in Prior. He knew that for these philosophers an expression like "It's raining now" is complete, being sometimes true and other times false. But he does not agree with this. He thinks that the truth-value of an expression cannot change with the passage of time. And this is one of his key ideas.

We find in his papers the germ of almost all temporal systems that are currently under study.[5] His most important works are Prior [22–24].[6] Prior's most important idea (inspired by his studies of Diodorus Cronus) is that our language and our thoughts have an "internal perspective of time", in which time is represented by setting a past and a future for a changing now.

This is deeply related to our idea that being a person "like us" entails that we have a temporal perspective with a minimum of correction, even if this temporal point of view is only internal to our points of view.

A logic of time must respect this inner sense of temporality that requires a "flowing now." One of the most important challenges for Prior's approach was to undo the feeling of paradox (rightly denounced by McTaggart) that seems to imply this idea that time itself "flows". Another important challenge was to find an argument against determinism.

In order to build his systems, Prior keeps the truth-functional operators and introduces many new ones, being the most important the temporal operators F, G, P and H. These operators allow writing propositions which refer to the past, present and future. While F and G are operators about the future, P and H are about the past. F and P are "diamond type" modal operators (weak tense operators), and G and H "box type" (strong tense operators). So, the intended meanings are as follows:

- FA "It will, at least once, be the case that A"
- GA "It will always be the case that A"
- PA "It has, at least once, been the case that A"
- HA "It has always been the case that A"

For example, to express that time flows, Prior introduces an axiom as A → HFA, that means "if A is given, then it has always been that A will be at least once". Thus, the events are ordered as past or future with respect to a present. The similarity of these operators with McTaggart's A-series is clear.

---

[5]A good history of temporal logic is Ohrstrom and Hasle [19].
[6]See also, the new ediion of one of his books, Prior [25].

With respects to the semantics, Prior's systems define, in general, a flow of time as a structure composed of a set of moments, ..., $t - 3$, $t - 2$, $t - 1$, $t0$, $t + 1$, $t + 2$, $t + 3$,..., a before/after relation on moments, which has different properties depending on the flow of time[7] and an evaluation function.

In this semantics, the formula FA will be true at the time $t0$, if and only if there is a later stage where A is true. In order to evaluate a formula, a before/after relation is used, as in McTaggart's B-series. We can see then that, while the syntax of the system has similarities with the A-series (we speak in terms of past, present and future), the semantics has similarities with the B-series. And while, in the syntax, time has a subjective character, in the semantics it seems more objective.

We have to take into account the two classical perspectives of understanding time, external and internal. From an external, objective perspective, time is described as a sequence of moments or instants, which could be enumerated. From an internal, subjective perspective, there is the perception of the human (or whatever) subject of the passing of time, of the moving now. Without this perspective we cannot understand what time is. And this is just what we have said is part of our idea of being a person "like us".

So, while the syntactic aspects of Prior's temporal logic are related to A-series and our internal perspective, they are deeply interrelated with the semantic aspects that call for B-series and an external perspective.

3 The Multimodality and Bidimensionality of Temporal Points of View

We have seen that our points of view are full of indexical ingredients, and that many of them involve an emplacement in time and changes of time. To understand how to represent emplacement in time, we have analyzed Prior's temporal logic and we have found that internal and external aspects are deeply related.

But this is not enough to understand changes in time. Change seems to involve a certain notion of indeterminism, as Prior clearly saw. He found that Aristotle speaks of propositions about events that are not already predetermined (the famous sea-battle). This was the germ of his branching time system.[8]

So, in order to give an account of changes of time, we need at least two types of indexical ingredients, one for the emplacement and one for the possibilities of change. It is precisely this kind of indexicality, the one that can provide a two-dimensional semantics.

This two-dimensional semantics was also developed by Prior in his Ockhamist system. He defines "an Ockhamist model as a line without beginning or end which may break up into branches as it moves from past to future, though not the other way; so that from any point on it there is only one route into the past, but possibly a number of alternative routes into the future".

---

[7]In any case, it has the irreflexive property, as constitutive of the flow of time.

[8]Although the idea of this system was suggested to him by Saul Kripke.

Many other authors[9] have used this idea of representing temporal indeterminist models (or even time) as a tree whose trunk, unique and solid, would be a symbol of the past, and whose branches would be the future. As I have said in other places,[10] I disagree with this image.

I prefer to understand indeterminist models as a collection of complete histories. I have used the image of a cable. Halfway through the cable, if we remove the outer cover, we have many smaller cables, which we can separate from each other. The half of the cable with the outer covering may represent the past and the other half uncovered, open, represents the future at a given point (this particular one). The small cables, that are covered, cannot be separated, but each one is the first part of one of those which is separated. Each one of them is a whole history.

What do I mean by this image? I do not see a common past and a future full of possibilities. I see a future full of possibilities, each one with its own past. Without each one of these pasts, we cannot understand each one of the futures. The pasts, the histories, are different and thus may lead to different futures, different possibilities. Understanding this one can understand the possibilities of change.

The image of the cable is not static, at each point we get other smaller cables, becoming a fractal-like structure. We could understand the brute facts, or the external facts to the PoV, as propositions (p, q, r,...) that share their truth value in the same point within the cable (independently of which history they belong to).

That is, if it's raining at a moment of the past, that is true in all the cables, in all the histories inside a PoV (and is even shared with other PoV). But the relations that this event (to have rained) has with earlier or subsequent events are different.

While the p (raining in a moment of the past) is true in all these histories and cables, and there is an equivalence between those moments in which p is given, we cannot say there is a logical identity between all those moments.

Several versions of Prior's Ockhamist system use axioms that represent time that is branching in the future but linear in the past. They add the generalization rule for the modal necessity and an axiom that imposes a restriction that avoids saying: "if A will be the case, then it is necessary that in the past A will be the case in the future". That is, avoiding A to be a proposition about the future. This restriction was in Prior's original system.

There have been other different axiomatizations of Ockhamist time logic.[11] The last paper I know of is one by Mark Reynolds that appeared in 2003,[12] in which he presented a complete axiomatization of Prior's Ockhamist branching time logic.

Hirokazu Nishimura[13] formulated a new temporal model which turned out to be slightly different from the Ockham model that Prior had considered. Although Prior's Ockhamist model is a good accurate representation of our intuitions

---

[9]See, for example, Álvarez [1].

[10]Vázquez [28].

[11]A good presentation is Thomason [27], in Gabbay and Guenthner [8].

[12]Reynolds [26].

[13]NIshimura [18].

concerning valid temporal reasoning, Nishimura shows some rare examples in which the Ockham theory is not sufficient.

Nishimura's model involved not only times, but also histories defined as linear subsets of the set of times. It is natural to view Nishimura's model of time as a union of disjointed histories. The tenses (past, present, future) are always relative to a history. Relative to one possible history, there is going to be a sea-battle tomorrow, and relative to another possible history there is not going to be a sea-battle. Nishimura's model involves a relation of identity of histories before certain events.

The axiomatization presented in 1994 by Gabbay, Hodkinson and Reynolds[14] takes into account the considerations made by Nishimura. They take an axiomatization for modal system S5, an axiomatization for linear time and some specific axioms for indeterministic time. They add the usual rules of linear time and an irreflexive rule.

These are the axioms for indeterministic time, where L is the necessity and M the possibility in modal logic:

$$\text{Ax1. } A \rightarrow LA, A \text{ not containing } F.$$
$$\text{Ax2. } (\neg p \wedge Hp \wedge LA) \rightarrow GLH((\neg p \wedge Hp) \rightarrow A)$$
$$\text{Ax3. } A \rightarrow GLPMA$$

The system is multimodal and bidimensional. It is multimodal because it has operators for several "modalities": time (past and future) and modality. It is bidimensional as it has two relations involving moments of time (an equivalence relation for the modal dimension and an irreflexive linear order for the temporal one) and the evaluation is always at two indices, time and history. When evaluating simple temporal operators such as F or P, the evaluation function examines moments before or after that one in which the evaluation is being done, but within the same history.

When evaluating modal operators, such as M or L, the evaluation function examines equivalent moments of different histories. The before/after relation is between moments of the same history. By contrast, the equivalence relation is given among moments of different histories.

When making evaluations, one must take into account several important things:

1. If two histories are different, they do not share any moment.
2. If two points are equivalent, there is an isomorphism between moments that precede them in their respective histories. This means that the present and the past have their equivalent moments in the other histories.
3. If we assign a value to a propositional variable at a time m, we assign the same value to that propositional variable at every equivalent moment.

---

[14]Gabbay et al. [9]. An axiomatization for a similar semantics is given in Zanardo [30].

We can use a method of semantic diagrams for checking the validity inside this semantics. This method is similar to that presented by Hughes and Cresswell for modal logic in their classic 1968 book.[15] It is an extension of the validity check for the logic of propositions, trying to find a model that makes false a formula. If such a model is possible, the method will allow us to build it. This would mean that the formula is invalid. In the other case, the formula is valid.

The rules of this method are:

1. Rules to put signs:

   - + above G 1 and F 0.
   - + under G 0 and F 1.
   - − above H 1 and P 0.
   - − under H 0 and P 1.
   - * above L 1 and M 0.
   - * under L 0 and M 1.

2. Rules to place a moment and history:

   - + under GA, we have to place a moment in the same history, after this, where A receives the value 0.
   - + under FA, we have to place a moment in the same history, after this, where A receives the value 1.
   - − under HA, we have to place a moment in the same history before this, where A receives the value 0.
   - − under PA, we have to place a moment in the same history before this, where A receives the value 1.
   - * under LA, we have to place a moment, equivalent to the actual moment, in another history, where A receives the value 0.
   - * under MA, we have to place a time, equivalent to the actual, in another history, where A receives the value 1.

3. Rules for assigning truth-values to moments:

   - + above GA, we have to assign the value 1 to A at every moment after the actual moment in this history.
   - + above FA, we have to assign the value 0 to A at every moment after the actual moment in this history.
   - − above HA, we have to assign the value 1 to A at every moment before the actual moment in this history.
   - − above PA, we have to assign the value 0 to A at every moment before the actual moment in this history.

---

[15]Hughes and Cresswell [12].

- * above LA, we have to assign the value 1 to A at every equivalent moment to the actual moment in other histories.
- * above MA, we have to assign the value 0 to A at every equivalent moment to the actual moment in other histories.

We are going to see an example: A → GLPMA (Ax. 3). This example allows us to build a model that captures the image of parallel pasts.

We give, in history 1 (*h1*) and moment 0 (*m0*), the value 0 to the formula. That is, we put 0 under →. Then, the antecedent, A, is 1 and the consequent, GLPMA, is 0. So, we put + under G.

As we have a + under GA, we have to place a moment in the same history (*h1*), after *m0*, where LPMA receives the value 0. We choose any moment, for example *m3*.

So, in *m3* of h1, LPMA is 0. Then, we put * under L and we have to place a moment, equivalent to the actual moment, in another history, where PMA receives the value 0. This could be the moment *m3'* of the history 2 (*h2*).

In *m3'* of *h2* we put a - above P. Then we have to assign the value 0 to MA at every moment before the *m3'* in this history, *h2*. Do we have any moment before *m3'* in *h2*? Yes. We know that if two points are equivalent, there is an isomorphism between the moments that precede them in their respective histories. As *m3'* and *m3* are equivalent, there is an isomorphism between moments that precede them in *h1* and *h2*. So, we have *m0* in *h1* and *m0'* in *h2*. *m0* and *m0'* are equivalent.

Then, in *m0'* of *h2* MA receives the value 0 and we put * above M. We have to assign the value 0 to A at every equivalent moment to *m0'* in other histories. The equivalent moment we have is *m0* in *h1*. In *m0 h1*, A had received the value 1. So, A is 1 and A is 0. We could not find a model that makes this formula false, so this formula is valid.

Let us see another example with which we can visualise the open possibilities in the future: MGA → GA. Imagine you are in the moment *m0* of the history *h1* and, to make this formula false, you have, at the same time that MGA is 1 and GA is 0. How can this be?

If MGA is 1, we put * under M in *m0 h1*. If GA is 0, we put + under G in *m0 h1*. As we have + under G, we have to place a moment in *h1*, after *m0*, where A receives the value 0. For example, in *m2 h1* (being *m2* after *m0*) we give to A the value 0.

If we have * under M in *m0 h1*, we have to place a time, equivalent to *m0 h1*, in another history, where GA receives the value 1. This could be *m0' h2*. So, in *m0' h2* GA receives the value 1 and we put + above G and we have to assign the value 1 to A at every moment after *m0'* in *h2*. But we have no moment after this one. So, we finish here and MGA → GA is false. We have found a model that makes this formula false. And, in this model, at a moment and history (*m0 h1*) MGA is true and GA is false at the same time. That is because the future is open, and, while in a future moment in *h1* A is false, there is another history *h2* in which A would be always true in the future. But, from *m0 h1* we don't know which path are we going to take. There is no contradiction here, so the formula MGA → GA is false.

These two examples show us how an indeterminist temporal logic and its bidimensional semantics with evaluations at two indices, is adequate to understand changes in time. So, they help to understand the existence and structure of temporal points of view.

4 Branching Time Structures, Temporal Points of View and Simulation Models

The importance of TPoV structures, understood as branching time structures, independent histories with parallel pasts, becomes very clear with the help of the concept of simulation and the idea of the scenario of a simulation. What is this?

A scenario of a simulation model is a series of results we get based on certain initial conditions. If we change the conditions, with the same simulation model, we have another scenario that could be very different.

Two different scenarios can be the same at the beginning and after that diverge. Do they share the same past? I think that they don't. Since their evolutions have been different, there must be a difference in the simulation (a delay, for example). At the same time, two scenarios that seem equal, could be produced by completely different structures. They would not be the same scenario, but equivalent scenarios.

So, we can have:

- two different scenarios from the same simulation model, and
- two equivalent scenarios from different simulation models.

Imagine we want to build a model of a past event, for example a sea battle. We can make two completely different models and obtain the same result, the sea battle. Would we say that we have produced the same simulation? No. The correct answer should be that these two simulations (and their resultant scenarios) were equivalent.

Moreover, using one of these two simulation models, and changing the initial conditions or some data, we could obtain a scenario where the sea battle does not happen. How is this possible? If the model is well built, it should allow us to give an account of a contingent event, that in some scenarios happens and in other scenarios does not.

5 Temporal Points of View and Beyond

To conclude, I would like to emphasize that with a bidimensional evaluation at two indices, as the moment and the history we have seen, we can represent the emplacement in time and the change in time as they are present in the philosophical notion of TPoV. Moreover, the notion of scenario also leads directly to the notion of change in time in a TPoV.

Even so, I will show that there is a lot of work to be done. On the one hand, it is important to refer to particular moments of time (hybridization) and, on the other hand, sometimes we need to talk about epistemic states.

As we have analyzed when talking about TPoV, our points of view are full of indexical ingredients and many of them involve an emplacement in time.

It is precisely this kind of indexicality, which we have explained in previous chapters, that can provide a two-dimensional semantics. And this is the kind of indexicality involved in the concept of scenario.

In Vázquez and Liz [29] we introduced the notion of a scenario of actions linked to a point of view. We defined a scenario of actions as a structure of possible actions, with different weights, related to a point of view. The actions of one scenario do not need to be carried out. It is sufficient that they are possible actions. In any case, each point of view will determine a particular structure of actions.

This brings us to the branching time structure. These possible actions are the futures that remain open. Each point of view would be represented by a complete history or branch.

When exploring the dynamics of the points of view, the underlying temporal structures, using a two-dimensional temporal logic seems a good starting point. There have been other approaches, such as the logic of Antti Hautamäki.[16] He proposes a number of logical systems for points of view, and based on them, also defines temporal logics. This has similarities with some branched-time systems. What happens is that Hautamäki's systems do not put any restrictions on the combinations of W and R, so that alternatives proliferate everywhere and there would be no difference, regarding the truth, between past events and future ones.

Therefore, it would perhaps be more appropriate to follow the reverse path for the development of a logic of points of view. We should get a bidimensional temporal logic and define operators similar to the ones of Hautamäki.

When we analyze the points of view and their structure, it seems clear that we must take into account the time t in which a point of view is attributed to a subject. A two-dimensional temporal logic, such as the one we have seen, which combines a modal dimension for possibilities and a temporal one for the flow of time, offers a clear view of the temporary location of a point of view. In this logic, we have histories, thanks to the temporal dimension, and the evaluation is made in two indices, time and history. These histories can be seen as different scenarios, providing a clear advantage when applied to the analysis of the notion of point of view.

But perhaps this is not enough, and we should go beyond. As we have seen, when we analyze the notion of PoV and its structure, it seems clear that we must take into account the time t in which a PoV is attributed to a subject. This allocation could be the attribution to a subject of a certain set of propositions, or of objects and properties, at a time t under certain conditions. But, it is not possible to do this with standard temporal logic. It has important limitations.

---

[16]Hautamäki [10].

The most important limitation is that our analysis does not allow us to make reference to particular moments in time. I mean, you cannot set the exact time t in which a proposition (or set of propositions) attributed to a subject is true. This limitation is a problem common to most one-dimensional or bidimensional temporal logics.

This is a serious limitation, since in natural language we make constant use of the temporality. One way to overcome this limitation could be hybridizing temporal logic.[17] This is, in fact, one of the aspects of temporal logic that hybrid logic seems to solve: extending the basic logic with a mechanism to refer to points in time.

Hybrid logics are modal logics that allow referring to the points in the model. In the case of temporal logic, they allow us to refer to a particular point of time, an instant. The principal ideas related to hybrid logics were introduced by Prior in 1967.[18] After him, it was developed by Bull and reinvented by a group of logicians from the Sofia School. In the 1990s, the research papers on this topic increased, and the principal authors are Blackburn, Areces and other researchers linked to the University of Amsterdam.[19] It is very interesting that hybridization is the fundamental idea used by Prior in his reduction of B-series to A-series.

Hybrid logic introduces "nominals" as a tool for naming, or reasoning about, the points in the set of semantic possible worlds. The new thing is that these nominals appear in the syntax, and we can build well-formed formulas with them (as something different from propositional variables). Nominals are true at only one point in any model. They "name" this unique point by being true there and nowhere else.

In Tense Logic,[20] nominals have been used as a mechanism for referring to times, solving a limitation of this kind of logic systems. When we say something such as "It was raining", it is important to have a reference point before the current one where it is true that it is raining. Syntactically, it is possible to define properties of frames that are not definable in ordinary modal and temporal logic.

There have been some interesting attempts at combining hybrid logic and temporal logic, the most important one for our purposes is the hybrid Ockhamist temporal logic (HOT) developed by Blackburn and Goranko [5].[21] In this paper, Blackburn and Goranko try to introduce, axiomatize and study the basic hybrid Ockhamist temporal logic. The semantics uses structures of bundled trees. A bundled tree is said to be an equivalent structure to an Ockhamist frame (as the one we saw previously). The elements of an Ockhamist frame can be thought of as branches in a bundle on a tree obtained by identifying the equivalent points. So, the Ockhamist truth of a formula is defined in a similar way to that of Gabbay, Hodkinson and Reynolds, using branches instead of moments and histories.

---

[17]See, Ponte and Vázquez [20] and Ponte and Vázquez [21].

[18]Prior [23].

[19]See: Blackburn and Seligman [6]; Blackburn [4]; and, Areces et al. [2].

[20]Blackburn [3].

[21]Blackburn and Goranko [5].

If we have a branching time logic in which we can refer to moments of time, we can solve the limitation of being incapable of making reference to particular moments of time.

Going further, Rafael Herrera presented in the last World Congress of Philosophy[22] a Temporal-Epistemic Logic (TEL).[23] In TEL, we need to evaluate sentences regarding both times and epistemic states (or possible worlds). Herrera considers the dynamic aspects of reasoning processes:

- How knowledge changes over time.
- The different kinds of knowledge an agent can have about past and future.

In order to achieve the first goal, he adds a temporal dimension to the epistemic one.[24] But, the second goal is much more difficult to achieve, as temporal operators come under the scope of epistemic ones.

In Herrera's words: "the most notable aspect of TEL systems is that we have to combine two different points of view in them, so that, on one hand, temporal points (instants) are determined from the point of view of an observer placed outside the world, and, on the other hand, the epistemic alternatives of each agent (in each instant) are relative to that agent".

So, this paper is only the beginning of a lot of work that needs to be done, we have lots of paths to explore.

Acknowledgments This work has been granted by Spanish Government, "Ministerio de Economía y Competividad", Research Projects FFI2008-01205 (*Points of View. A Philosophical Investigation*), FFI2011-24549 (*Points of View and Temporal Structures*), and FFI2014-57409-R (*Points of View, Dispositons, and Time. Perspectives in a World of Dispositions*).

References

1. Álvarez, S. (1994). Tiempo, cambios e indeterminismo. *Análisis Filosófico, 14*(2).
2. Areces, C., Blackburn, P., & Marx, M. (2001). Hybrid logics: characterization, interpolation and complexity. *The Journal of Symbolic Logic, 66*(3), 977–1009.
3. Blackburn, P. (1994). Tense, temporal reference and tense logic. *Journal of Semantics, 11*, 83–101.
4. Blackburn, P. (2000). Representation, reasoning, and relational structures: a hybrid logic manifesto. *Logic Journal of the IGPL, 8*(3), 339–365.
5. Blackburn, P., & Goranko, V. (2001). Hybrid Ockhamist temporal logic. In C. Bettini & A. Montanari (Eds.), *Proceedings of the Eigth International Symposium on Temporal Representation and Reasoning. TIME-01*, Cividale del Friuli, Italy, June 14–16.
6. Blackburn, & Seligman, J. (1998). What are hybrid languages? In M. Kracht, M. de Rijke, H. Wansing & M. Zakharyaschev (Eds.), *Advances in modal logic* (Vol. 1). Stanford: CSLI Publications.

---

[22]Celebrated in Athens in August 2013.

[23]Related with this topic, Herrera and Vázquez [11].

[24]More about temporal epistemic logic in Engelfriet [7].

7. Engelfriet, J. (1996). Minimal temporal epistemic logic. *Notre Dame Journal of Formal Logic, 37*(2), 113–130.
8. Gabbay, D., & Guenthner, F. (Eds.). (1984). *Handbook of philosophical logic* (Vol. 2). Dordrecht: Kluwer Academic Publishers.
9. Gabbay, D., Hodkinson, I. & Reynolds, M. (1994). *Temporal logic. Mathematical foundations and computational aspects* (Vol. 1). Oxford: Oxford University Press.
10. Hautamäki, A. (1983). The logic of viewpoints. *Studia Logic: An International Journal for Symbolic Logic, 42*(2/3), 187–196.
11. Herrera González, R., & Vázquez Campos, M. (2011). Hacia una lógica temporal-epistémica basada en legunajes híbridos. *Análisis filosófico, XXXI*(1), 33–46.
12. Hughes, G. E. & Cresswell, M. J. (1968). *An introduction to modal logic*. London: Methuen.
13. Liz, M. (Ed.). (2009). *La realidad sin velos*. Barcelona: Laertes.
14. Liz, M. (Ed.). (2013). *Puntos de Vista*. Barcelona, Laertes: Una investigación filosófica.
15. Liz, M. & Vázquez, M. (2010). Subjects and points of view. Where privileged access meets with intersubjectivity. In *DISCOS International Conference in Intersubjectivity and the Self*, Budapest.
16. Liz, M. & Vázquez, M. (2011). Two approaches to the notion of point of view. In *14th Congress of Logic, Methodology and Philosophy of Science*, Nancy, Francia.
17. McTaggart, J. M. E. (1927). In C. D. Broad. *The nature of existence* (Vol. II). Massachusetts: Cambridge University Press.
18. Nishimura, H. (1979). Is the semantics of branching structures adequate for non-metric Ockhamist tense logics. *Journal of Philosophical Logic, 8*, 477–478.
19. Ohrstrom, P., & Hasle, P. (1995). *Temporal logic. From ancient ideas to artificial intelligence*. Norwell: Kluwer Academic Press.
20. Ponte, M. & Vázquez, M. (2008). An approach to the notion of point of view through hybrid temporal logic. *ESPP 2008*, Utrech.
21. Ponte, M. & Vázquez, M. (2012). Tense and temporal reference. Hybrid temporal logic. *Logique et Analyse, 55*(220), 555–578.
22. Prior, A. (1957). *Time and modality*. Oxford: Clarendon Press.
23. Prior, A. (1967). *Past, present and future*. Oxford: Clarendon Press.
24. Prior, A. (1968). *Papers on time and tense*. Oxford: Clarendon Press.
25. Prior, A. (2003). *Papers in time and tense*. New Edition, editado por Hasle, Ohrstrom, Braüner y Copeland, Oxford University Press.
26. Reynolds, M. (2003). An axiomatization of prior's Ockhamist logic of historical necessity. *Advances in Modal Logic, 4*, 355–370.
27. Thomason, R. (1984). Combinations of tense and modality. In D. Gabbay & F. Guenthner (Eds.), (pp. 135–165).
28. Vázquez, M. (2013). El cable del tiempo. In Liz (Ed.), (pp. 249–262).
29. Vázquez, M. & Liz, M. (2011). Models as points of view. The case of system dynamics. *Foundations of Science, 16*(4), 383–391.
30. Zanardo, A. (1985). A finite axiomatization of the set of strongly valid Ockhamist logic. *Journal of Philosophical Logic, 14*, 447–468.

Chapter 6
Change, Event, and Temporal Points of View

Antti Hautamäki

Abstract A "conceptual spaces" approach is used to formalize Aristotle's main intuitions about time and change, and other ideas about temporal points of view. That approach has been used in earlier studies about points of view. Properties of entities are represented by locations in multidimensional conceptual spaces; and concepts of entities are identified with subsets or regions of conceptual spaces. The dimensions of the spaces, called "determinables", are qualities in a very general sense. A temporal element is introduced by adding a time variable to state functions that map entities into conceptual spaces. That way, states may have some permanency or stability around time instances. Following Aristotle's intuitions, changes and events will not be necessarily instant phenomena, instead they could be processual and interval dependent. Change is defined relatively to the interval during which the change is taking place. Time intervals themselves are taken to represent points of view. To have a point of view is to look at the world as it is in the selected interval. Many important concepts are relativized to intervals, for instance change, events, identity, ontology, potentiality, etc. The definition of points of view as intervals allows to compare points of view in relation to all these concepts. The conceptual space approach has an immediate semantic and structural character, but it is tempting to develop also logics to describe them. A formal language is introduced to show how this could be done.

1 Introduction

The interest of studying points of view has grown during the last decades. The idea of point of view is intuitively clear; it is a way to see the world. But when we want to explore the function and meaning of points of view in our cognition, we face a serious dilemma: To define exactly the concept of point of view, we have to select a

A. Hautamäki (✉)
University of Helsinki, Helsinki, Finland
e-mail: Antti.hautamaki@kolumbus.fi

A. Hautamäki
University of Jyväskylä, Jyväskylä, Finland

© Springer International Publishing Switzerland 2015   197
M. Vázquez Campos and A.M. Liz Gutiérrez (eds.), *Temporal Points of View*,
Studies in Applied Philosophy, Epistemology and Rational Ethics 23,
DOI 10.1007/978-3-319-19815-6_6

specific field where points of view are relevant and applied. My personal observation is that the discussion about points of view is quite vain without formalizing basic concepts. Good formalism helps us to clarify our intuitions and in fact opens new research questions which are hard to see outside the formalism. This is especially evident in studying temporal aspects of points of view, the target of this article.

In my mind, the time aspect is underscrutinized in logic. A large majority of logical studies is related to a "static" aspect of logic; that is, temporal features of language are neglected and models are standard structures without a time component (cf. [16]). Another bias is that time is taken to be discrete leading to unnatural concepts like "the next moment" which are in contradiction with the intuition of the continuity of time. A related bias is to define the truth of formulas relative to single points in time (see [25, 27]), losing the process nature of change and motion.

My topic in this article is to study the concepts of change and event that are temporal and time related. The issue of time is philosophically sensitive. I consider it impossible to study temporal aspects of entities and changes without a philosophical framework. My strategy is to take Aristotle's philosophy of nature as a starting point. In his book *Physics*, Aristotle treated such important concepts like time, place, motion, change, and substance and tried to solve complex problems related to describing motion and change. What is interesting is that Aristotle adopted quite "modern" conceptions, such as continuity of time, space, and motion. Aristotle also developed a typology of changes, which is relevant also today. Although, Aristotle's conceptions are not defined in a clear, unambiguous way, there is enough consensus about its central points and suppositions, making it possible to discuss the Aristotelian conception of time and change.

To study temporal points of view, I use conceptual space approach. This approach permits me to formalize Aristotelian intuitions about time and change in a way that allows me to offer a new insight into them. I do not follow Aristotle's solutions in all aspects of my analysis. Although Aristotle's conceptions of continuity of time and the processual character of change are central for my presentation, I introduce modern analogues of the time related concepts of Aristotle that contain features not found in Aristotle's texts.

I have used the conceptual space approach in my earlier studies about points of view [12, 13, 15, 18, 19]. In that approach, properties of entities are represented by the location of entities in multidimensional conceptual spaces. The dimensions of spaces, called *determinables*, are qualities of entities in a very general sense. Concepts of entities are subsets or regions of conceptual spaces (see [10, 13]). Points of view could be defined in many ways with this approach. A clear way is to take a subset of determinables to represent points of view [13]. A more elaborated approach is to add theories or suppositions to points of view [15].

In this article, I adopt a slightly different strategy. First, I introduce a time element into the conceptual space approach by adding a time variable to state functions that map entities into conceptual spaces. In my notation, $S(x, t)$ is the state of an entity x at a time instance t. In basic structure of temporal conceptual space I do not make any assumption about change. In this framework I explicate an Aristotelian concept of change according to which states have some permanency or stability around time instances. Thus changes and events are not instant phenomena, instead they are processual and interval dependent—change is defined relative to the interval during which the change is taking place.

What about points of view? My approach here is to take time intervals to represent points of view. To have a point of view is to look at the world as it is in the selected interval. Many concepts are relativized to intervals, for example, change, events, identity, ontology, and even potentiality. Therefore the definition of points of view as intervals allows me to compare points of view in relation to these concepts. For comparison I will apply also an interval algebra developed by Allen [1].

The conceptual space approach is semantic and structural in nature. It starts from sets of entities and from their aspects or qualities and considers their relations. But it is tempting to develop also logic to describe temporal conceptual spaces. I do that by introducing a language to talk about determinables and changes of entities. In this language, there is no direct reference to time instants. Instead, formulas are interpreted relative to conceptual spaces and time. The logic is not axiomatized in this article; its sole purpose is to show how a temporal logic could be developed in the framework of conceptual spaces.

2   Time and Change in Aristotle

In Aristotle's philosophy of nature, the concept of change (*kinêsis*) is a central one. Aristotle defines change as "the actuality of that which exists potentially, in so far as it is potentially this actuality (*Physica*. III, 1, 201a10–11, 201a27–29, 201b4–5)." That is, change rests in the potential of one thing to become another. In the change, the potentiality is in the process of becoming actual. In other words, change is the process of actualization of potentiality. When the change is complete, the potentiality has become actual.

In general, there are three kinds of change: *generation*, where something comes into being; *perishing*, where something is destroyed; and *transformation*, where some attribute of a thing is changed while the thing itself remains constant. The permanent form of a thing is called its *essence* and changing attributes are called *accidental properties* (*Physica*. V).

All change or process involves something coming to be from out of its opposite. Something becomes what it is by acquiring its distinctive form—for example, a baby becomes an adult. According to Aristotle there are two general kinds of change defined by logical concepts contradictory and contrary determinations (*Physica*. V, 1–2.)

Contradictory determinations hold between properties A and not-A, and contrary determinations hold between "different" properties, say A and B such that A and B cannot be attributed to same entities at the same time. In Book V of *Physics*, Aristotle makes a distinction between substantial change and change as *kinêsis*. Substantial changes are the generational (birth, genesis) and the perishing (phthora) of entities, that is, appearance and disappearance of entities. Substantial change is instant, happening at a certain moment of time.

Table 1 Types of changes in Aristotle (physics)

Categories	Type of change	Contradictory change	Contrary change
Substance	Birth (genesis)	Non-S → S (x starts to exist)	–
	Perishing (phthora)	S →· non-S (x ceases to exist)	–
Kinêsis (real change) Quality Quantity Place	Qualitative change	–	F → F* (F and F* incompatible)
	Quantitative change	–	F is quantity with values n →· m (n < m or n > m)
	Changing place	–	F is a place p → p* (p ≠ p*)

There are three kinds of *kinêsis* type change: qualitative change, quantitative change and change of place. A change in color of an entity (e.g., from green to yellow) is a qualitative change, and a change in size (growth) of an entity is a quantitative change. The change of place is easy to understand; in modern terms, the space coordinates of a moving entity change. *Kinêsis* is a processual change, taking an interval to complete.

Aristotle counts quantity and quality among the accidents that modify a subject or substance entirely and directly.[1] In *Metaphysics,* Aristotle defines quantity as follows: "We call a quantity that which is divided into constituent parts, each or every one of which is by nature something one and individual. Thus plurality, if it is numerically calculable, is a kind of quantity." (*Metaphysica.* 1020 a 7–10).

On the other hand, quality is anything apart from quantity that belongs to a substance (*Metaphysica.* 1020 b 7). According to Aristotle, in its primary sense, quality is that which distinguishes a thing in its essence. A second kind of quality are the properties of changing things insofar as its change is detectable, namely the properties in view of which changes are distinguished (*Metaphysica.* 1020 b 14). This distinction refers to the difference between essential and accidental qualities.

It is important to know that Aristotle used in *Physics* the concept of accidental change to refer to external changes of entities, exemplifying it with, a scholar walks (*Physica.* V, 1). If, on the other hand, something in a thing is changing, then the change is called *simple* by Aristotle. The simple change is often the change of some part of a thing, for example, a diseased eye becomes healthy. Aristotle says clearly in Book V of *Physics* that we can set aside accidental changes (*Physica.* V, 1, 224 b 25).

We can classify the concepts of change (Table 1) by the using the schema

"x which is S is F"

---

[1]http://peenef2.republika.pl/angielski/hasla/a/accident.html, accessible 6.10.2014.

where x is an individual, S is a substantial genus, and F is a quality, quantity or location. S is a form of x and it is unchangeable in Aristotle's system, whereas F is a changing accident.

Aristotle bound time and movement together; there is no time without motion and no motion without time (*Physica.* V). In the Book V, Aristotle made statements concerning the continuum of time and motion. Motion and time are continuous and they cannot be composed of indivisible atoms (of movement or time). An "atom" of time would be an instant of time with no duration and therefore it would be impossible to build an interval of time starting from these atoms, Aristotle argues. Similar arguments are given for the continuity of motion. According to Aristotle, changes in quantity and location are continuous. This means in modern terms that change in quantity is taking place as a process where the value of quantity is approaching the target values, the new state of entity.

Aristotle's conception of change of quality is not so clear. If a person is changing from illness to health, is that a discrete change: a person was ill and, after some moment of time, he is healthy. Aristotle seems to make an assumption, that there is a continuum of degrees of illness (or health), and similar for the change of color and of many other qualities. My interpretation of the Aristotelian conception of change is that quantities and qualities are both continuums and there is no jump or gap in change. Aristotle's theory of continuum is central for his theory of change. Change is a process concept. According to continuum theory, there is no way to say what is the "next state". Therefore it is not possible to define the change as referring to differences between one instant and the next. Also is not possible to define a change by referring to a single instant. Aristotle's solution is to accept the first moment of change (*Physica.* VI, 5). So the new state, after a change, begins at some moment and continues for some time (see Fig. 1). The first moment or state is the beginning of the actuality of some potentiality of an entity.

The background of this discussion about the "first moment" is the paradoxical character of change and motion when concentrating on single moments. If we say that x changed color from blue to yellow at the moment t, then it seems that x is blue and yellow at the same time. To avoid this logical contradiction, change must be analyzed in another way (see also [33]). Aristotle's solution is to split change into two parts: *Before t,* x is not yellow, and *from t onward,* x is yellow. This solution is based on the continuity of time. Zeno's paradox is related to the same issue: the motion is impossible because the moving object must be in rest and in motion at each moment (*Physica.* V).

I have described Aristotelian notion of change by pointing out several philosophical principles:

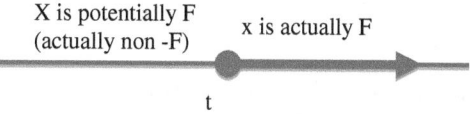

Fig. 1 Change identified by the starting point

- Time and motion are interdependent concepts.
- Time and motion are continuous, containing no jumps.
- There are changes related to substance (genesis and perishing) and changes related to accidents (*kinêsis*: quality, quantity and location).
- Substantial changes are instant (coming to be or perishing).
- Changes (*kinêsis*) in accidents are processual, continuous, and interval related phenomena.
- There is a first moment when change has taken place (the first moment of the actuality of the target state of change).
- Change is the actualization of potentiality.

I take these features of Aristotelian change theory as my starting point when developing temporal points of view. But I will make also some important deviations from them. Of course, each of these features has fueled extensive philosophical discussions [21, 22, 24], and my comments in this article are not intended to be scholarly.

3 Conceptual Spaces and Determinables

Conceptual space is an effective way to present properties of entities. Its idea is to map entities into quality space, where qualities are dimensions of the space. Johnson, in his book *Logic*, called these qualities *determinables*. Johnson's definition of determinables is the following:

> I propose to call such terms as colour and shape determinables in relations to such terms as red and circular which will be called determinates. (Johnson, Part I p. 174)

A determinable is something that could be further specified and this specification gives determinates (determinate values). Determinables could be seen from a linguistic viewpoint as general terms or from an ontological viewpoint as qualities, properties, or dimensions of things. Johnson's definition is linguistic, referring to *terms*. Johnson prefers to say that determinables are abstract names, which stand for adjectives [17]. According to A.N. Prior, "'red' is the proper name of an individual universal, if we may speak so; while 'colour' is the name of the class of universals to which this individual one belongs" ([28], 8).

My interpretation of determinables is that they are functors associating entities to their values. In schematic forms, determinables could be presented as follows: "the $\_\_1$ of $\_\_2$". For example, the phrase "the color of the table" presents the determination of "color" to the entity "table" and the value could be presented by the sentence:

> "The color of the table is brown."
> Using function notation the sentence could be coded
> [color](table) = brown.

I suppose that there is no absolute set of determinables describing the world as a whole. Determinables are first of all related to situations in which we live and act. We have to construct concepts to talk about our situations. Some determinables are physical and some mental, some are concrete and some abstracts, some are personal and some social, and so forth. In a traditional context (e.g. Aristotle), a distinction is made between qualities and quantities. In Johnson, determinables are intended to be qualities. But in a mathematical context, quantities are more natural and qualitative variables are transformed into quantities. So one can consider values of determinables to be just real numbers, but we do not assume that. So determinables could be qualities as well as quantities. It is interesting in this context that Aristotle seems to suppose that qualities are somehow continuous and have an infinite number of degrees; like degrees of colors or illness.

The number of determinables is a complex issue. In some earlier studies, like Hautamäki [13], it is supposed that the number of determinables for each set of entities is countable (see also [5]). Now it seems that it is better to assume that there are finite numbers of determinables needed to describe and identify entities in relevant applications. This is not a thesis concerning the absolute number of determinables. I suppose in this article that, in each context, a useful finite set of determinables is selected. This supposition reflects my conviction that the human (brain) capacity to identify entities is limited and so "ontology" in my sense is always finite in the sense of the fundamental variety of entities.

With the concept of determinable the concept of conceptual space is defined. Conceptual spaces are linked to entities by state functions, which "locate" entities in spaces.

For the definition, I use the notation B^A for the set of functions from the set A to the set B:

$$B^A := \{f \mid f\colon A \to B\}.$$

If $f \in B^A$ and $C \subseteq A$, then the *restriction of f into C*, denoted by f/C, is defined by

$$f/C := \{\langle x, y \rangle \in f \mid x \in C\}.$$

Let I be a finite set of determinables and let D be a set of (possible) values for determinables. We call the set

$$D^I = \{f \mid f\colon I \to D\}$$

a *conceptual space*. To talk about properties or qualities of entities, we just need a state function S from the set of entities E into the conceptual space D^I:

$$S\colon E \to D^I.$$

$S(x)$ is the state of x. If $S(x)(i) = a$, we say that "i of x is a", for example

$$\text{"the color of the pen is yellow"}$$

when i is the determinable color, a is the color yellow and x is the pen. The state S(x) tells the "complete story" about x, because it specifies the values of all determinables for x.

As an example, let I = {color, form} and let S(x) = {⟨color, blue⟩, ⟨form, oval⟩} then the color of x is blue and the form of x is oval.

For clarity of terminology let us use the following terms to describe conceptual spaces:

Determinables, such as color, are aspects of entities and dimensions of space; we call them also functions of entities.

Determinates, such as red, are attributes of entities and values of determinables.

If $S(x)(i) = a$ then a is a *property* of x and the value of the determinable i for x.

The term conceptual space comes from the fact that concepts could be presented as subsets of conceptual space. Let C be a subset of the conceptual space D^I then the concept C applies to all entities whose states belonging to C:

$$\text{"x is C" if and only if } S(x) \in C.$$

Note that in my presentation I "collect" all values to a single set D, which might be the set of real numbers. Although this is little bit artificial, in practice it does not set any restrictions on the applicability of determinables or conceptual spaces: it is a normal practice, say in statistics, to quantify qualities. In Hautamäki [13] all determinables have their own set of values.

The terms conceptual space or quality space are used in van Fraassen [31, 32], Stalnaker [30], Hautamäki [13, 14], Gärdenfors [10], and Clark [6], among others.

4   Time Aspect

The idea behind time is that entities change in time. Taking this as a starting point, we have to express change with a time variable. There are at least two options. Either we take time as one of the determinables or we connect time to the entities. I prefer the last one, because time is not considered a similar kind of aspect as determinables. Time is not a quality in the sense of Johnson's logic, where the determinables are principles of "fundamentum divisionis". Time is something to which changes of determinates are internally linked. On the other hand, both options make possible to express different aspects of changing entities. When time variable is a component of state function, as we suppose in this presentation, we could express continuity of entities as supposed in *endurantism*. It one take time to be one of determinables, then one could talk about entities with different temporal parts, like animals with relict parts (perdurantism).[2]

Let T be a set of times. For lucid presentation, we take T to be the set of real numbers $\mathbb{R}$. The elements of T are called (time) *moments* or *instants*.

---

[2]The concepts of enduratism and perdurantism was introduce to me by Manuel Liz. I would like to thank him for many valuable comments.

We extend the state function S to include the time beside the set of entities:

$$S: E \times T \rightarrow D^I, S(x, t) \in D^I, \quad x \in E, t \in T.$$

For convenience we shall use the expression i[x] for the phrases "the i of x" or "the value of determinable i for x". If $S(x, t)(i) = a$ we write often "i[x] = a at t".

But further elaboration is needed, because if we add this concept of time to the state function without qualifications, changes of entities start to seem arbitrary and unnatural. For example, it might be that a determinable i has a value a at t, but not a before and after t. Intuitively entities have permanence over some period of time (cf. Aristotle). To express that aspect we use intervals to guaranty temporal unchangeability. The idea of our treatment of continuity is to suppose that, if an entity x has some property at t, it has the same property also near t, either before or after t.

Another crucial issue is the existence of entities. When time is under consideration, entities are changing in time but their existence might also be temporal. Entities appear or disappear. Between their beginning and ending, their existence is unbreakable, that is, there is no gap of existence between their appearance and disappearance. So intuition seems to propose that there must be an interval of existence for all entities. Before or after that interval the entity does not exist. There are no properties of entities outside their interval of existence, either. I take these ideas into my definition of temporal determination base.

For notation, the basic definitions for intervals of real numbers are

$[t, t^*] := \{x \in \mathbb{R} | t \leq x \leq t^*\}$, closed interval;
$(t, t^*) := \{x \in \mathbb{R} | t < x < t^*\}$, open interval.

Definition 1 A *(temporal) determination base* B is the structure

$$B := \langle I, F, D, E, T, S \rangle$$

where I, F, D and E are non-empty sets, I is finite and F is a subset of I, T is the set of real numbers $\mathbb{R}$, and S is a partial function from $E \times T$ into the set D^I such that

1. S is an injection for all time moments: if $S(x, t) = S(y, t)$ then $x = y$;
2. if $S(x, t)/F = S(y, t)/F$ for all $t \in T$, then $x = y$;
3. if $S(x, t)$ is defined, then there is a closed interval $\pi_x = [m, n]$ containing t such that

   if $t^* \in \pi_x$ then $S(x, t^*)$ is defined;
   if $t^* < m$ or $t^* > n$, then $S(x, t^*)$ is not defined;

The elements of I are called *determinables*, the elements of F are called *fundamental determinables*, the elements of D are called *determinates* or *determinate values*, the elements of E are called *entities*, the elements of T are call time *instants* or *moments*, and the function S is called a *state function*. Other determinables are called *supplementary*.

The set D^I of functions is called a *conceptual space* for entities in E. For all $x \in E$ and $t \in T$, $S(x, t) \in D^I$ is called the *state of x* at t. The notation $S(x, t)(i) = a$ and $S(x, t) = s$ implies that $S(x, t)$ is defined. When we write $S(x, t) = S(y, t)$ or $S(x, t)(i) = S(y, t)(i)$ we mean that both are defined and identical or both are undefined.

The introduction of fundamental determinables is a novel feature in the theory of points of view as compared to Hautamäki [13, 15]. They are fundamental in the sense of providing a constraint for identity. In many applications of conceptual spaces, fundamental determinables include three space coordinates (x, y, and z axes).

Condition 1 above means, that whenever two entities are in the same state they are identical. It is a variant of the principle of Identity of Indiscernibles by Leibniz in his *Discourse on Metaphysics*, Sect. 9 [23]. But it is possible that two entities are in the same state but at a different time.

Condition 2 above is also a variant of Leibniz's principle: If two entities have the same values for all fundamental determinables at all moments then they are the same entity. Note that the condition permits the case that two different entities have same fundamental properties at some moment. Of course, then they have some differences in other determinables, otherwise they are identical by Condition 1. In this Condition 2, I make a deviation from Aristotelian conception of essence, because for him essence is stable and permanent (in form), whereas here fundamental determinables could change their values but still define identity over time. So my concept of "essence" is dynamic.

Condition 3 means that x has "properties" only in the interval $\pi_x = [t, t^*]$. We interpret that so that *x exists only in the interval* π_x. Before or after π_x, x is non-existent (or x has a virtual, empty existence). The interval π_x is called the *lifecycle* of x. At the instant *t, x appears* and, at *t*, x disappears*.

It is useful to generalize temporal state function to intervals, denoted by π, as follows:

$S(x, \pi)(i) = a$ if and only if for all $t \in \pi$: $S(x, t)(i) = a$,
$S(x, \pi) = s$ if and only if for all $t \in \pi$: $S(x, t) = s$.

Then I will write also that "i[x] = a in π" (Fig. 2).

Because a major interest for the time aspect comes from changes of entities, we have to define what change is in the context of conceptual space. As a basic case we take the change of the value of a determinable for an entity in some instant of time. It is quite complicated to define exactly the change point, because in a continuous

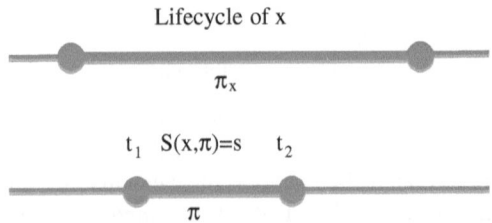

Fig. 2 The lifecycle of x and the interval condition for $S(x, \pi) = s$

time system there is no "next moment" of time. Of course, we could recognize the existence of change in the interval [t, t*] by observing that S(x, t) ≠ S(x, t*). But I want to study the process of change and therefore I use intervals to see where the change starts or ends in the Aristotelian sense. Note that my definition 1 of temporal determination base is neutral in relation to the nature of change.

Definition 2 Let B be a determination base B = ⟨I, F, D, E, T, S⟩ and let i ∈ I, x ∈ E, t ∈ T,

$$s \in D^I, a \in D, S(x,t) = s \text{ and } S(x,t)(i) = a.$$

x is *stabile in the interval* $\pi \subseteq \pi_x$ in respect of i if S(x, t)(i) is constant in π.

x is *stabile at t* in respect of i if there is an open interval π containing t in which x is stabile.

x *A-changes at t* in respect of i if there is an interval $\pi = (t^1, t^2)$ such that $t^1 < t < t^2$ and

1. S(x, t*)(i) = a for all t* such that $t \leq t* < t^2$
   S(x, t*)(i) ≠ a for all t* such that $t^1 < t* < t$

Or

2. S(x, t*)(i) = a for all t* such that $t^1 < t* \leq t$
   S(x, t*)(i) ≠ a for all t* such that $t < t* < t^2$

   A-change is a nick name for Aristotelian change.

In the case 1 we say that i[x] *starts (begins)* to be a at t.
In the case 2 we say that i[x] *ceases (ends)* to be a at t.

If i ∈ F is a fundamental determinable, the change of x in respect of i is called a *fundamental change.*[3]

The original Aristotelian model of change accepts only changes with Condition 1, that is, changes are always beginnings. I deviate here from Aristotle by accepting also endings as changes.

Aristotelian changes are "regular" in the sense that in them something is stabile either before or after of the change. But are all changes so nice? In theory not; entities can change all time following numerous different patterns. Instead of classifying them in any imaginable way, I just define a concept of "irregular change" as follows

---

[3]The concept of change is relevant when one studies the traditional doctrine of essentialism. There is the distinction between essential properties and accidental properties, where essential properties are those that survive change and accidental properties are those that do not (see Aristotle *De Int* 4a10, *Met.* 1028a31–33; [7]. In my system, fundamental determinables correspond to essential properties, whereas supplementary determinables might be called *accidental*.

An entity x *changes irregularly at t* in respect of the determinable i if for all intervals π containing t there is a moment of time t* such that

$$S(x, t^*)(i) \neq S(x, t)(i).$$

Note that if x changes irregularly at t in respect of i then x is not changing in Aristotelian sense nor it is stabile at t.

The other interesting case of change is the change without starting or ending point. In fact, even in the Aristotelian change one have accept this king of change. Namely, when i[x] ceases to be a at, it might be that i[x] = b in an open interval (t, t*) with b ≠ a. Then i[x] changes its value at t so that there is no starting point for the new value b.

Because the definition 1 of determination base is quite general and allows many patterns of change, it might be worthwhile to restrict possible patterns in very definition of determination base. For example, if we think that Aristotelian concept of change is a "natural one", we could adopt the following definition.

Definition 1A   Let B = ⟨I, F, D, E, T, S⟩ be the determination base. Then it is *Aristotelian* if the state function S satisfies the following condition:

Whenever S(x, t) is defined and i ∈ I there is a closed interval π containing t such that S(x, t)(i) is constant in the interval π.

This definition means, that entities keep their determinate values always during some interval. It is easy to define Aristotelian change in this setting: if S(x, t)(i) = a and t is the first point of the interval where i[x] = a and i[x] ≠ a before t in some interval then i[x] start to be a at t, and similarly for the ending case.

An important question is the nature of continuity in change. This issue could be studied by exploring "graphs" of states of entities. I use for that the concept of world-lines, which are functions from the set T of time moments into the conceptual space. This allows me also define time-related concepts.

If we think of conceptual space as multidimensional space and track how the state of an entity x changes in time, we will reach a "world-line" or curve of x in the space. So the life cycle of x could be represented by its world-line in conceptual space.

Definition 3   The elements of the set $(D^I)^T$ are called *world-lines* and they are functions from T into D^I. The subsets of $(D^I)^T$ are called (time-related) *concepts*.

The *world-line of an entity x* is the function w(x) defined by

$$w(x)(t) = S(x, t) \quad \text{for all } t \in \pi_x.$$

The *extension* of a concept C is the set

$$EXT(C) := \{x \in E \mid \text{there exist } f \in C \text{ such that } w(x) \subseteq f\}.$$

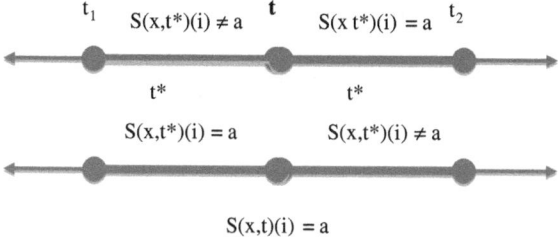

Fig. 3 There is an A-change: in the *upper line* i[x] starts to be a and in the *lower line* i[x] ceases to be a at t

w(x) is a function of the time mapping time moments in x's lifecycle into the state space reaching all states of x. In mathematical terms, a world-line is a *vector* of time. The extension of a concept is defined so that an entity satisfies the concept C, if its world-line is a part of some element of C. Note that we do not presuppose, that the world-line w(x) is member of C, because entities have lifecycles not covering the whole time span in T. The presented notion of concepts is interesting, because concepts do not only express the properties of entities but also they specify how properties of entities change in time.[4]

It is a difficult question, whether world-lines must be continuous. As we observed above, Aristotle seems to suppose that all changes are continuous. On the other hand, he hesitates to extend the continuity assumption to the case of qualitative change. I have defined determination base so that continuity over change moments is not supposed. So a jump is possible, as in the case A in Fig. 3. My intuition differs here from Aristotelian intuition. I see it as an empirical question whether a determinable is changing in a continuous way or not. A separate problem is that in genesis entities start to exist suddenly with a bundle of properties (qualities and quantities), and similarly they lose all properties when perishing. So genesis and perishing are not continuous changes; they are not *kinêsis* in Aristotle's terms.

In Fig. 4, the world-line A contains a jump or gap, the world-line B is a continuous curve, and the world-line C is a straight line, where the i1-coordinate is constant. Note that this kind of representation of world-lines does not tell how entities move in time in the conceptual space; for that a time dimension must be added to the coordinate system.

To reach a workable definition of the continuity of state function, one need to introduce a topology into conceptual spaces. Gärdenfors [10] has done this by supposing that there is a distance measure d defined between states of the space and using it to define a betweenness relation. Gärdenfors uses that relation to define the

---

[4]Many *scientific theories* could be presented as time-related concepts (e.g., [4]. Let Th be a time-related concept and x an entity. Then $Th \subseteq (D^I)^T$ and the proposition "$x \in EXT(Th)$" is an empirical claim stating that the entity x (a model of Th) "obeys the laws" of theory Th (see [15].

Fig. 4 A set of world-lines in
a two dimensional space

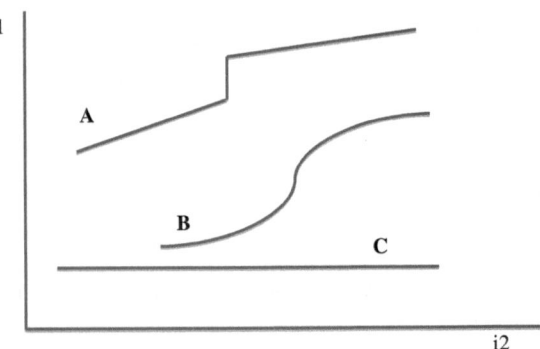

convexity of subsets (regions) of the conceptual space. A subset C of the conceptual space is *convex* if for all states s and s* in C all states between s and s* are also in C [11]. I use convexity to define the continuity of state function S as follows:

The state function $S: E \times T \to D^I$ is *continuous* if sets S_x are convex for all $x \in E$ where

$$S_x := \{S(x, t) | t \in \pi_x\}.$$

This definition implies that if s and s* are states of an entity x then all states between s and s* are also x's states. Note that this definition says nothing about times when states in S_x are reached.

If we have an total order $\leq_D$ in the set D of values, then continuity could be define for determinables as follows:

The state function S is *continuous in respect of i* if for all entities x and for all intervals $\pi = [t_1, t_2]$ it holds that

if $S(x, t_1)(i) \leq_D a \leq_D S(x, t_2)(i)$ then there exists $t^* \in \pi$ such that $S(x, t^*)(i) = a$.

If S is continuous in respect of i then S(x, t)(i) reaches all values between any couple of values x reaches in any intervals during its lifecycle. Continuity implies that there is no gaps in world-lines of entities.

5 Temporal Points of View

A point of view is a special way to see the objects in cognition and in discourse. With point of view, some aspects of objects are selected as the focus of attention, leaving some other aspects out of consideration or awareness [26]. In conceptual space framework, several methods are used to approach points of view. In Hautamäki [13], points of view were defined to be a finite selection of determinables. In Hautamäki [15], another element was added to the definition of points of view: that of a theory. Thus a point of view is a structure $V = \langle B, K, T \rangle$, where B is a determination base, K is a (finite) set of determinables and T is a subset of subspace D^K called a theory (T is not time there).

The time aspect of objects leads naturally to consider time as an element of points of view. Instead of adding time to previous concepts of points of view, we take time as a sole element of points of view. Single moments of time are not relevant to define points of view, however. Instead, intervals of time are how we conceive time in cognition. A temporal point of view is an interval into which we concentrate our attention.

Definition 4 A *temporal point of view* over the determination base $B = \langle I, F, D, E, T, S \rangle$ is the structure $TPV = \langle \pi, B \rangle$, where $\pi \subseteq T$ is a closed interval.

When the determination base B is known, I identify a point of view $\langle \pi, B \rangle$ with the interval π. The interval π is closed, because it is natural to suppose that there are starting and ending instances in temporal points of view.

Identity of entities could be relativized to points of view. Of course, if entities are in the same state at some time they are identical. But if we consider the fundamental determinables of entities in some interval, we get an interesting concept of relative identity (or "incidence" as per [29]).

$$x \approx_\pi y \text{ if and only if for all } t \in \pi : \ S(x,t)/F = S(y,t)/F.$$

If $x \approx_\pi y$ I will say that x and y are *fundamentally identical in* π. This means that x and y identical in respect of all fundamental determinables in π. It might be that there is a moment t before or after π such that $S(x, t)/F \neq S(y, t)/F$ and therefore $x \approx_\pi y$ does not imply $x = y$.

Ontology is related to what exists and does not exist. π-*ontology* is the set of entities existing in π. Entities have their own lifecycles. Their appearance or disappearance could be relativized also to points of view. Entities could change in some respects and remain the same in other respects in π. It is interesting to recognize in which respects an entity is changing.

Definition 5 Let π be a point of view and $x \in E$. I will say that

x exists in π if $\pi \cap \pi_x \neq \emptyset$;
π-*ontology* is the set $O(\pi) := \{x \in E | \pi \cap \pi_x \neq \emptyset\}$;
x appears in π if $\pi_x = [t, t^*]$ and $t \in \pi$;
x disappears in π if $\pi_x = [t, t^*]$ and $t^* \in \pi$;
x A-changes in π *in respect of i* if $i[x]$ A-changes at some $t \in \pi$;
x A-changes in π when x A-changes in π in respect of some $i \in I$;
x is stabile in π *in respect of i* if $i[x]$ is constant in $\pi \cap \pi_x$;
The *extension of a concept C relative to* π is the set

$$\mathrm{EXT}(C, \pi) := \{x \in E | w(x)/\pi {\subseteq} f \ \text{for some} \ f \in C\}.$$

6 Time Direction and Accessibility

So far all my definitions are in fact neutral to the direction of time, only the phrases "start" and "beginning" must be change to phrases "cease" and "ending" and vice verse. In fact all above concepts related to time and change are *symmetric*.

But intuitively and from the experience of every day life it seems plausible to suppose that time is "forward" directed. Our memory is building a picture of time flowing from past to present and now-moments turn to be passed away. One solution to this challenge is to bound the direction of time to temporal points of view by adopting the actual moment of time: the "now". A *temporal point of view with present time over the determination base B* is the structure $\mathrm{VPT}(n) = \langle \pi, n, B \rangle$ where π is closed interval and $n \in T$ is the *now-moment*.

Then it is natural to call all moments $t < n$ *past time* and all moments $t > n$ *coming time*. Note that it is not supposed that n belongs to π. This means that the interval π under consideration in $\mathrm{VPT}(n)$ might be a passed interval or a future interval.

In the Aristotelian tradition the process of actualization of potentials is deeply bounded to the direction of time. We can even state that the process of actualization sets the direction to time. Behind the change there are causes, whether material, formal, efficient or final (*Physica* II, 3). These causes are answers to the question "why" something is moving or changing. According to Aristotle an event's **"final cause"** is the end toward which it directs; say the final cause of a seed is the full-grown plant. In summary, the whole theory of change and causes contains implicitly the direction of time, from causes to results.

One way to clarify the potentiality and the direction of time is to take into account of the possible transitions of states in conceptual space. I define the potentiality to be a relation between consecutive states on the basis of world-lines of entities.

Definition 6 The *accessibility* relation $A \subseteq D^I \times D^I$ is the set

$$\mathrm{ACC} := \{\langle s, s^* \rangle | S(x, t) = s, S(x, t^*) = s^* \ \text{and} \ t < t^* \ \text{for some} \ x \ \text{and} \ t, t^*\}.$$

If ACC(s, s*), we say that the state s* is *accessible from or possible after* the state s.

The *potentials of an entity x* at t is the set

$$\mathrm{POT}(x, t) := \{s \in D^I | \mathrm{ACC}(S(x, t), s)\}.$$

The *potentials of an entity x at t from the point of view* π is the set

$$POT(x, t, \pi) := \{s \in D^I | s = S(x, t^*) \text{ and } ACC(S(x, t), s)) \text{ for some } t^* \in \pi\}.$$

If s* is accessible from s then they are in the same world-line for some entity x, s* being a later state than s. Potentials of entities at t are their consecutive states after t. In POT(x, t, π) the set of potentials is restricted to states x owns in π.

Another interpretation of the accessibility relation ACC is that it expresses the *disposition of entities to change*. If $\langle s, s^* \rangle \in A$ and $S(x, t) = s$ then x has the disposition to reach the state s*.

According to the definition of x's potentials at t, the state s is potential for x if it is accessible from it's state at t. We say that a potential s of x is *actualized* at t* if S (x, t*) = s. When potentiality is relativized to a point of view π, then only those states are considered to be potential which are actualized in π.

We can develop a numerical measurement or probability to express transitional potentiality:

$$P(s^*/s) = r \quad \text{and } r \in [0, 1]$$

P(s*/s) is the probability that an entity is in the state s* in the condition that it has been in the state s. To define this probability we take sets EV(X), $X \subseteq D^I$, to be *events* in the sense of probability theory:

$$EV(X) := \{\langle x, t \rangle | S(x, t) \in X\}.$$

Note that EV({s}) = {$\langle x, t \rangle$} when S(x, t) = s.

7   Comparison of Points of View

A temporal point of view is the structure TPV = $\langle \pi, B \rangle$, where $\pi \subseteq T$ is a closed interval. Let B be fixed. I will compare intervals of T. Relations of intervals could be classified by applying interval algebra developed by Allen [1].

Let $\pi = [t_1, t_2]$ and $\pi^* = [t_3, t_4]$, then there are basically seven different relations between π and π* as shown in Table 2 and Fig. 5.

Table 2 Basic relations of intervals according to Allen [1]

Relation	Interpretation $\pi \rightarrow \pi^*$	Converse $\pi^* \rightarrow \pi$
(A) $t_2 < t_3$	(π) proceeds (π*)	(π* is) proceed by (π)
(B) $t_2 = t_3$	Meets	Met by
(C) $t_1 < t_3 < t_2 < t_4$	Overlaps	Overlapped by
(D) $t_1 < t_3 < t_2 = t_4$	Finished by	Finishes
(E) $t_1 < t_3 < t_4 < t_2$	Contains	During
(F) $t_1 = t_3 < t_2 < t_4$	Starts	Started by
$t_1 = t_3$ and $t_2 = t_4$	Equals	Equals

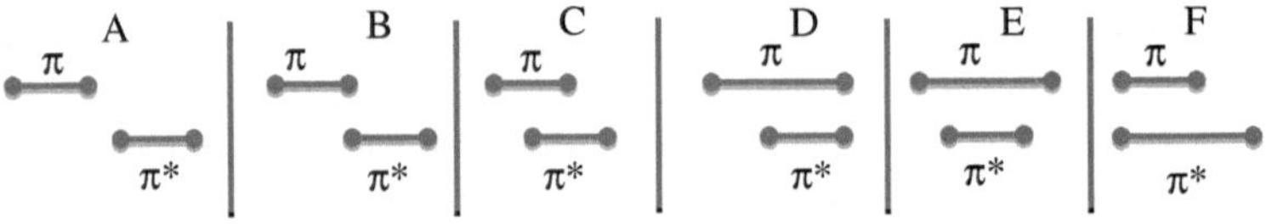

Fig. 5 Graphical presentation of relations between intervals (following [1]). *Note* The case of equality is a trivial identity

In Fig. 5 these concepts are presented by intervals.
Some observations about these cases:

In cases B, C, D, E and F, $\pi \cap \pi^* \neq \emptyset$ and $\pi \cup \pi^*$ is also an interval.
In cases B, C and F, $\pi \cup \pi^*$ contains π, thus being an enlargement of π.
In cases D and E, π contains π^*.
In the case A, $\pi \cap \pi^* = \emptyset$ and $\pi \cup \pi^*$ is not an interval.
In the case B, π and π^* meet or join each other.

The concepts of existence, ontology, change, permanence, and potentiality are all relativized to points of view. I point out only some interesting cases, leaving others to the reader.

If $\pi \subseteq \pi^*$ and $x \in E$ and $y \in E$ then:

$O(\pi) \subseteq O(\pi^*)$, more instances imply more existing entities;
$POT(x, t, \pi) \subseteq POT(x, t, \pi^*)$, more instances imply more potentiality;
$EXT(C, \pi^*) \subseteq EXT(C, \pi)$, because if $w(x)/\pi^* \subseteq f$ then $w(x)/\pi \subseteq f$;
if $x \approx_{\pi^*} y$ then $x \approx_\pi y$, if x and y are identical in an interval they are also identical in its subintervals.
The converses of these relationships do not hold in general.

8 Events

The nature of events is one of the most exiting topics in philosophy; discussed by such contemporary thinkers than Alain Badiou, Donald Davidson, Nelson Goodman, Jakerow Kim, David Lewis and W.V.O. Quine among others.[5] But the event is also important topic in Hellenistic philosophy, especially for Aristotle.

As an example we consider Kim's [20] account of events. According to him events are instantiations of properties. Events are composed of three things: Object (s) [x], a property [P] and time or a temporal interval [t].

Events are defined using the operator [x, P, t]. A unique event is defined by two principles:

---

[5]http://plato.stanford.edu/entries/events/, read on 9.9.2014.

(a) the existence condition: [x, P, t] exists if and only if object x exemplifies the
     n-adic P at time t;
(b) the identity condition: [x, P, t] is [y, P*, t*] if and only if x = y, P = P* and
     t = t*.

Kim does not specify what kind of properties P's are, instead he states that only
constructive properties create distinct events. Also events are related to time
intervals, but not their spatial situations. A conceptual space approach opens new
opportunities to handle events. I take Kim's proposal seriously and elaborate it in
my framework.

Intuitively, events are associated to important changes in things. The "important"
part is difficult to define without priorities of determinables and for that I use the
concept of fundamental determinables. There are two kinds of events, the generation of a new thing or the appearance of new fundamental properties of existing
things. Both changes can be treated by states of entities. Because changes are
relative to points of view as time intervals, the concept of event is defined by
referring to intervals. Roughly, a state s is an event relative to point of view π, if it is
a new state inside π. My definition of the concept of event is in the spirit of
Aristotle.

Definition 7 Let $B = \langle I, F, D, E, T, S \rangle$ be a determination base, let $\pi = [t_1, t_2]$ and let
$TPV = \langle \pi, B \rangle$ be a temporal point of view. A triple $e = \langle x, s, t \rangle \in E \times D^I \times T$ is an *event
from the point of view π* if

$x \in O(\pi)$, $s = S(x, t)$, $t \in (t_1, t_2)$, and there exists $i \in F$ such that
$S(x, t^*)(i) = a$ for all t^* such that $t \leq t^* \leq t_2$ and
$S(x, t^*)(i) \neq a$ for all t^* such that $t_1 \leq t^* < t$.

We say also that e is a *fundamental change of x in respect of i*. If there are
several determinables changing at t, an event is said to be a *composite fundamental
change*. The time moment t of the event is called *event time* and we can say that *e
happens at t*.

According to the definition of event, the state s of an event $\langle x, s, t \rangle$ is new in the
interval π. If the event e is a fundamental change of x in respect of i then i[x] will
keep its value at t to the end of the interval π. A fundamental change is an
appearance of a new value of a fundamental determinable. So the appearance of a
new value of a supplementary determinable is not, as such, an event.

The definition allows there to be several changes taking place in the same event.
For example the color and form could both change, thus being a composite fundamental change. So both changes are counted as parts of the same event (cf. [8].

To specify an event e at t one needs to pick up an entity x and a state s of x such
that the combination $e = \langle x, s, t \rangle$ is an event in π. The entity x is going through a
fundamental change at t and i[x] gets a new value for some fundamental deter-
minable $i \in F$.

This definition for events is a clear advance in comparison to Kim's concept.
Properties are specified to be actual values of determinables in events. The entity x

exemplifies the properties presented by the state s. An event is also directed toward the future, that is, something new has happened and this will continue for some time. Identity condition for events is basically:

$$\text{if}\langle x, s, t \rangle = \langle y, s, t \rangle \quad \text{then } x = y.$$

The condition is satisfied. It follows from Condition 1 of the definition of determinable (Definition 1) that, if $S(x, t) = s$ and $S(y, t) = s$ then $x = y$ (S is an injection). Of course, Kim's account is more complicated if one problematizes the identity conditions of time moments and states.

9   A Temporal Modal Logic of Viewpoints (TL)

There are many ways to elaborate a temporal modal logic over temporal determination bases. For example, formulas could be valuated in relation to instants of time or intervals of time. Here I use instants, but in truth relation I refer to intervals, as well. Time element could be added also to the language, but my choice is to treat time in semantics: Formulas are evaluated relative to conceptual spaces and time instances. For treating points of view, it is possible to introduce an interval operator. Scott [29] proposed to use *progressive tense* operator [↔] defined by the stipulation (following my notation):

[↔]φ is true at t iff there is an open interval $J \subseteq T$ with $t \in J$ such that φ is true at all $t^* \in J$.

According to Scott, McKinsey, and Tarski, the propositional logic of this operator is S4 (referred to in Scott [29], p. 160). I make some modifications to Scott's proposal.

9.1   Language of TL

The language TL is an extended first order language with two kinds of variables, positive and negative predicates for determinables and concepts, and with three modal operators for intervals. Variables for entities are x, y, x*, x**, and so forth and for determinates v, v*, v**, and so forth. Constants for entities are c, c*, c**, and so forth and for determinates a, b, a*, a**, and so forth. Variables and constants are terms, denoted by t, t*, t**, and so forth.

In the language TL there is an identity relation $t = t^*$ and a predicate EX(x) for existence of entities. To express determinables, I introduce a finite (non-empty) set Δ of determinable indexes and a two-place predicate P_δ for each $\delta \in \Delta$. Then I have also a set of one-place predicates C, C*, C**, and so forth for concepts.

Primitive formulas: $\mid t = t^* \mid EX(t) \mid [P_\delta](t, t^*) \mid [-P_\delta](t, t^*) \mid [C](t) \mid [-C](t) \mid$

$t = t^*$	means "t is t^*", t and t^* are both constants for entities or for determinates;
$EX(x)$	means "x exists";
$[P_\delta](x,v)$	means "the value of the determinable δ for x is v";
$[-P_\delta](x,v)$	means "the value of the determinable δ for x is *not* v";
$[C](x)$	means "x is C";
$[-C](x)$	means "x is not-C"

Time operators:

$[\leftarrow]\varphi$	backwards φ holds;
$[\mapsto]\varphi$	forwards φ holds;
$PoV\varphi$	from the point of view φ holds.

Formulas: $\mid$ primitive formulas $\mid \neg\phi \mid \phi \wedge \psi \mid \forall x\phi \mid \forall v\phi \mid [\leftarrow]\phi \mid [\mapsto]\phi \mid PoV\phi \mid$

A *sentence* is a formula without free variables.

9.2   Semantics of TL

The semantics of TL is a standard semantics for first order time logic except that modal operators are not interpreted by referring to single moments of time as are the standard operators now, past, and future (see [25]).

Definition 8   A *model* M of TL is a structure $M = \langle B, V, \pi \rangle$, where

1. $B^6 = \langle I, D, E, T, S \rangle$ is a determination base with a set I of determinables $(I = \Delta)$, a (non-empty) set D of determinate values, a (non-empty) set E of entities, a set T of time moments $(T = \mathbb{R})$ and a state function S from E into D^I (with condition 1 and 3 of the Definition 1);
2. $\pi \subseteq T$ is a closed (non-empty) interval;
3. V is a valuation function such that:

   - $V(c) \in E$, for all entity constants;
   - $V(a) \in D$, for all determinate constant;
   - $V(C) \subseteq D^I$, for all concept predicates.

---

[6]I leave the set of fundamental determinables away in this definition.

Let $M = \langle B, V, \pi \rangle$ be a model of TL, $t \in T$, and φ a sentence. The *truth relation* $M \vDash_t \varphi$, "φ is true in M at t", is defined as follows:

$M \vDash_t t = t^*$	iff $V(t) = V(t^*)$ (t and t^* are constants);
$M \vDash_t [P_\delta](c, a)$	iff $S(V(c), t)(\delta) = V(a)$ and $S(V(c), t)(\delta)$ is defined;
$M \vDash_t [-P_\delta](c, a)$	iff $S(V(c), t)(\delta) \neq V(a)$ and $S(V(c), t)(\delta)$ is defined;
$M \vDash_t [C](c)$	iff $S(V(c), t) \in V(C)$ and $S(V(c), t)$ is defined;
$M \vDash_t [-C](c)$	iff $S(V(c), t) \notin V(C)$ and $S(V(c), t)$ is defined;
$M \vDash_t EX(c)$	iff $t \in \pi_x$, π_x is the closed interval where x exists;
$M \vDash_t \neg\varphi$	iff not $M \vDash_t \varphi$;
$M \vDash_t \varphi \wedge \psi$	iff $M \vDash_t \varphi$ and $M \vDash_t \psi$;
$M \vDash_t \forall x \varphi$	iff $M \vDash_t \varphi[\underline{e}/x]$ for all $e \in E$;
$M \vDash_t \forall v \varphi$	iff $M \vDash_t \varphi[\underline{a}/v]$ for all $a \in D$;
$M \vDash_t [\leftarrow\!\dashv]\varphi$	iff if there exists a moment $t^* < t$ such that $M \vDash_{t^{**}} \varphi$ for all t^{**} with $t^* < t^{**} < t$;
$M \vDash_t [\mapsto]\varphi$	iff if there exists a moment $t^* > t$ such that $M \vDash_{t^{**}} \varphi$ for all t^{**} with $t < t^{**} < t^*$;
$M \vDash_t PoV\varphi$	iff $M \vDash_{t^*} \varphi$ for some $t^* \in \pi$

$\varphi[\underline{e}/x]$ ($\varphi[\underline{a}/v]$) denotes the closed formula which we get after each free occurrence of x(v) in φ is replaced by the name $\underline{e}$ of the entity e in E (by the name $\underline{a}$ of the entity a in D); I stipulate also that $V(\underline{e}) = e$ ($V(\underline{a}) = a$). The interval π is the *point of view* of the model $M = \langle B, V, \pi \rangle$.

In this definition, I take the set I of determinables to be just Δ to simplify the treatment. A more accurate way would be to map Δ into I by a bijection. My definition of the truth condition for the point of view PoV operator differs from that of progressive tense operator in the sense of not demanding that the "actual state" is in the point of view.

Note that the state function S is partial causing truth-value caps. I solve this problem by taking negative predicates $[-P_\delta]$ or $[-C]$ as well as positive predicates $[P_\delta]$ or $[C]$, thus avoiding introducing three-valued logic in which the third value is "undefined". This device is used in Hautamäki ([13]; see also [9]. For axiomatization of TL, one should include formulas to tell how $[P_\delta]$, $[-P_\delta]$, $[C]$ and $[-C]$ behave for all $\delta \in \Delta$:

$\forall x \forall v \neg([P_\delta](x, v) \wedge [-P_\delta](x, v));$
$\forall x \forall v (EX(x) \rightarrow ([P_\delta](x, v) \vee [-P_\delta](x, v)));$
$\forall x \forall v \forall v^*(([P_\delta](x, v) \wedge [P_\delta](x, v^*)) \rightarrow v = v^*)$ (functionality);
$\forall x \neg([C](x) \wedge [-C](x));$
$\forall x (EX(x) \rightarrow ([C](x) \vee [-C](x))).$

The concept "x starts to be a in respect of δ change in the respect of the determinable δ" is defined by the formula

$$\text{Start}(x,\delta,a) \leftrightarrow EX(x) \wedge [P\delta](x,a) \wedge [\mapsto][P\delta](x,a) \wedge \forall v([\leftarrow\!\dashv]([P\delta](x,v) \rightarrow \neg v=a)$$

The predicate "entity x appears" is defines by the formula:
App(x) ↔ EX(x) ∧ [↦]EX(x) ∧ [↤]¬EX(x).
The predicate "entity x disappears" is defined by the formula:

Dis(x) ↔ EX(x) ∧ [↤]EX(x) ∧ [↦]¬EX(x).

Unfortunately, the larger development of this temporal logic is not possible in the context of this article. The fragments show, regardless, that a relevant modal logic for temporal points of view, based on conceptual space framework, could be built. I made similar proposal for non-temporal points of view in my article [15].

10 Conclusions

I have shown that interesting time related concepts like change, event, and potentiality could be represented and analyzed in the conceptual space framework. My elaboration was inspired by Aristotle's philosophy of nature. Aristotle's analysis of time and change in his *Physics* was shown to be interesting also from a "modern" viewpoint. Particularly his view that time and motion are continuous is still relevant and superior as compared to some recent attempts to develop time logic based on discrete treatment of time. The Aristotelian approach is to emphasize the processual character of motion and change. Therefore we have to consider intervals along with single moments of time. I used that insight in developing a temporal concept of conceptual space.

In my presentation of temporal conceptual space, time is not a determinable (or a dimension of entities). My device is to add time to state functions, mapping entities and moments of time into conceptual spaces. This works well, but the real challenge was to incorporate continuity. In Aristotelian model an entity x is in the state s at the moment t then it is also in some interval containing t as an end point. This means that entities are permanent in some period and it is not possible that they are in a state only in a single moment but not just before and after it. Still there are difficult problems in continuity. Aristotle thinks that states of entities are also continuous. This is clear in the case of quantities, but what about qualities? Aristotle was tempted to suppose that qualities are also continuous, for example a continuum of degrees of illness or redness. Some topology in conceptual spaces is needed before we can define the continuity of determinables. In this article, I just gave some hints how to define continuity of determinables and state function.

One aim of the article was to introduce a time-related concept of point of view in conceptual space framework. My choice is to take closed intervals of time to represent points of view. This means that to have a temporal point of view is to look at world from some period of time. So a point of view is a kind of "time window". Many time-related notions could be relativized to points of view, like existence, change, and event. This makes it possible to compare points of view.

I show also in this article how to build a temporal logic. In the logic TL three modal operators were introduced: PoV (point of view), [⊢→] (forwards) and [←⊣] (backwards). These operators allow one to express, in TL, interesting time-related concepts like event, appearance, and disappearance. I also used positive and negative predicates for determinables: $[P_\delta](x, v)$ (the value of δ for x is v) and $[-P_\delta]$ (x, v) (the value of δ for x is not v). By this device I could avoid introducing three-valued logic, the third value being that of undefined. There is a special predicate EX(x) to express the existence of entities.

The extension of the conceptual space approach to include time provided it with new dynamic features. It has been a common bias of logic that it has applied static conceptual models. It is perfect for describing ahistorical, synchronic structures, but if we want to apply logic to events and processes, new kinds of dynamic concepts have to use. Temporal conceptual spaces provide a promising framework to explore the world, changing in time.

References

1. Allen, J. F. (1983). Maintaining knowledge about temporal intervals. *Communications of the ACM, 26*(11), 832–843.
2. Aristotle. *Metaphysica.*
3. Aristotle. *Physica.*
4. Bunge, M. (1977). *Treatise on basic philosophy. Vol. 3. Ontology I: The furniture of the world.* Dordrecht-Holland/Boston, USA: D. Reidel Publishing Company.
5. Charro, F., & Colomina, J. J. (2014). Points of view beyond models: towards a formal approach to points of view as access to the world. *Foundations of Science, 19*(2), 137–151.
6. Clark, A. (1993). *Sensory qualities.* Oxford: Clarendon Press.
7. Copi, I. M. (1954). Essence and accident. *The Journal of Philosophy 51*(23), 706–719.
8. Davidson, D. (1969). The Individuation of Events. In N. Rescher (Ed.) *Essays in honor of Carl G. Hempel* (pp. 216–234). Dordrecht: Reidel.
9. Feferman, S. (1984). Towards useful type-free theories. *Journal of Symbolic Logic, 49*, 75–111.
10. Gärdenfors, P. (2000). *Conceptual spaces: On the geometry of thought.* Cambridge: The MIT Press.
11. Gärdenfors, P., & Williams, M. A. (2001). Reasoning about categories in conceptual spaces. In *Proceedings of IJCAI 2001*, pp. 358–392. Palo Alto: Morgan Kaufmann.
12. Hautamäki, A. (1983). The logic of viewpoints. *Studia Logica, XLII*(2/3), 187–196.
13. Hautamäki, A. (1986). *Points of view and their logical analysis.* Acta Philosophica Fennica 41. Helsinki: University of Helsinki.
14. Hautamäki, A. 1992. A Conceptual space approach to semantic networks. *Computers & Mathematics with Applications 23*(6–9), 517–525 [Also in *Semantic networks in artificial intelligence*, ed. F. Lehmann, pp. 517–525. Oxford: Pergamon Press].
15. Hautamäki, A. (2015). Points of view—A conceptual space approach. *Foundations of Science* (Forthcoming).
16. Jacquette, D. (Ed.). (2006). *A companion to philosophical logic.* London: Blackwell Publishing.
17. Johnson, W. (1964). *Logic, part I, (orig. 1921) and part II (orig. 1922).* New York: Dover.

18. Kaipainen, M., & Hautamäki, A. (2011). Epistemic pluralism and multi-perspective knowledge organization, explorative conceptualization of topical content domains. *Knowledge Organization, 38*(6), 503–514.
19. Kaipainen, M., & Hautamäki, A. (2015). A perspectivist approach to conceptual spaces. In P. Gärdenfors & F. Zenker (Eds.), *Conceptual spaces at work*. Berlin: Springer.
20. Kim, J. (1976). Events as Property Exemplifications. In M. Brand & D. Walton (Eds.), *Action theory* (pp. 159–177). Dordrecht: Reidel.
21. Knuuttila, S. (1986). Remarks on the background of the fourteenth century limit decision controversy. In M. Asztalos (Ed.), *The Editing of Theological and Philosophical Texts from the Middle Ages* (pp. 245–266). Stockholm: Almqvist & Wiksell.
22. Kretzmann, N. (Ed.). (1982). *Infinity and continuity on ancient and medieval thought*. Ithaca, London: Cornell University Press.
23. Loemker, L. (Ed.). (1969). *G. W. Leibniz: Philosophical papers and letters* (2nd Ed.). Dordrecht: D. Reidel.
24. Morrow, G. R. 1966. Qualitative Change in Aristotle's Physics. In *Naturphilosophie bei Aristoteles*, ed. Düring, 154–167. Heidelberg: Carl Winter Universitätsverlag.
25. Needham, P. (1975). *Temporal perspectives. A logical analysis of temporal reference in English*. Uppsala: University of Uppsala.
26. Polanyi, M. (1958). *Personal knowledge, Towards a Post-Critical Philosophy*. Chicago: University of Chicago Press.
27. Prior, A. (1967). *Past, present and future*. Oxford: Clarendon Press.
28. Prior, A. (1949). Determinables, determinates and determinants, I and II. *Mind, 58*, 1–20, 178–194.
29. Scott, D. (1970). Advice on modal logic. In K. Lambert *Philosophical Problems in Logic, Some Recent Developments*, ed. , 143–173. Dordrecht: Reidel Publishing Company.
30. Stalnaker, R. (1979). Anti-essentialism. In P. French, T. Uehling, & H. Wettstein (Eds.) *Midwest studies in philosophy, IV studies in metaphysics* (pp. 343–355). Minneapolis: University of Minnesota Press.
31. Van Fraassen, B. (1967). Meaning relations among predicates. *Noûs, 1*, 161–179.
32. Van Fraassen, B. (1969). Meaning relations and modalities. *Noûs, 3*, 155–167.
33. Von Wright, G. H. (1986). Truth, negation, and contradiction. *Synthese, 66*, 3–14.

Chapter 7
Grounding Qualitative Dimensions

Juan J. Colomina

Abstract The purpose of this chapter is further complete an approach to Point of View (*PoV*) that could help us to set up a framework of evaluation and comparison. First, I introduce the notion of *PoV* as access in order to distinguish the model of accessing from other ways to understand *PoV* and then taking distance from purely relativist approaches. Second, I provide an explanation and development of some important notions introduced for the applicability of *PoV*. Third, since *PoV* allow setting a reference frame from where to evaluate the different objects accessed, I offer a comparison within *PoV* (independently of their different possible bearers). Since objects and states can be differently evaluated from different *PoV*, these will be considered as different qualitative dimensions of the world and its objects. They will serve as crucial elements of translation between all possible *PoV*, establishing degrees of comparison when the translation is not completely at hand.

1 Introduction

Some contemporary debates in philosophy of science and metaphysics include a principle of relativization of the truth-conditions of events and states of affairs. According to this principle, the feasibility of the properties of the world is relative to the adopted evaluative dimension to access to them. Then, the features of the

This work has been granted by Spanish Government, "Ministerio de Economía y Competitividad", Research Projects FFI2008-01205 (*Points of View. A Philosophical Investigation*), FFI2011-24549, (*Points of View and Temporal Structures*), and FFI2014-57409-R (*Points of View, Dispositions, and Time. Perspectives in a World of Dispositions*).

J.J. Colomina (✉)
The University of Texas at Austin, Austin, USA
e-mail: colomina-alminana_juan@austin.utexas.edu

© Springer International Publishing Switzerland 2015
M. Vázquez Campos and A.M. Liz Gutiérrez (eds.), *Temporal Points of View*,
Studies in Applied Philosophy, Epistemology and Rational Ethics 23,
DOI 10.1007/978-3-319-19815-6_7

world would be differently addressed depending on the particular value that the perspective applies to objects and states of the world.

I have previously developed a novel theory about these perspectives under the notion of Point of View (*PoV*).[1] According to this theory, *PoV* should be analyzed "taking as a model the notions of location and access… [T]he internal structure of a *PoV* is not directly addressed, and the emphasized features of *PoV* are related to the function that *PoV* are intended to have. That is, *PoV* are directly identified by their role and they can solely be understood as ways of accessing the world that bring some kind of perspective about it" [4, 137].

Since this theory accepts the existence of objects and events of the world as independent of the perspective adopted, the theory of *PoV* as access accepts realism about the world and its objects. Since the way that objects and events appear would vary depending on the qualitative dimension from the world is accessed, the feasible characteristics and properties will be intensionally interpreted, and only possibly evaluated from the inside of the same *PoV*. In other words, it analyzes the objects of the world as grounded in *PoV* but, unlike purely relativist theories, it considers *PoV* as primitive ontological structures.

The purpose of this chapter is further developing some notions introduced in the mentioned previous account in order to complete an approach to *PoV* that could help us to set up a framework in which *PoV* could be achieved. First, I introduce the notion of *PoV* as access using some examples, in order to distinguish the model of accessing from other ways to understand *PoV* and then taking distance from purely relativist approaches (Sects. 1 and 2). Second, I provide an explanation and development of some important notions introduced for the applicability of *PoV*, mainly the notions of system of generators, remainder, and scenario (Sect. 3). Given the importance of distinguishing between the different objects that inhabit the world, I use the determinate/determinable distinction to generate finer-grained properties inside *PoV*. These new generated characteristics will determine the objects differently, distinguishing between the different particulars of the world that a certain *PoV* has access to, generating different perspectives on the same objects. Third, since *PoV* allow setting a reference frame from where to evaluate the different objects accessed, I offer a comparison within *PoV* (independently of their different possible bearers) (Sect. 4). Since objects and states can be differently evaluated from different *PoV*, these will be considered as different qualitative dimensions of the world and its objects. They will serve as crucial elements of translation between all possible *PoV*, establishing degrees of comparison when the translation is not completely at hand.

---

[1] I use *PoV* to refer to Point of View in singular as well as Points of View in plural, as it is usual and has conveyed in the introduction and other chapters of this volume.

2 Points of View as Access to the World

It has assumed that, since it is intuitive to think of the things of the world as related or been held one to another, there is some kind of metaphysical ground. By metaphysical ground is usually meant the existence of some kind of constitutive form of determination [5]. The theory of *PoV* accepts this view. Since "*PoV* are ontologically primitive entities" [14, 386], they are very idiosyncratic primary ontological entities that offer some kind of perspective to the reality. Then, given that everything is approached from a certain *PoV*, they are fundamental in the constitution of the world and its objects. Literally, *PoV* are the ground from where other things are determined.

In their seminal analysis on *PoV*, Vázquez and Liz [14] have established two different ways to account for *PoV*. On one hand, *PoV* can be analyzed following the usual approach to propositional attitudes. This assumes the inner structure of *PoV* as constituted by a set of contents and relations between the bearer of the *PoV* and these contents. On the other hand, we can analyze *PoV* taking as model the notions of location and access. According to this second approach, the internal structure of a *PoV* is not directly addressed, and the emphasized features of *PoV* are related to the function that *PoV* are intended to have. That is to say, under this second approach, *PoV* are just different models from where to see, think, or act on the world and its objects. In other words, *PoV* are "a particular way to conceptualize the world independently of the precise content that constitutes the particular *PoV* under consideration" [4, 139].

To exemplify the differences between the two positions, suppose a devise employed to look forward the natural scope of human cognition, a telescope for instance. (A microscope, a radar, a sonar, a periscope, or a binoculars will also serve for this purpose). Since the psychological life of the bearer is not relevant for the account provided within the scope of the device, the way to approach to the view offered by the telescope is from the certain angle and distance the devise is placed, and this will depend on the intrinsic capabilities of the device itself. In other words, the limitations of the device will sanction the way that the world and its objects can be accessed within the telescope, independently of the particular capacity of discrimination of the observer.[2]

I will adopt the second way of analysis of *PoV*, the model offered by the notions of location and access, and then they should be understood as objective structures of access to the objects of the world. Under this approach, to summarize, *PoV* are primitive ontological realities that have some causal powers (if I may use the terminology that Magnani [10] employs to analyze models), the non-deterministic capacity to act according to the particular rules (the logical form) that structure *PoV*.

---

[2]It is very important to realize that here the particular *what it is like* to be a telescope, or a bat, or whatever other thing phenomenally based is not relevant. Since the particular content of the *PoV* neither is the basis of evaluation nor is addressed, the subjectivity of the bearer of the *PoV* is out of the equation. Then, the theory of *PoV* as access is not a relativist position in the usual way since *PoV* have an objective and metaphysically grounded reality independently of their contents.

3 Previous Analysis to the Notion of Point of View as Access

In the Section IX of *An Enquiry Concerning of the Principles of Morals*, Hume states the following:

> When a man refers to someone else as 'my enemy,' 'my rival,' 'my antagonist,' 'my adversary,' he is understood to be speaking the language of self-love; he is expressing sentiments that are specifically *his*, and arise from *his* particular circumstances and situation. But when he characterizes someone as 'vicious' or 'odious' or 'depraved,' he is speaking a different language, and expressing sentiments that he expects to be shared by all who hear him. In this second case, therefore, he must depart from his private and particular situation and choose a point of view that is common to him and the others; he must move some universal principle of the human frame, and touch a string to which all mankind have an accord and symphony.
>
> If what he means to express is that this man has qualities whose tendency is harmful to society, then he has done what is needed for this to be proper moral speech. That is, he has chosen a common point of view and has touched the principle of humanity that is found in some degree in everyone. For as long as the human heart is made out of the same elements as at present, it won't ever be wholly indifferent to public good, or entirely unaffected by the likely consequences of characters and manners. [9, 56–57]

It seems that here two different senses of the notion of *PoV* as access could be distinguished. According to the first sense, the notion of *PoV* can define nothing else that a particular spatial (or egotistic) location, but this is a non-philosophically relevant (and highly problematic) sense. The philosophically relevant sense is the second one, which claims that "a point of view is not a place from which one views things and events, but a way of viewing them" [11, 192].

According to Moline, the first philosopher in the contemporary era to take seriously the notion of *PoV*,[3] we should adopt a "behaviorist" approach to *PoV*, since to adopt a *PoV* implies to overtly take a particular pattern of action or to try to understand it. Thus, "to adopt a particular *PoV* implies sharing the same assumptions, criteria, interests, and other relevant elements with everybody else that has this particular *PoV*" [4, 141], a view that highly resembles contemporary contextualism. Therefore, unlike Kantian and neo-Kantian approaches that emphasized the practical answer to questions by reference to the possession of certain principles, Moline defends a view of *PoV* that is independent from the possession of a mental life.[4]

Even though I think that this approach has a huge potential, and given the issues of this position (highlighted in Charro and Colomina [4, 142]), I will prefer to keep closer to the more formal approach to *PoV* as access proposed by A. Hautamäki.

---

[3]Even though the notion of determination is analyzed several times in the history of philosophy, being of key importance, the notion of *PoV* as being taken aside, and there is no a systematic analysis of what it is and what structure it has.

[4]Moline's is not the only behaviorist approach to *PoV*. Brandom [1], for instance, also develops an account of *PoV* as referred to moral practical reasoning. See Charro and Colomina [4, Sect. 2] for more on this matter.

His purpose was "to create a logics in which the truth-value of a proposition depends also on ways to conceptualize the world... [then] viewpoint means a way to conceptualize the world" [6, 187]. His idea is thus "to identify (extensionally) a point of view with the set of the worlds which the point of view selects" [7, 226].

As I have defined it before, Hautamäki develops a propositional logic of *PoV* that expresses statements about entities, their properties, and the available information about those entities. Since, according to him, properties and relations "depend on the nature of entities which are defined in terms of determinables" [8, 83], in this logic we can define a determination basis, $D = (D(i))_I$, that partitions a set of entities E so that the state function of E is an injection from E to the logical space XD [8]. In other words, the Cartesian product of determinables XD obtained, that can be expressed as a vector of determinates in a particular framework, forms a logical space whose elements represent states of a given entity of the world and contains all the available information about such entity that belongs into this logical space.

Therefore, it seems that *PoV* produce a selection of relevant aspects depending of the interests, aims, backgrounds, and so on. Since the available information from a certain *PoV* will be always partial in contrast with the complete information offered by the reality, there is no such a thing as an absolute *PoV*. A particular *PoV* then is nothing else than a particular selection of aspects. And these aspects are determinables as indices from the particular determination basis adopted by the particular *PoV* under consideration. So a *PoV* is, according to Hautamäki [8, 65], a proper finite subset of I where I is the set of determinable-indices available according to the appropriate projection function in the particular logical space XD associated to the *PoV* in question.

4 Determinables, Determinates, and Determinants

Since the theory here defended accepts *PoV* as primitive ontological entities able to consider different objects of the world, I will construct my argumentation on the assumption that the world has a determinate/determinable structure. Once the particular *PoV* from we access the world is placed, this distinction will allow us to view the same properties of the same objects from different qualitative perspectives according to the particular characteristics enlighten by the dimension at hand.

Because I have supposed that the world has a determinate/determinable structure, I also accept that the world objectively has properties that can be further fine-grained into determinates if necessary. And *PoV*, as entities belonging to the world, inherit this peculiar structure. Therefore, as I have defended elsewhere, "a particular *PoV* will structure the world through a number of classes that will refer to characteristics of objects of the real world" [4, 144].

Since a determinable is a current feature of one of the objects of the physical world, its existence does not dependent of *PoV*. But at the same time, the *PoV* generates such a determinable in the very same act of accessing to the object. Thus, "the set of determinables of the world that the *PoV* can access serves the purpose of a model of the physical world (PW)" (Ibid.). So considered, determinables have a dual status. On one hand, they are physical properties of the objects that populate the world. On the other hand, they are the way in which a *PoV* gets access to the objects of the world and their characteristics.

I introduced the notion of filtered world to bring together these two different aspects of determinables. A filtered world is the result of applying a filter function to a particular *PoV*. For instance, suppose that O is a *PoV*. The filter function for it will be f_o. The physical world will act as the input for the filter function, providing the material to generate the set of the feasible determinables that O has access to as an output. Then, the filtered world $f_o(PW)$ will be generated.

I will borrow Hautamäki's notion of state function [8, 83]. According to Hautamäki, the state function S maps an entity of the world into the state space,[5] a Cartesian product of all possible determinables that are feasible from the *PoV*. Since the notion of state space can be identified with the notion of filtered world, the state function will filter the entities of the physical world from the particular access perspective, and will generate different vectors of determinates accordingly.

Consider the following example to illustrate how this process works (borrowed from Charro and Colomina [4, Sect. 3]):

$$S(\text{Ulysses}) = (40\,\text{A}, 180\,\text{W}, \text{Ulysses R})$$
$$S(\text{Penelope}) = (20\,\text{A}, 125\,\text{W}, \text{Penelope R})$$
$$S(\text{Jason}) = (40\,\text{A}, 180\,\text{W}, \text{Jason R})$$

I am assuming that our *PoV* O's filtered world $f_o(PW)$ consists of the characteristics 'age in years' (A) and 'weight in pounds' (W), so that the state space is the Cartesian product A × W. When the world that O has access to consists only of two individuals, say Ulysses and Penelope, no problem arises since both individuals can be perfectly distinguished with the properties at hand. But suppose that the individuals O has access to are Ulysses and Jason. Both individuals appear as identical within the *PoV*. How to distinguish between them?

This situation is particularly idiosyncratic. Since in this situation Ulysses and Jason are undistinguishable from the *PoV*, they should appear as indiscernibles. But, at the same time, given that from the *PoV* both individuals are recognized as different elements of the filtered world, then the state space cannot consist only of the mentioned characteristics A and W. A third property would be implicitly working in the fact that Ulysses and Jason are different elements of $f_o(PW)$. Since

---

[5]Carnap employed this notion to similarly apply it to families of related properties [2]. Posteriorly, he classified primitive attributes also into families such that the attributes of a family are related to each other by belonging to the same general kind according to the state space [3].

some of the characteristics of the filtered world could be expressed in terms of other properties without notice, I will call this additional characteristic remainder (R). In our example, R is described in terms of {A, W}, the property that is not A, nor W. And one could even measure the difference between objects of the world using the notion of remainder as a metric, as a distance function (with a numeric value that subsumes the attributes-pair), as introduced by Carnap [3, 50–51 and 78–79] in connection to families of related properties.

The notion of remainder has the advantage of notational flexibility, given that it can be further fine-grained if required to guaranty the access to new features. In other words, one can split the remainder into several finer-grained determinables and let R be a vector itself. And this can happen because Ulysses and Jason are both feasible objects of the physical world.

Thought this way, determinables are identified from sets of determinates only when such a set of determinates is available in two or more objects of the world (or they are "in the same line of business", as Searle [12, 154] says. Such a determinate can be perfectly described from an object of the world that has not the characteristics such and such, and this process could become more complex when the number of feasible determinables grows and somehow defines determinates. I will call scenario to this complete set of determinables, and the determination of the minimum number of elements that constitute a system of generators in a scenario will depend on the particular structure of the filtered world.

So described, the theory of *PoV* that I have presented allows for a comparison between different *PoV*.

5   Frame of Translation and Degrees of Comparison

Let x be an object of the world. The state function of x is given by

$$S(x) = (D_1(x), D_2(x), \ldots, D_n(x))$$

where n is the total number of accessible determinables that is given by the filter function. The functors $D_i(x)$ are the determinables (literally, the complete set of all possible values of the dimension from where to evaluate the properties of the objects of the world), and assign to x one and only one value $d \in Im(D_i) = \{d_1, d_2, d_3, \ldots\}$, where $Im(D_i)$ denotes the image of the functor D_i, and is the set of all feasible determinates.

Alternatively,

$$Im(D_i) = \{\text{Coordinate i of } S(x) \text{ for every x in the accessible world}\}$$

Since the main purpose of determinables is to allow comparison within a *PoV*, thus we can set a reference frame that situates such a *PoV* in the world and serves the purpose of a point of access to the world. In other words, since the feasible

Table 1 Possible *PoV* for a
world with 6 objects and 5
perceptible features

x1	n/a	d12	n/a	n/a	d15	S(x1)
x2	d11	d22	n/a	d14	n/a	S(x2)
x3	d21	d23	d13	d24	d25	S(x3)
x4	d31	d24	d23	d34	d35	S(x4)
x5	d41	n/a	n/a	n/a	n/a	S(x5)
x6	n/a	n/a	d41	n/a	d45	S(x6)
	Im(D_1)	Im(D_2)	Im(D_3)	Im(D_4)	Im(D_5)	

determinables depend on the particular structured world and the available determinates are fine-grained according to their resemblance, we can fix a particular framework that serve us as a point of translation and comparison within a particular *PoV*. Because, in this sense, it is clear that determinables allow comparing the different objects we can have access from a certain *PoV*. As a natural consequence, the question of how to compare different *PoV* (independently of their bearers) arises.

To illustrate this point, imagine a world composed by six objects and five perceptible features (Table 1). We can generate a table that applies a different value for each characteristic and each object from the particular *PoV* we access to them. This allows different approaches that will apply different qualitative aspects to the different objects depending on the particular characteristic that is enlighten in each moment (including slots of time). And the set of enlighten characteristics will conform the feasible properties available from that particular *PoV*. Since the values can vary, the properties accessible can also vary. Then, what we obtain is a way to calculate the different qualitative dimensions available for the objects of the world depending on the way that the feasible features can be accessed from the particular *PoV* adopted. For example, given the restrictions applied in our frame from the perspective offered by Im(D_2), objects x1 − x4 will appear in a particular way, been x5 − x6 inaccessible. In the same fashion, if we will like to give the complete possible description of, for instance, the object x4 within this *PoV*, its complete description will be reduced to the totality of feasible characteristics as accessed from the different qualitative dimensions within the same *PoV*, expressed by S(x4).

Since we can obtain a frame that is going to generate different perspectives within a *PoV*, or qualitative dimensions, for the different objects of the world depending on the different ways that the different characteristics can be accessed, we can also distinguish at least four levels of comparison between different *PoV*.

1. Impossibility
   Informal explanation: No comparison is possible because from the available *PoV* there is no access to the features of other *PoV*. (If the determinant here would be the bearer of the *PoV*, we could say that "from the *PoV* that you are talking about, nothing make sense to me because I have a completely different *PoV*. I cannot see what you are talking about.")
   Formal explanation: Assume the determinable D ∈ PoV1 and D ∉ PoV2. Then, D is meaningless to PoV2, as it cannot even exist, or be feasible, in there.

2. Mismatch:

   Informal explanation: Even though the relevant particular feature is available from the relevant *PoV*, or some of the features are available from the relevant *PoV*, the feasibility of that characteristic in the relevant available objects is not possible from the particular qualitative dimension that the world is approached. (If the bearer of the *PoV* would be the relevant determiner, then this position can be summarize as: "I don't know what you are talking about, or I know what you are talking about but still don't understand.")

   Formal explanation: In this case $D \in PoV1$ and $D \in PoV2$ and, given an object x, $D(x) = d_1$ for B_1. Now, assume that $d_1 \notin Im_2(D)$, where $Im_2(D)$ denotes the image of the functor D under PoV2. In this case, for B_2 it might not be able to infer D from d_1 alone, even if D would be available. In the best possible case, B_2 will be able to infer D from context but still will not have access to d_1. (B_1 and B_2 are two different moments of accessing the world).

3. Disagreement:

   Informal explanation: Some (or the majority) of objects of the world that can be accessed, even that can be accessed from the different *PoV*, coincide. Even coincide the features that these objects can show, but the dimensions that are generated into the frames of these *PoV* are different. Then, the features that could be applied to the same objects as accessed from the same dimensions differ. (If bearers were important, we can say that "we are talking about the same thing but have different opinions about the characteristics that can be attributed to it.")

4. Agreement:

   Informal explanation: Some (or the majority) of the objects of the world and their features that can be accessed, even that can be accessed from different *PoV*, coincide. Even coincide the features that these objects can show. But, unlike the previous degree, the dimensions generated into the frames of these *PoV* also coincide. Then, when the features of these objects are accessed from different *PoV*, they are the same. ("We are talking about the same thing, and are of the same opinion about the features that can be attributed to it" is something that could be informally said, if the bearers of a *PoV* would be the relevant determinant.)

The formal explanation of cases 3 and 4 require a subtler analysis. Assume that for a given object x, $D(x) = d_1$ and $D(x) = d_2$ are the two different qualitative dimensions of access from two different *PoV*. This difference does not necessarily mean that there is a disagreement between these *PoV*, say P_1 and P_2, as both of them can coincide, or not, and even though be perfectly and independently correct. Imagine for example that D is the determinable "size" and $D(x)$ is "large" and "small" for P_1 and P_2, respectively. Apparently, an instance like this is contradictory, as in absolute terms x cannot be "large" and "small" simultaneously. In this case, we say that there is a comparison in a strong sense.

However, since the world can only be accessed through only a *PoV* at time, it is not reasonable to consider absolute frameworks that, by definition, would necessarily be exterior to any given *PoV* (and hence, unfeasible). Any reasonable approach, then, should instead consider notions that are intrinsic, that is, that do not depend on exterior considerations but only on the information accessible to the relevant *PoV*.

The important point here is to notice that P_1 and P_2 have their own (probably different) reference to qualitatively measure x. Hence, the natural way to find an intrinsic quantity is to use another property y as a reference. In this way, we discover that, even if $D(x)$ and $D(y)$ both depend strongly on the *PoV*, the way they are ordered does not. This can be understood as a comparison in a weak sense. In practice, repeated tests with several different objects will allow both P_1 and P_2 to identify correspondences between their sets of determinates $Im_1(D)$ and $Im_2(D)$.

There is a key point here. Both $Im_1(D)$ and $Im_2(D)$ have a certain internal consistency,[6] in the sense that there is a partial order relation among the elements of $Im_1(D)$ and $Im_2(D)$. This will lead to a new notion that I will like to introduce. It is the notion of partially ordered set (POSET), and is described by a graph that summarizes the relationships between the elements that belong to it. (Notice that in a POSET some elements might not be ordered with respect to each other, but this is not important here.)

The description of the POSET is inductive and can be obtained confronting pairs of objects. If no logic contradiction is obtained, then P_1 and P_2 will be able to transfer each other's set of determinables, and a translation is then possible. But then, two different scenarios arise. First, the determinable will be wrongly identified, because for some reason the feature applied to the relevant object is not the same, and we have a disagreement (case 3). Second, it is possible that the complete set of features for that particular object coincides in both *PoV*. Then, we have a case of agreement (case 4). (Realize that the complete set of features for all the possible objects accessible from different *PoV* cannot coincide, because then, for Leibniz's Law, we would be talking about one and the same *PoV*.)

An instance of construction of the POSET in our example will serve to point out this distinction. Let us assume the role of P_1 and consider a collection of objects x_i (for our purpose $D(x_i)$ does not need to be all different, the ones that are equal will serve the purpose of a coherence test). Then P_1 will only need to show pairs of objects to P_2 and keep the track of the answers to learn about the order that the relation P_2 is considering, and how to translate between the two different POSETs.

In our example, size is a total order, in the sense that if two different objects, A and B, are related then either $A \geq B$ or $A \leq B$. This is not necessarily true in a POSET, as two elements might not be related. If a logic inconsistency appears in the process of translating the POSET, then we will find a mismatch. (In informal terms, and if bearers would be relevant determinants, "maybe she didn't mean D,

---

[6]For reasons of space, I will leave open here the question of the coherence of different *PoV*, which it will be analyze in future research.

since she has used a determinate that is not logically consistent with her other answers. She must refer to something I'm missing.")

Incidentally, in our example, the reference property/object could be the one employed to define the standards of measurement, for instance any object of length equal to 1 meter. Comparison in a strong sense is possible within the same determinable in different *PoV*. It allows ruling out levels of comparison 3 and 4, but neither 1 nor 2. Weak comparison, however, allows distinguishing between levels of comparison 1 and 2. For the reasons above indicated, we should distinguish then between strong and weak degrees of comparison.

Realize that this inductive construction also allows noticing if the two *PoV* have totally different POSETs (totally different graphs), or partly different POSETs. However, in principle, only exhaustive testing would allow to totally identify the POSET, as we would need to test over all possible objects in the world. Incidentally, the number of tests growths exponentially with the number of objects (it has the cardinal 2^N, where N is the number of objects) making it unpractical. Also, actually only a projection of the POSET into our *PoV* could be available, since there could be parts of the other POSET that escape the generated determinables and determinates.

Another interesting problem is that there is a level of uncertainty in this approach, since we can only know other *PoV* up to certain extent through projection into our own *PoV*, but in general we cannot be sure that we totally know it. A modal logic that uses probability of certainty is in order to clarify this point, but this is matter for another article.[7]

References

1. Brandom, R. (1982). Points of view and practical reasoning. *Canadian Journal of Philosophy, 12*(2), 321–333.
2. Carnap, R. (1950). *Logical foundations of probability*. Chicago: The University of Chicago Press.
3. Carnap, R. (1971). A basis system of inductive logic. In R. Carnap & R. C. Jeffrey (Eds.), *Studies in inductive logic and probability* (Vol. I, pp. 33–165). Berkeley, CA: University of California Press.
4. Charro, F., & Colomina, J. (2013). Points of view beyond models. Towards a formal approach to points of view as access to the world. *Foundations of Science, 19*(2), 137–151.
5. Fine, K. (2012). Guide to Ground. In F. Correia & B. Schneider (Eds.), *Metaphysical grounding* (pp. 8–25). Cambridge: Cambridge University Press.
6. Hautamäki, A. (1983). The logic of viewpoints. *Studia Logica, 42*(2–3), 187–196.
7. Hautamäki, A. (1983). *Dialectics and points of view. Ajatus, 39*(218), 231.
8. Hautamäki, A. (1986). *Points of view and their logical analysis* (Vol. 41). Helsinki: Acta Philosophica Fennica.

---

[7]I have also intentionally left aside another way to compare determinables: the very interesting distinction between physical and mental determinables introduced by Yablo [13]. The discussion of this point here will unnecessarily extend the length of the chapter.

9. Hume, D. (1751). *An enquiry into the sources of morals* (J. Bennett (Ed.), *An enquiry concerning the principles of morals*, Trans.). Available in http://www.earlymoderntexts.com/pdfs/hume1751.pdf.

10. Magnani, (2012). Scientific models are not fictions. Model-based science as epistemic warfare. In L. Magnani & P. Li (Eds.), *Philosophy and cognitive science. Western and eastern studies* (Vol. 2, pp. 1–38). Heidelberg: Springer.

11. Moline, J. (1968). On points of view. *American Philosophical Quarterly, 5*(3), 191–198.

12. Searle, J. (1959). Determinables and the notion of resemblance. *Proceedings of the Aristotelian Society, Suplemmentary, 33*, 141–158.

13. Yablo, S. (1992). Mental causation. *The Philosophical Review, 101*(2), 245–280.

14. Vázquez, M., & Liz, M. (2011). Models as points of view: The case of system dynamics. *Foundations of Science, 16*(4), 383–391.

Chapter 8
Kinds, Laws and Perspectives

Sebastián Álvarez Toledo

Abstract This chapter deals with the main characteristics of natural kinds, and analyzes three approaches to them. The first approach argues that natural kinds are characterized by their essential properties (in a modern, scientific sense), but encounter difficulties even on the physico-chemical level, which is where it seems to be better implemented. On the other hand, the constructivist stance, much more liberal, does not explain why certain kinds are inductively useful and not others. Third, an introduction, with comments, is provided on the approach of Richard Boyd, among others, which understands natural kinds as homeostatic property clusters that accommodate to the causal structure of the world. In this view, natural kinds are usually fuzzy sets with no clear boundaries, subject to time and space limitations, and relative to some perspective. However, it solves the problems of the other approaches mentioned without forgoing a realistic conception of natural kinds. Finally, a proposal is made on how an application of Boyd's ideas to the analysis of laws of nature can help to solve the old chestnut about the distinction between scientific laws and accidentally true generalizations.

1 Introduction

An important cognitive activity is to classify, form kinds; in other words, bring different things together according to certain shared similarities, and whenever possible, link several kinds in an order or hierarchy. Yet not all the groups we can

This work has been granted by Spanish Government, "Ministerio de Economía y Competividad", Research Projects FFI2008-01205 (*Points of View. A Philosophical Investigation*) and FFI2011-24549, (*Points of View and Temporal Structures*).

S. Álvarez Toledo (✉)
University of Salamanca, Salamanca, Spain
e-mail: sat@usal.es

© Springer International Publishing Switzerland 2015
M. Vázquez Campos and A.M. Liz Gutiérrez (eds.), *Temporal Points of View*,
Studies in Applied Philosophy, Epistemology and Rational Ethics 23,
DOI 10.1007/978-3-319-19815-6_8

form have the same cognitive interest or utility; hence the traditional division between natural and non-natural kinds. Examples of natural kinds are chemical elements and compounds, such as oxygen, carbon, water, and salt; biological species, such as oaks and whales, or groups such as birds, reptiles, mammals, fishes, and amphibians. However, we would not use the term of natural kinds to refer to groups of a simple anthropocentric character (poisonous or edible fruits, pets), or groups whose members have very little in common (white animals, star constellations, or decades).

We can see there is an important feature in natural kinds: they allow us to make inductive inferences and predictions because their members share many interesting properties. According to Whewell, Mill, and Quine, this is the main feature of natural kinds. Yet let us look more closely at what a natural kind is, and what differentiates it from an arbitrary grouping.

The topic of natural kinds has raised two philosophical questions, which although related are quite different. One question is metaphysical: what is a natural kind? The other one is semantic: what do natural kind terms mean? What do they refer to? My focus here will be on the first question, on the metaphysical question, which is very closely related to realism in the philosophy of science and in metaphysics. From a realist approach, one may think, for example, that a natural kind, far from being a mere convention, accurately reflects the very structure of nature. A good classification would be one that carves nature at its joints (using the Platonic metaphor). However, can we gain so much in our classifications? What's more, does nature have joints?

This chapter will analyse a number of approaches to natural kinds. I am going to start with the more realistic approaches, according to which each natural kind is defined by an essential, necessary and sufficient property, whereby discovering such properties would amount to detecting the joints in nature. These approaches have the undeniable appeal of offering a specific criterion for the definition of natural kind based on a modern and scientific idea of essence. However, this criterion faces serious difficulties within and without the group of its paradigmatic examples: chemical elements and compounds. Then I shall dwell briefly on the conventionalist or constructivist approach, for which nature does not have a structure or set of joints that allows certain classifications and prohibits others: each classification is purely anthropocentric, depending on our practical interests. The problem with this approach is that it fails to explain why our classifications are useful predictive tools.

Under the conviction that the solution to this issue must come from the broad intermediate area between essentialist and constructivist approaches, I shall be defending the proposal made by Richard Boyd, among others, according to which a natural kind is defined as a "homeostatic property cluster". Boyd's position is realistic about the constitution of natural kinds because these must respect some basic features of nature, what he calls the causal structure of the world. It is not a mere constructivist position. However, unlike essentialist realism, it conceives natural kinds as entities with fuzzy boundaries, being both temporal, since they are subject to evolution and change, and also relative, because their relevance depends on other scientific disciplines and perspectives. This concept therefore allows for ontological pluralism, according to which there is no single way to classify objects and events, as there are different possible natural classifications of the same domain.

In this analysis of different approaches to natural kinds an issue emerges that seems crucial, namely, the relationship between these and the laws of nature. So the last paragraph briefly presents an understanding of the laws of nature that I think may help to complement Boyd's standpoint on natural kinds.

2   A New Essentialism

A strongly realist position about natural kinds is one that seeks to define them by their very *essence*, understanding this concept in a modern, scientific sense. It can be said that the early proponents of this position were Kripke [17] and Putnam [21], who formulated it to address the meaning of natural kind terms.[1] Later, Ellis and Lowe, among others, have continued to develop this essentialist approach.

The core of this approach is the notion that the elements of a natural kind share a set of properties, one of which, or perhaps a small number, is its essence (its *real essence*, in Locke's terms). Usually, the essential property of a natural kind is not superficial or readily observable, such as the colour, weight or size of its elements, but a property that requires a deep understanding of these; a microstructural property in many cases. So according to Kripke, the essence of gold is its atomic number, the number of protons in its nucleus, 79; the essence of water is its composition, H_2O; and the essence of light is a flux of photons. Therefore, from this modern essentialism, sciences are our best access to the essence of things [17, 330]. Chemical elements and compounds become paradigmatic examples of natural kinds. Ellis seems to restrict his essentialist realism to the physico-chemical level, and is very sceptical about natural kinds in higher levels. In fact, he is reluctant to accept that biological species are natural kinds. Kripke and Putnam, however, accept species, and Kripke even speaks about the possibility of historical (non-micro-structural) essences.

The essential property of a kind is a necessary and sufficient condition to belong to it. So for a group of elements to be a natural kind, all its members, and only them, have to have the corresponding essential property.[2] Hence, if the essence of, for example, gold is its atomic number, it follows that something is gold if, and only if, it has that atomic number.

---

[1]In this chapter I will consider the metaphysical aspect of natural kinds, and not the semantic one; these are two sufficiently distinct aspects, because the position one adopts on one of them does not determine the answer to the other.

[2]"... some things ... hold some or all of their intrinsic properties necessarily in the sense that they could not lose any of these properties without ceasing to be things of the kind they are, and nothing could acquire any set of kind-identifying properties without becoming a thing of this kind. These kind-identifying sets of intrinsic properties are the ones I call the real essences of the natural kinds" [12, 237–238].

Although the essential property of a natural kind explains the presence of surface or manifest properties typically associated with it, such as colour, size, shape, etc. (what Locke called *nominal essence*), their connection is loose enough so as not to constitute sufficient or necessary conditions for membership of that natural kind. They are not sufficient conditions because, in principle, they could occur regardless of the essential property. As Putnam argues when referring to his "Twin Earth", a liquid that is colourless, tasteless, good to drink, etc., would not be water if its composition were not H_2O, and according to Kripke, fruits that are superficially similar to lemons but which have a different genetic structure are not lemons [17]. On the other hand the manifest properties are not necessary because the essential property does not invariably produce them. One can only say that members of a kind exhibit a tendency to have them. Therefore, the explanation the essential property of a natural kind can offer of surface properties is not deductive, but rather an inductive explanation.

So defined, natural kinds are the suitable subjects for universal and exceptionless laws of science. According to Ellis, laws of nature are the descriptions of essences of natural kinds, hence their metaphysical necessity [12, 145–150]. Lowe maintains that natural kinds are ontologically basic entities that enable the formulation of laws of nature. Kepler's first law, stating that planets move in an ellipse, actually states that the kind of the planets is characterized by the property of moving elliptically [19, 144]. We could say that according to this new essentialist approach the reference to a natural kind is fundamental to distinguish between authentic laws of nature and so-called accidental or accidentally true generalizations.

Therefore, this modern essentialism is an extreme, strong form of realism with two elements or theses: an ontological thesis, according to which the world has a determined structure (its own joints); and an epistemological one, by which science allows us to unveil this structure by discovering its component kinds and the relations among them. In this sense, science manages to carve nature at its joints.

3   The Problems of New Essentialism

This modern or scientific essentialism has been the subject of many and severe criticisms.[3] As we have seen, this approach to natural kinds finds its paradigmatic cases on the physic-chemical level. Yet can it be extended to accommodate cases of higher levels? Can it accommodate biological kinds, as Kripke and Putnam thought? Here some problems arise. First, it should be borne in mind that Biology does not offer a single definition or concept of species. Perhaps the so-called *biological concept* of species is the one most widely accepted: it is the criterion of species based on the idea of reproductive isolation; but there are many others, such as the ecological, the evolutionary or the phenetic. They correspond to different

---

[3]For example: Mellor [20], Dupré [10, 11], Sober [24, Chap. 6], Bird [1, Chap. 3], Williams [25].

conceptions, and divide organisms in different ways. This variety of criteria does not seem to be consistent with the essentialist claim to define species in a single way by essential properties.

On the other hand, species are local and dynamic entities that change over time. They are not immutable and universal, as modern essentialism seems to assume. If the essential property of a chemical element is its atomic number, it is understood that this property is timeless and should always characterize all samples of this element anywhere in the universe. However, as we know, biological species evolve, and even branch out into different species. They are historical, and confined to certain places. Species we know belong only to our planet Earth, because if we discover animals on another planet, for example, lions, and they were very similar to those we know, they would not belong to the same species as our lions, because they do not share the same ancestors and the same story. No wonder there are biologists who propose seeing species as individuals rather than kinds. And no wonder, therefore, that an essentialist such as Ellis argues that biological species cannot be natural kinds. Isn't this exclusion too high a price to pay?

Still in the field of Biology, it seems sensible to think that diseases, or at least some of them, are natural kinds. Ellis, for example, admits that processes can be natural kinds too,[4] and a disease is a process. However, is it possible to define a disease in terms of an essential property or, more generally, in an essentialist style? Williams [25] has examined this question and concludes that, in the case of diseases, the essentialist claims that the properties in question are etiological: the essence of a disease kind is whatever underlying physical condition that causes the instances of the disease and the associated symptoms. There are well-known examples of diseases for which the essentialist picture appears to work just fine, such as tuberculosis, cholera, meningitis, botulism, syphilis, and a number of other diseases. However, there are many other diseases (e.g., the case of rheumatoid arthritis, which Williams analyzes in that paper) that do not appear to satisfy the essentialist desiderata. They are diseases that fail to have a neat causal structure, or have multiple causes.

Moving now into Physics, it is not difficult to imagine the problems that arise when one tries to define natural kinds in terms of essential properties, especially in the case of theoretical entities, such as neutrinos or electrons: directly unobservable entities that have been conjectured to explain certain phenomena [1, 109–110]. Let us now focus briefly on a specific case. As we saw, Lowe said that Kepler's first law means that the elliptical movement is the proper or characteristic movement of planets. However, the Newtonian explanation for the elliptical movement of planets does not need to talk about planets, but about masses moving around a much larger mass. We would not say that "masses" form a natural kind [2].

Nevertheless, essentialist realism has difficulties even in Chemistry, its favoured hunting ground, because neither atomic number nor chemical composition

---

[4]Ellis identifies three types of natural kinds: *substantive* (elements, particles, gases, salts), *dynamic* (interactions, processes), and *properties* (mass, load, shape) [13, 141–142].

accurately defines the corresponding elements and substances. First, as is known, the spatial arrangement of the atoms in the same chemical element can result in different allotropic varieties that could very well be considered different natural kinds, as with diamonds and graphites, whose atomic composition is the same (carbon). Furthermore, in the case of isomers, we find substances with the same chemical composition, but which may however display very different properties. For example, ethyl alcohol and dimethyl ether are different substances with the same composition: C_2H_6. Furthermore, we know there are no chemically pure substances. Impurities are unavoidable. Any water we may find in nature has ingredients other than oxygen and hydrogen, to the extent that impurities may result in a different kind. Ruby and sapphire, for example, are two types of corundum (Al_2O_2), which, at least in gemmology, form two distinct kinds. Finally, it should be noted that the chemical composition of a substance does not suffice to account for its observable properties. A water molecule, whose formula is H_2O, lacks important properties of water, such as being liquid and having a temperature or boiling point [1, 105–109].

After considering all these problems, we might apply Mellor's critique of the works by Kripke and Putnam about natural kinds to essentialist realism in general. He concluded that "their essences can go back in their Aristotelian bottles, where they belong" [20, 311].

4   A World Without Kinds

A quite different approach to natural kinds is the constructivist or conventionalist one. Its weak version would admit that nature is endowed with a profound structure of distinct kinds, but we are unable to understand them, so our classifications, both in science and in everyday life, are hopelessly artificial. They are constructions based on appearances, and accepted by their utility for certain purposes or interests, in particular, inductive inferences and predictions. It is a type of epistemological constructivism, a form of scepticism, given that it is not about nature itself, but about our ability to suitably understand it. To a great extent, this was Locke's position, who although he believed in the existence of unobservable real essences of things, which would allow us to classify them in a natural way, was convinced that our classifications can only attend to nominal essences, that is, the surface properties of things.

The strong version of constructivism denies that nature is really structured in any way. It is an ontological constructivism, according to which there are only unique and unrepeatable individuals, so what we mean by natural kinds, far from being discoveries, are merely constructions made for practical purposes, and inevitably depend on our different interests and conceptual frameworks. From this point of view, there would be no significant difference between the classifications of scientists and the ones of laypersons, only that interests would be different; so the classifications made by a gardener or a cook can be as natural as those of a botanist.

A qualified mouthpiece for this approach is Nelson Goodman [16], for whom classifying is equivalent to "making worlds", so worlds (or domains of individuals), like their components, may be very different from each other depending on how these are organized in kinds, taking into account that the various "modes of organization...are not `found in the world´ but *built into a world*" (p. 14). So the world of the Inuit who do not have the unitary concept of "snow" is very different from the world of the inhabitants of Samoa or New England who have not grasped the various concepts or kinds the Inuit use to refer to snow (pp. 8–9). Furthermore, the different worlds constructed need not necessarily be translatable among each other: "our passion for *one* world is satisfied, at different times and for different purpose, in *many* different ways". Nevertheless, according to Goodman, not all alternatives are equally good. We would not accept a kind that combines tomatoes and triangles and typewriters and tyrants and tornadoes (pp. 20–21). This is not because natural kinds may be classified as true or false, categories that are totally alien to them, but because what really matters is that they rely on "projectible" properties (pp. 126–127); in other words, they allow us to make inductive inferences and form expectations and predictions regarding their elements. This pragmatic criterion is, according to Goodman, the only one we have for distinguishing between right and wrong classifications, not between natural and artificial ones, as he believes that the correctness of classification affects both those we make in Biology and those we use with musical works or artefacts.

I find some very interesting aspects in the constructivist view and in Goodman's approach that I have just outlined. I am referring to his pluralist perspective on classifications, to his affirmation that the truth has no regulatory capacity in the justification of a natural kind, and to his defence of projectability as a practical criterion for the acceptability of natural kinds. It is nonetheless inevitable to wonder how certain groupings of elements we have formed artificially and freely may be "projectible", while others are not, without nature having any kind of structure that is responsible for that difference. This is a question to which the constructivist perspective has no answer. There follows an attempt to find a solution in a realist approach that is far removed from the essentialism we have seen in the first section, and which is close to the more appealing aspects of constructivism.

5 Realism Without Essences

It is easy to assume that many more sensible approaches are possible between essentialist realism and this extreme conventionalism: approaches that characterize natural kinds as a set of properties occurring truly intertwined in nature, but without any essential property among them, and which give rise to certain natural constraints to our freedom to form natural kinds. A good example of this sort of approach is the proposal by Richard Boyd and Ruth Millikan, among others, according to which a natural kind is characterized by a *homeostatic property cluster*. Let us now look at the main features of this approach according to Boyd.

A natural kind would be characterised by a series of properties without an essential subset or property. What really matters is the relationship or connection among them. Hence the reason for positing that the typical properties of a natural kind form a homeostatic cluster. As we know, homeostasis is a feature of some systems by which they maintain their properties and even a certain balance or equilibrium among them, offsetting changes from the outside with a rearrangement of their values. We find the clearest examples of homeostatic systems in plants and animals, and in their internal reactions to changes in their environment.

Therefore, such clusters are not accidental or strange coincidences of properties, but are guaranteed by some nomological relation of interdependence among them. Boyd stresses this idea by saying that these property clusters are based on causally sustained regularities, and that natural kinds therefore *accommodate* to the causal structure of the world. I consider that what he understands by the causal structure of the world may well be interpreted as the *nomological* structure of the world, that is, the network of causal and non-causal relationships between events (e.g., coexistence relationships of properties). The kinds we consider natural need to accommodate themselves to this network. Through this idea of accommodation, as fuzzy as it is interesting, Boyd expresses the type of realism he defends in his version of natural kinds.

Boyd does not say that natural kinds reveal or describe groupings existing in nature, or that they represent the real distribution or organization of kinds that are independent of human beings and their interests, or that they correspond with that structure (which are commonplace expressions in the philosophy of realist orientation), but rather that they accommodate to it, which is tantamount to saying they exploit it or use it for a specific end, which in this case is none other than the formulation of inductive inferences and, therefore, predicting and explaining. The complex structure of what is real would be alien to our classification interests, but would allow inferences that we could use or select for forming natural kinds.

There follows a description of the ramifications of this notion of accommodation applied to natural kinds, although it is already easy to see that Boyd's position is clearly far removed from the constructivism we have just mentioned, as natural kinds would not be arbitrary creations we have made, but instead, in the final instance they depend on the existence of a nomological structure of the world. On the other hand, the realism that stems from Boyd's concept of accommodation is very different to the realism associated with the essentialist version. As we shall explain in what follows, natural kinds, without being merely our own inventions, are not discoveries or reproductions of divisions that appear in nature. The expression "cut the world at its joints" appears to be devoid of meaning in Boyd's realism, because the nomological structure of the world would not be actually classificatory, it would lack joints.[5]

---

[5]Although Boyd states that "kinds useful for induction and explanation must always 'cut the world at its joints'" [3, 139], he does so simply to stress that natural kinds are not merely arbitrary and conventionally accepted constructions, but instead their inductive usefulness resides in their accommodation to the causal structure of the world.

We shall now consider certain basic aspects of Boyd's version. Firstly, the type of relationship with the nomological structure of the world that the notion of accommodation suggests means that the notion of homeostasis, applied to natural kinds, is not perfect. What it seeks to express is the following:

> the presence of some of the properties tends (under appropriate conditions) to favour the presence of the others, or there are underlying mechanisms or processes that tend to maintain the properties. [4, 143]

It therefore follows that none of the typical properties of a natural kind is necessary or sufficient to belong to it. Two individuals may belong to the same natural kind, albeit having certain different properties. However, they need to share a sufficient number of those making up the homeostatic property cluster. Thus, for example, although the platypus lays eggs, has a bill and is poisonous, it shares a series of characteristics with mammals (its body is covered in fur, it has lungs to breathe, it is warm-blooded, etc.) which, while none of them is in itself "essential", allow classifying it as a mammal. In short, the particular properties of a homeostatic property cluster are not a necessary or sufficient condition, but only symptoms of membership of a natural kind.

It is easy to deduce from the above that this approach introduces an unavoidable vagueness in natural kinds. Even the concept of natural kind becomes a comparative concept, at least in some cases, because it allows varying degrees of kind membership. So, for example, we could say that a platypus is less mammal than a cow, and an eagle is more bird than a chicken. Yet according to Boyd, this vagueness would not be a default of this approach to natural kinds, but a consequence of the naturalness of these ones, in other words, of their accommodation to the structure of the world:

> The resulting 'vagueness' in the extension of the associated kind term reflects not an inappropriate imprecision but a precise accommodation of classificatory practices to relevant causal phenomena... Biological species are paradigmatic HPC [homeostatic property clusters] natural kinds. It follows from evolutionary theory that they will ordinarily lack completely determinate boundaries, so any precisification of a definition of a species would misrepresent biological reality and thus undermine accommodation. [5, 216–217]

Therefore, any attempt to make natural kinds clearer and sharper runs the risk of denaturing them and turning them into something artificial. It is curious that, from this approach, vagueness is not justified as a matter of method, but instead ontologically.

Nevertheless, the imprecision or vagueness of the natural kind has its limits. The notion of accommodating to the nomological structure of the world, together with the demand for inductive usefulness, does not allow using natural kinds to define accidental groupings of properties (e.g., the kind of white animals). In addition, excessively broad groupings would not merit the name of natural kind. Boyd says he is "against a conception of natural kinds according to which they... must figure in eternal, ahistorical, exceptionless laws" [4, 164], which allows us to highlight other aspects of his concept of natural kinds. It is understandable that a principle such as Newton's law of universal gravitation, which deals with relations between masses, does not refer to a natural kind: mass would not be a natural kind because it is not a homeostatic property cluster. What variety of properties do masses share? And the same may be said of the concepts of force or energy, above all if we compare them to more specific ones, albeit ones that are more varied in their content, such as metal, mammal, or even planet. On the other hand, it seems desirable for a natural kind to have other contrasting kinds, which is what often permits the formation of classifications or hierarchies. Indeed, natural kinds are generally defined by their difference or opposition to other kinds. Thus, for example, the concept of solid is opposed to that of liquid or gas, and that of oak to those of beech or pine; but this is not the case with the aforementioned concepts of mass, force or energy. In sum, natural kinds should not be too broad, as this will lead to a reduction in shared properties and the ability to be juxtaposed with others.

According to Boyd, this understanding of natural kinds enjoys a wide field of application. Firstly, it is easy to imagine that the imprecision and vagueness we have seen in this notion of natural kind would be enough to remove the problems facing some chemical elements and substances when trying to subject them to the rigid essentialist criteria of kind membership. Moreover, Boyd believes that the concept of homeostatic property cluster also applies to groups of physical entities. So, for example, he argues that metals, besides sharing a certain set of properties (thermal and electrical conductivity, ductility, malleability, etc.) also have in common certain homeostatic correlations among the values of some of them (e.g., conductivity and temperature) [4, 83–84]. On the other hand, and as we have seen, biological species for Boyd are paradigmatic natural kinds. Yet even higher taxa (such as genus or family) can be homeostatic property clusters, too. He also admits natural kinds in social sciences and the humanities, and extends the concept of natural kind to many groupings in our everyday language, provided they accommodate to some causal structure. According to him, ordinary kinds and scientific natural kinds lie along a continuum [4, 162].

Yet what's more, a major consequence of the concept of accommodation applied to the formation of natural kinds is that these are not absolutely so, but instead always depend on some point of view or perspective. Boyd says:

> It follows from the accommodation theory that the naturalness of a natural kind is discipline relative. There are not kinds which are natural simpliciter, but instead kinds that are natural with respect to the inferential architectures of particular disciplinary matrices. [5, 217]

> A kind may be natural "from the point of view of some discipline or disciplinary matrix, but not "from the point of view of another"... This relativity to a discipline or disciplinary matrix does not compromise the naturalness or the reality of a natural kind. Natural kinds simply are kinds defined by the ways of satisfying the accommodation demands of particular disciplinary matrices. [4, 160]

These affirmations are consistent with our comments on Boyd's notion of accommodation, and further highlight the relative nature of the concept of natural kind. Boyd is referring here to disciplinary matrices, which initially leads us to the field of scientific theories or paradigms, but the meaning of this expression is very broad here, as we can infer from the many fields (not only in sciences) in which according to Boyd natural kinds can be formed.

> Although my choosing the term disciplinary matrix undoubtedly betrays my special con-
> cern with the issue of kinds in the theoretical sciences, everyday life provides disciplines or
> at any rate regimes of inferential and practical activity in which accommodation of practice
> to causal structure is central. [4, 160–161]

The examples he provides of these practices include gardening, flower arranging, landscaping, interior decoration, and so on.

The constitution or construction of a natural kind, and the accommodation to nature that this requires, is always undertaken from some point of view. Nature does not impose specific kinds or classifications, but instead permits a diversity of classifications and perspectives. This has an immediate consequence. If an element belongs to a specific kind from one point of view, it may well belong to a different kind from another point of view. In other words, the same element may belong to different natural kinds according to different points of view. Given that both points of view may be equally valid, there is an unavoidable acceptance of pluralism in the notion of natural kind. In this regard Boyd agrees with those philosophers (e.g., Dupré, Mishler, Brandon, Donoghue, Kitcher, and Ereshefsky) who defend the pluralist view according to which there are different but equally legitimate strategies for classifying the same things, for example, "for sorting organisms into species" [4, 169].

An example of Dupré's pluralism is his comment on the classification of whales as mammals. He argues [11, Chap. 2] that even if we do well to include whales in the kind of mammals, it is not that they are mammals in some ontological sense, but this classification is helpful in Biology. However, we can imagine a situation in which an ecological point of view acquires paradigmatic character in Biology, and whales are best classified as fish. It is therefore a methodological question: what is the most helpful classification in making predictions and inductions?

This type of dependency of the concept of natural kind on a point of view or perspective is what Boyd seeks to express when referring to the discredit of the concept of human race in biological classifications.

> The critic who denies the 'reality' of races would then be understood to be denying that
> races, as currently understood, play an epistemically legitimate role in biology. She would
> not then need to deny that those very categories are natural kinds in the social sciences that
> study stratification, poverty, and political oppression. [5, 222]

Based on what we have seen as regards Boyd's approach to natural kinds, and especially in terms of the relative and plural character he assigns them, another of their major features becomes apparent: they are entities that are subject to limitations in time and space. Philosophical studies on natural kinds traditionally tend to consider them to be eternal, universal and immutable. The essentialist approach to which we referred earlier is a good example of this; therein, as we saw, lie the difficulties for accepting biological species as natural kinds. Nevertheless, from Boyd's approach, an element of a specific kind may subsequently be placed in a different kind, a natural kind may cease to be one with the passage of time, and the rightness of a classification may be restricted to a certain space. The aforementioned comment by Dupré about the classification of whales, with which Boyd would agree, shows how a change in paradigm or disciplinary matrix may lead to a relocation of the elements of a domain under classificatory concepts that are different to the previous ones (or incommensurable, according to Kuhn). Furthermore, I believe that Boyd's opinion on the relative natural character of human races requires a comment to vouch for the temporal nature of natural kinds. According to Boyd, although the classificatory concept of race lacks any basis in Biology, races would be natural kinds in social sciences for the study of the cultural, economic or political situation of a people. Yet I understand that even in social sciences, the concept of race has ceased to play the role of a natural kind. When each one of the various human groups classified as races lived in the same geographical, economic, political and religious setting, for example, the concept of race might have had a certain value from an inductive and predictive standpoint, but in a world with the level of globalisation we have today, the morphological traits that define races do not provide any socially interesting information on individuals. The concept of race may be a good example of the temporal nature or precariousness of certain natural kinds. On the other hand, I believe that Boyd's approach perfectly fits Goodman's earlier reference to the different ways of classifying snow between the Inuit and the inhabitants of Samoa or New England, a difference of classificatory criteria that reflects the local nature that some natural kinds may have.

This understanding of natural kinds as dynamic entities that can adapt to the characteristics of a restricted spatial setting plays a key role in explaining the broad field of application that Boyd assigns them, especially in Biology, social sciences and the humanities. Based on this understanding, biological species may be considered natural kinds despite their spatiotemporal limitations; and the historical character and geographical restrictions of concepts such as "feudalism or capitalism, or monarchy and parliamentary democracy" [4, 154–157] cease to be an obstacle for their status as natural kinds.

Although this involves a version of natural kinds in which these appear as fuzzy groupings, constructs undertaken from specific perspectives and subject to spatiotemporal limitations, as we have seen, Boyd constantly champions the realism of his version, understanding that the issue of realism is not a matter of "metaphysical fundamentality or anything of that sort", but rather of accommodation to a structure of the world that is objective and independent of us. Nevertheless, given the relevance Boyd attaches to the relativist character of natural kinds about points of view or perspectives, it is easy to understand that his realism is curiously very similar to the internal realism that Putnam proposed later [22, 23]:

> Natural kinds are features not of the world outside our practice, but of the ways in which that practice engages with the rest of the world!... Still, natural kinds are social artefacts. That's why asking whether a kind exists independently of our practice is the wrong way to inquire about its reality. No natural kinds exist independently of practice. [4, 174–175]

Finally, I would like to draw attention to a matter in which, however, Boyd and the essentialist philosophers we have considered earlier coincide, at least partially. I am referring to the relationship between natural kinds and the laws of nature. Boyd denies that the naturalness of kinds depends on their presence in the laws of nature as subject of them, and states that, instead, reference to natural kinds is central to the formulation of an important number of laws of nature. As we have seen, he does not refer to eternal, ahistorical and exceptionless laws, that is to say, to the fundamental laws or principles of theories, because the subjects of them (mass, energy, force) are too abstract and general to constitute a natural kind. Boyd refers to "laws with more specific subject matters", those laws that are often called secondary or derivative, and which explicitly or implicitly incorporate certain restrictions of space, time or individuals [4, 157–158]. Regardless of these details, however, Boyd seems to agree with essentialist approaches, such as those taken by Ellis and Lowes, whereby natural kinds are ontologically more basic than the laws of nature.

In short, Boyd's version of natural kinds manages to successfully circumvent the difficulties of application that the essentialist approach encounters, rejecting its strict demands in the definition of natural kind and tolerating a certain margin of imprecision or vagueness for them (required by the very nature of the system to which they are applied), a relative value to the point of view from which they are constructed or formulated and, in certain cases, a temporal and local nature. Nevertheless, although Boyd maintains that natural kinds are our own constructions and not discoveries in a world previously divided into disjointed kinds, it avoids the stumbling block of mere conventionalism or constructivism by insisting on the necessary accommodation of natural kinds to the nomological structure of the world, an accommodation that would justify their inductive, predictive and explanatory utility.

Notwithstanding, one may well question the ontological priority that Boyd seems to give natural kinds over the laws of nature.[6] It might be worth considering that this is how Boyd (as well as Ellis and Lowe) is seeking to resolve the old chestnut of the distinction between the laws of nature and accidentally true generalizations. Yet in his case, and from what we have seen, we find that the principles of the theories would not be actual laws of nature, as they do not refer to natural kinds. On the other hand, is it enough for a general and well-confirmed statement to refer to a natural kind to be considered a law? The statement that there is no gold sphere with a diameter of more than one mile should therefore be a law of nature because "gold" is a natural kind, but we know that this is a typical example of accidental generalization. Boyd might add that although we know gold is a natural kind, such a restriction on volume is not part of the homeostatic property cluster that defines the kind "gold". To put it another way, the causal structure of the world does not provide a nomological connection between being made of gold and having some form of size limitation. Yet in such a case, the sensible option is not to think that the laws of nature should refer to natural kinds, but instead that both natural kinds and such laws depend on a causal structure of the world, as they are different ways of accommodating to that structure.

---

[6]There are those who contend, in contrast, that the laws of nature are ontologically more basic than natural kinds. See, for example, Bird [1, Chap. 3].

Nevertheless, I understand that Boyd's approach to natural kinds is a very interesting conceptual framework for analysing the laws of nature. In what follows, I shall provide a summary of the basic aspects of the problem of defining the notion of law of nature (and to do so, I shall focus on the laws of science) and then I shall propose, very schematically, a way of resolving it. As I develop these points, I will be briefly commenting on how the concept of scientific law is liable to be interpreted using the characteristics that Boyd attributes to natural kinds.

6 Natural Laws and the Concept of Accommodation

Firstly, it might not be out of place to remember that on some of the occasions when the expression "laws of nature" has appeared here, its meaning is that of *statements* that affirm certain regular (or necessary, according to some) relationships between events. Clear examples of these would be the laws of science, such as those of Snell, Newton, Ohms, Coulomb or Mendel (in these cases, the author's name reveals their human origin), but in this sense laws of everyday knowledge, such as "All human beings are mortal" or "snow is melted by heat", would also be laws of nature. This would not however be the case, as we have seen, with accidentally true generalizations. Let's call the laws of nature in this sense LN^1. Yet there is another different meaning of "laws of nature" according to which the expression does not refer to statements, but instead to regular relationships and patterns of behaviour that, independently of us, exist in nature and whose complex network forms the nomological structure of reality (or what Boyd calls the causal structure of the world).[7] Let's call the laws of nature in this other sense LN^2.

An ingenuous realism seeks to establish a one-to-one relationship between these two types of laws, assuming that each of the LN^1 laws is the precise discovery of an LN^2 law and corresponds to it. Yet there are no grounds for this assumption. At least in the cases of scientific laws (the LN^1 laws which I shall focus on henceforth), the contrived process of abstraction and idealisation that enables them to be formulated converts them into conditional statements whose antecedent describes certain highly idealised conditions that never actually appear in nature. This is the case to such an extent that there is a long tradition in the philosophy of science that even denies that some scientific laws (especially fundamental laws of Physics) are true statements (e. g., [7, Essay 3] and [15, 90–91]). This is not the occasion to discuss whether the degree of abstraction typical of the laws of science may deprive them of their truth value, but it is enough to allude to this matter to remember how the reference the laws of science make to the nomological structure of nature is much more complicated, indirect, and selective than an ingenuous realism may suggest.

---

[7]It is obvious that, according to Boyd, the natural laws in this second sense would be ontologically more basic than natural kinds.

For this reason, amongst others, I believe that Boyd's concept of accommodation may be recovered for the laws of science. Just as Boyd affirmed that natural kinds are not discoveries or descriptions of mutually excluding groupings of things or events that appear in nature, neither should a scientific law be interpreted as representing the corresponding fragment of the nomological or causal structure of the world. What can be said is that those laws accommodate to that structure, using it to become rules that predict and explain phenomena. The complex structure of reality would be beyond our practical and theoretical interests, but from its network we can extract the type of persistent relations that are stated in the laws of science. Accordingly, the idea of accommodation serves as the basis for a realistic conception of the laws of science. We shall now focus briefly on the main difficulties facing the definition of the concept of scientific law, and the divide between such laws and accidental generalizations.

The philosophy of science usually takes Physics as its science model; this is not unwarranted, given that it is the branch of science with the longest track record, the highest level of development, and boasts the largest number of theoretical and practical successes. It is no surprise, therefore, that the fundamental laws or principles of Physics are considered models of scientific laws. Indeed, many handbooks on the philosophy of science begin by defining scientific laws as general (universal or probabilistic) statements without limitations of space, time or individuals. This definition is perhaps applicable to the principles of the main theories of Physics, but it is too exigent: it leaves out those laws of a lower level, which generally arise from the application of theoretical principles to more specific circumstances.

Carroll [6] contends that the genuine laws of science depend solely *on* nature itself (as in the case of the principle of inertia) and such status is not merited by those generalizations that without being casual or accidental are due to specific circumstances or situations that arise *in* nature. Thus, for example, even if it were true that "there is no gold sphere with a diameter of more than a mile", if the reason for this is that there is not enough gold in the entire universe to make one, that generalization, while true and in no way accidental, would not be a law of science because it does not depend on nature itself but instead on certain specific conditions that arise in it. However, this criterion does not solve the problem. There are many laws that depend on specific circumstances or events that have occurred in nature. Kepler's laws depend on the masses of the Sun and the planets, while Galileo's law of free fall depends on the Earth's mass. And what about the laws of geology or Biology? According to Carroll's criterion, only theoretical principles can be referred to as laws. Yet even so, to what extent do theoretical principles depend solely on nature itself? We know that, given their level of abstraction, they are fulfilled solely in ideal situations that only occur, and to a certain extent, in rigorous laboratory experiments.

Some versions of the laws of science do not define them according to the characteristics they have as separate entities, but instead by virtue of their membership or link to scientific theories, distinguishing between fundamental laws or principles and the laws derived from them. It is understood that the derivation of a law entails the explicit or implicit introduction in it of certain restrictions of space, time or individuals. Thus, laws such as those of Kepler and Galileo mentioned earlier, the law of pendular movement, the coefficients of thermal expansion of different metals, or the laws that permit the design of optical devices, all would owe their consideration as a law to the fact they are the derivations or applications of theoretical principles or more general laws. This distinction between fundamental and derived laws introduces a hierarchy among scientific laws. According to this criterion, however, the consideration of laws would not apply to those generalizations, which while admittedly true, lack a theoretical explanation or coverage. Thus, and staying with the previous example, although there is no gold sphere with a diameter of more than a mile, this generalization does not hold as a law because it has no theoretical explanation. By contrast, the status of scientific law would apply to the generalization that no uranium sphere has that diameter, as it is deduced from quantum theory and certain data on the nature of uranium [18, 112].

This definition of the laws of science, while convincing in many cases, poses two major problems. Firstly, there are many true generalizations that we would not consider laws of science even though they may have a theoretical explanation or coverage. Staying with the same example (so often used in theses discussions), the fact that here are no gold spheres of such size because, as Carroll assumed, there is not enough gold in the universe to make one, is open to a scientific explanation. It would be a geological explanation, albeit an overly complex one, above all in comparison to the example of uranium. Therefore, the theoretical explanation or coverage is not, in itself, a clear criterion for determining the nature of a law of science. On the other hand, it is important to note that this approach to laws leaves out the so-called experimental laws. I am referring to important regularities discovered in nature that do not (as yet) have a theory to explain them. In fact, the aforementioned laws formulated by Kepler and Galileo predated Newtonian Physics, the discovery of the constant speed of light came before the general theory of relativity, and the knowledge of the expansion of the universe preceded Big Bang theory.

Therefore, overcoming these difficulties in the characterisation of a scientific law should involve a double task. On the one hand, it should impose some form of limitation on the explanatory chain that runs from the principles of a theory to the law derived from them, and thereby impede the attribution of the status of scientific law to true generalizations whose explanation is, nonetheless, overly complex, and therefore calls for the introduction of highly specific initial conditions. And on the other hand, it will need to explain the existence of experimental laws, lacking any theoretical backing. On this point, and in a purely tentative manner, I suggest that the solution to these problems could involve defining a law of science as a general and suitably confirmed statement that *strikes a balance between the simplicity of its theoretical explanation and the breadth of its practical applications or predictions*; in other words, a statement that achieves a suitable combination of a small number of premises with a large number of consequences. This would rule out the consideration as laws of true generalizations whose explanation requires either the explicit or implicit inclusion of too many specific conditions, thereby restricting their field of application or their predictive interest (e.g., "all pencils thrown in the air fall to the ground" or, once again, "there is no gold sphere with a diameter of more than a mile"). On the other hand, experimental laws, devoid of a theoretical explanation, would justify their consideration as a law because of the breadth and interest of their predictions.

In a definition of this sort, the concept of scientific law would be a comparative concept, and necessarily vague and imprecise, as occurred with Boyd's concept of natural kind. A theoretical principle, for example, would be a higher degree law than the laws derived from it, and amongst these there would also be degree differences depending on the complexity of their derivation and the extent of their predictive capacity. At the same time, it would be an imprecise and fuzzy concept, because if a scientific law is characterised by achieving a balance or compensating between such diverse properties as the simplicity of their theoretical explanation and the breadth of their practical applications or predictions, one would not expect it to be possible to draw up a fixed boundary between scientific laws and generalizations that do not achieve that status.

Finally, let us now see how it is possible that akin to what happens with natural kinds according to Boyd's version, the concept of scientific law is relative, and what is understood by scientific law would depend on the science in question, which would permit us to refer to a certain pluralism and different points of view as regards the concept of law.

In light of what I have been affirming, it seems obvious that qualifying a general statement as a law of science is not a metaphysical issue, as both the laws of science and accidentally true generalizations accommodate to the causal structure of the world, but is instead a methodological issue, given that if the latter are not considered laws, this is merely for pragmatic reasons related to the organisation of scientific knowledge. Having arrived at this point, it is worth wondering whether requirements of the scientific method are the same no matter what science is involved or, conversely, it is possible to admit a methodological pluralism (i.e., a diversity of work styles, decision criteria, forms of expression, etc.). For many decades now, it has been commonplace in philosophy to defend a kind of methodological monism in science that has taken as its model the method followed in Physics, whereby this science has served as a yardstick in such matters as experimentation, the formulation and confirmation of hypotheses, the constitution of laws, the structure of theories, etc. This monism fit perfectly with the notion of the unification of science, which became one of the major topics in the philosophy of science in the first half of the 20th century. Nonetheless, there are many authors who, by critically analysing the old ideal of a unified science, defend a pluralist conception of it, which also encompasses methodological aspects, as being more in keeping with the true history of science and its current situation (Cf. [8, 10, 14]). Although this is not the place to present the details and arguments for this position, it seems obvious that a methodological pluralism implies a pluralism of perspectives on the laws of science. If methodology of science (or better, of sciences) should consider the history of every science, its stage of development and, especially, the characteristics of its object of study, it seems reasonable to conclude that the laws of the different sciences need not have the same degree of precision, of timelessness, of universality, of absence of exceptions, of experimental reproducibility, etc. So what is considered a scientific law depends in each case on the science or "disciplinary matrix" in question, and there would be no reason to deny on principle the existence of genuine scientific laws in the different branches of biology and social sciences.

These few considerations on the laws of nature need to be developed in more detail, but I think they may be an interesting starting point for an approach to these laws based on the main ideas used by Boyd, among others, on natural kinds, without seeking to establish any direct dependence between natural kinds and the laws of nature.

References

1. Bird, A. (1998). *Philosophy of science*. Montreal: McGill-Queen's University Press.
2. Bird, A., & Tobin, E. (2012). Natural Kinds. In E. N. Zalta (Ed.) *The stanford encyclopedia of philosophy* (Winter 2012 ed.). http://plato.stanford.edu/archives/win2012/entries/naturalkinds/.
3. Boyd, R. (1991). Realism, anti-foundationalism and the enthusiasm for natural kinds. *Philosophical Studies, 61*, 127–148.
4. Boyd, R. (1999). Homeostasis, species and higher taxes. In R. A. Wilson (Ed.). *Species. New interdisciplinary essays* (pp. 142–185). Cambridge, Mass.: MIT Press.
5. Boyd, R. (2010). Realism, natural kinds and philosophical methods. In H. Beebee & N. Sabbarton-Leary (Eds.), *The semantic and metaphysics of natural kinds* (pp. 212–234). New York: Routledge.
6. Carroll, J. (2008). Nailed to hume's cross? In J. Hawthorne, T. Sider, & D. Zimmerman (Eds.), *Contemporary debates in metaphysics* (pp. 67–81). Oxford: Basil Blackwell.
7. Cartwright, N. (1983). *How the laws of physics lie*. Oxford: Oxford University Press.
8. Cartwright, N. (1999). *The dappled world. A study of the boundaries of science*. Cambridge: Cambridge University Press.
9. Dupré, J. (1981). Natural kinds and biological taxa. *The Philosophical Review, 90*(1), 66–90.
10. Dupré, J. (1993). *The disorder of things: Metaphysical foundation of the disunity of science*. Cambridge, Mass: Harvard University Press.
11. Dupré, J. (2002). *Humans and other animals*. Oxford: Clarendon Press.
12. Ellis, B. (2001). *Scientific essentialism*. New York: Cambridge University Press.
13. Ellis, B. (2008). Essentialism and natural kinds. In S. Psillos & M. Curd (Eds.), *The routledge companion to philosophy of science* (pp. 138–148). London and New York: Routledge.
14. Galison, P., & Stump, D. (Eds.). (1996). *The disunity of science*. Stanford: Stanford University Press.
15. Giere, R. (1999). *Science without laws*. Chicago: University of Chicago Press.
16. Goodman, N. (1978). *Ways of worldmaking*. Indianapolis: Hackett Publishing Company.
17. Kripke, S. (1972). Naming and necessity. In G. Harman & D. Davidson (Eds.), *Semantics of natural language* (pp. 253–355). Dordrecht: Reidel.
18. Loewer, B. (1996). Humean supervenience. *Philosophical Topics, 24*, 101–126.
19. Lowe, E. J. (2006). *The four-category ontology: A metaphysical foundation for natural science*. Oxford: Clarendon Press.
20. Mellor, D. H. (1977). Natural kinds. *The British Journal for the Philosophy of Science, 28*, 299–312.
21. Putnam, H. (1975). The meaning of 'meaning'. *Minnesota Studies in the Philosophy of Science, 7*, 215–271.

22. Putnam, H. (1981). *Reason, truth and history*. Cambridge: Cambridge University Press.
23. Putnam, H. (1987). *The may faces of realism*. La Salle Ill: Open Court.
24. Sober, E. (1993). *Philosophy of biology*. Boulder Co: Westview Press.
25. Williams, N. E. (2011). Arthritis and Nature's Joints. In J. Keim Campbell, M. O'Rourke & M. H. Slater (Eds.), *Carving nature at its joints* (pp. 199–230). Cambridge, Mass: The MIT Press.

Chapter 9
Synchronic and Diachronic Luck

Steven D. Hales

Abstract In the present paper I argue that luck attributions are structured by points of view. In particular, whether one is prepared to say that an event or a person is lucky is partly determined by one's temporal perspective. If an event is seen in isolation, at a moment in time, it might not be a matter of luck at all, but when the same event is considered as an element in a temporal series, then it becomes either lucky or unlucky. Since neither temporal point of view enjoys any kind of logical priority or metaphysical privilege, it is not possible to make consistent assignments of luck without first assuming a synchronic or a diachronic point of view. Since no extant theory of luck acknowledges or incorporates such points of view, all fall short of adequacy. This failure matters broadly in philosophy, because understanding luck underwrites a number of philosophical projects.

1 Why Luck Matters

Appeals to luck play a role in epistemology, ethics, political philosophy, metaphysics, and the philosophy of science. Epistemologists worry about the problem of epistemic luck (cf. [14, 18]). Knowledge is something more than mere true belief—one could have stumbled upon the truth by accident, but a lucky guess or set of circumstances is not enough to rise up to knowledge. Epistemologists have struggled for 50 years (i.e. since Gettier) to analyze knowledge in such a way that no lucky possession of the truth could satisfy the analysis. No consensus has been reached, and some have despaired of the whole project.

This work has been granted by Spanish Government, "Ministerio de Economía y Competividad", Research Projects FFI2008-01205 (*Points of View. A Philosophical Investigation*), FFI2011-24549, (*Points of View and Temporal Structures*), and FFI2014-57409-R (*Points of View, Dispositions, and Time. Perspectives in a World of Dispositions*).

S.D. Hales (✉)
Bloomsburg University, Bloomsburg, USA
e-mail: shales@bloomu.edu

© Springer International Publishing Switzerland 2015
M. Vázquez Campos and A.M. Liz Gutiérrez (eds.), *Temporal Points of View*,
Studies in Applied Philosophy, Epistemology and Rational Ethics 23,
DOI 10.1007/978-3-319-19815-6_9

In ethics is the issue of moral luck (cf. [13, 21]). The praiseworthiness or blameworthiness of an agent is generally taken to depend solely on their intentional actions, and not on external circumstances over which they have no control. On the other hand, we tend to regard a drunk driver who killed a pedestrian as morally worse than a drunk driver who got home safely, even though the latter might only be because she was lucky that no pedestrians ran in front of her car. Yet it is extremely counterintuitive to think that moral judgment should turn on something so extrinsic to the agent, which was unpredictable or uncontrollable.

In political philosophy is the issue of luck egalitarianism (cf. [3, 19]). Some people have won the natural lottery: they are born intelligent, attractive, and innately talented. Others have lost the natural lottery: they are born with cognitive or physical defects. Likewise some have won the social lottery, and are born to wealthy, caring families in peaceful nations. It is easy to imagine losers in the social lottery as well. In this way, one's life prospects are strongly connected to luck. What obligations do we have as a society to overcome the effects of luck and level the playing field? Collaterally, the concept of privilege (in the sense of white privilege, or heterosexual privilege) seems to be grounded in the antecedent notion of luck egalitarianism.

Metaphysicians interested in free will are troubled by the possibility that if libertarianism is the correct theory of freedom, then some actions seem to solely the result of luck (cf. [10, 12]). If an agent performs an action A, but might have performed action B instead, even given the same past and the laws of nature, then there seems to be nothing that determines her performance of A. Performing A instead of B then looks arbitrary and outside of her control. If performing A had good effects, then the agent was lucky that she did A when might have easily done the less optimal B instead. For example, if the agent hit a five-iron to the green, she was lucky that she didn't pick the three-iron that her caddy recommended, which would have caused her to fly the green. But if she might have picked the three-iron, even given the same past and laws of nature, then that undermines the sense in which she is in control of which club she uses. It seems like mere luck which club winds up in her hand.

Philosophers of science care about explaining the logic of scientific discovery. However, there seems to be a large element of serendipity in discovery and advance—scientists who were looking for one thing and found something else, or who stumbled onto something great by accident (cf. [6, 7]). Such unexpected fortune is difficult to integrate with theories of discovery that understand science as a rigorous matter of nomological-deductive reasoning.

2   Theories of Luck

There are three theories of luck in the literature, each of which tends to appeal to philosophers pursuing different concerns. The first is the probability theory, according to which an occurrence is lucky (or unlucky) only if it was improbable to occur [1, 4, 17]. The second theory of luck is the modal theory, according to which an event is lucky only if it is fragile—had the world been very slightly different it would not have occurred [10, 14, 16, 20]. The third theory of luck is the control view, which states that if a fact was lucky or unlucky for a person, then that person had no control over whether it was a fact [9, 10, 12].

Theory 1: Probability. The consensus view among mathematicians and scientists, according to the probability theory an event is lucky or unlucky only if it is improbable. The idea that luck is to be explained by probability goes back at least to Abraham de Moivre's *The doctrine of chances, or, A method of calculating the probability of events in play* [5]. The ancients regarded luck as an occult property, one that was granted by the whim of the gods or might be harbored in a rabbit's foot or four-leaf clover. Instead of luck being an inexplicable turn of Fortuna's wheel to be harnessed by magic, de Moivre argues that it is fully explainable by mathematics. To that end his book is an original development of probability theory, including a partial proof of the central limit theorem. Bewersdorff [4] also describes the early development of probability theory as arising out of gamblers' need to explain bad luck (pp. 8–9). Mazur [11] argues that the gambler's fallacy of expecting good fortune to follow after a run of bad simply arises from a misunderstanding of Bernoulli's law of large numbers. "Luck," Masur writes, "can be cogently explained by the rules of probability" (p. xvii). Ambegaokar [1] concurs that probability theory is "why it is possible to reason quantitatively about luck" (p. 10).

The main philosophical defender of the probability theory is Nicholas Rescher. In Rescher [17] he argues that only improbable events can be lucky or unlucky, and that their degree of luck is a function of the event's improbability and its importance ($\Delta(E)$). He offers this formula [17, p. 211] to measure the amount of luck (λ) in an event E: $\lambda(E) = \Delta(E) \times pr(not\text{-}E)$. Thus the occurrence of a mildly improbable event that is very important might be just as lucky as a very improbable event that is only somewhat important. Very important, very unlikely events are the luckiest of all. No luck whatsoever attaches to events that are wholly unimportant or are certain to occur.

Theory 2: Modality. The modal view is common among epistemologists. According to this view, an event is lucky only if it is fragile—had the world been very slightly different it would not have occurred. The most prominent defender of the modality theory is Duncan Pritchard, who writes, "if an event is lucky, then it is an event that occurs in the actual world but which does not occur in a wide class of the nearest possible worlds where the relevant initial conditions for that event are the same as in the actual world" [14, p. 128]. Epistemologists like the modality approach because then epistemic luck involves "a true belief that could very easily have been false" [15, p. 272] and due to epistemic luck, "the fact that you could very easily have been deceived is a ground to deny you knowledge, even if in fact you were not deceived" [15, p. 275]. These ideas pave the way for requiring a safety condition on knowledge, which states that S knows that *p* only if S's true belief that *p* could not have easily been false. Safety may not be the whole story about knowledge—even Pritchard now thinks it must be supplemented with a virtue account of success from ability—but it is widely considered to be a major player in theories of knowledge.

Modally robust events, on the other hand, are not due to luck. A true belief that is false only in distant possible worlds is (or is at least a worthy candidate for) knowledge. It cannot be a matter of luck that a necessary truth is true, or that an inevitable event occurs. A proposition that remains stably true as one moves further and further away from the actual world is less and less attributable to luck.

Theory 3: Control. Philosophers interested in moral luck, luck egalitarianism, the luck problem in free will, and virtue epistemology tend to gravitate towards the control theory of luck: if a fact was lucky or unlucky for a person, then that person had no control over whether it was a fact. Al Mele writes, "Agent's control is the yardstick by which the bearing of luck on their freedom and moral responsibility is measured" [12, p. 7]. John Greco too is sympathetic to the control account, writing, "something is a matter of luck in relation to some agent just in case it is not the agent's doing. Put differently, something is a matter of luck just in case it is external to the agent's own thinking, choosing, and acting" [9, p. 130]. Events outside the control of an agent are properly attributable to luck, irrespective of how probable or modally robust those events are. Other philosophers opt for hybrid views. Neil Levy [10, p. 36], for example, votes for a disjunctive view: luck can be either the modal kind (which he calls chancy luck) or the control variety (which he calls non-chancy luck). He regards them as independent kinds of luck. Wayne Riggs, on the other hand, defends a conjunctive theory: luck is a combination of both the control and probability approaches [18, see esp. p. 340].

The problem of synchronic and diachronic points of view in luck attributions undermines all three formulations of luck, and combination views fare no better.

3   Synchronic and Diachronic Luck

The problem of diachronic luck is when an event is judged to be lucky as a part of series or streak that takes place over time, but is not regarded as lucky when the same event is considered synchronically, independently of its relations to other events. I will provide some examples of diachronic luck and then examine how the theories of luck on offer falter when confronted with it.

Example 1: slot machine. Suppose you are playing an old-fashioned mechanical slot machine (new ones are digital, computer-controlled, and randomized). Pull the lever, and three reels spin around independently of each other, each with the same probability to land on a lemon, cherry, apple, lime, grape, watermelon, etc. A common setup is to have 16 different images per reel. The reels do not stop all at once; the one on the furthest left stops first, then the middle reel, then the one on the right. You pull the lever. The first reel lands on a cherry. That's not luck; you don't care. It is irrelevant what the first symbol is. Then the second reel also stops on a cherry. You're still not feeling too lucky, because there's no payout for two cherries. But now you are certainly crossing your fingers for the third reel, hoping for a visit from Lady Luck. When it stops, it too lands on a cherry. Jackpot! You were very lucky that the 3rd reel came up cherry.

Told in that manner, it is perfectly sensible that the first cherry was not a matter of luck at all, the second cherry also not luck (or maybe a tiny bit lucky), but the third cherry was tremendously lucky. Viewed as an element of a diachronic series, the final cherry was lucky, since it secured the jackpot. However, the spins of the reels are independent trials and are not causally connected to each other. Furthermore, each wheel had to land on the same symbol in order to win; it was no more necessary that the 3rd reel land on cherry than it was for the first two. Given that the 3rd reel was cherry, the first two had to be as well. Viewed synchronically, no one wheel seems any luckier than any other. They all had to cooperate together to yield a payout.

Consider how our three theories of luck fare. The probability that the final symbol would be a cherry was no lower than the chance the first symbol was. The chance that all three would hit on the same fruit was low (0.02 %), so the probability theory gives the correct result that beating a slot machine is lucky. But whether it is lucky to get a streak of cherries is not the issue: *was getting a cherry on the 3rd reel luckier than getting one on the 1st or 2nd reel?* On the probability theory the answer is no. The chance was the same for each reel. The modal view gives the same result. All it takes is a small change in the world (the reel stops a few clicks later or a few before what it did in the actual world) and reel 3 would have not hit on a cherry. Yet the exact same thing is true of reels 1 and 2. The success of reel 3 in producing a cherry is no more modally fragile than the other two reels; therefore a cherry on reel 3 is no luckier than the other two. The control theory lines up with the others. A player has no control over where any of the reels stop spinning. One has no less control over the 3rd reel than over any of the others. Therefore under the control theory the relative luck assigned to each of the reels is exactly the same. It doesn't matter for our purposes here whether it was lucky or not lucky that a reel hit on cherry. The salient issue is whether it was luckier that the third reel did so.

Synchronically, all the theories get it right: landing on a cherry on reel three isn't any luckier than getting one on either of the other two reels. Yet none of the theories are able to accommodate the diachronic judgment that during play the successful spin of the third reel seems vastly luckier than the other two. Hitting a cherry on the third reel seems both luckier than a cherry on the first two reels (viewed diachronically) and also not luckier at all (viewed synchronically). While the extant theories of luck can accommodate synchronic luck, they cannot explain diachronic luck.

One might argue that hitting cherry on the third reel was more significant than it was for the first two reels. So even if the probability of, modal fragility of, or control over cherry on the third reel was no different from the first two, its importance was, and therefore it really was luckier for the third reel to come up cherry than it was for the first two reels. However, this rejoinder is mistaken, and in fact highlights the problem of synchronic versus diachronic luck. Considered synchronically, in isolation of its position in a temporal series, it is no more important that a cherry come up on one reel over any other. It was equally essential for the same fruit to appear on each reel to hit the jackpot. It was just as important for the first or second reels to come up cherry as it was for the third. However, what happens on the third reel does seem to matter more when it is considered diachronically, as an element of a series. Given that there were cherries on the first two reels, it is now of considerable significance that a cherry come up on the third. The fundamental phenomenon is the differing attributions of luck depending on the diachronic or synchronic perspectives.

One might wonder whether things work out differently in cases of skillful performance. It will be argued below that the reasoning is the same for cases involving agency, and not wholly impersonal chance.

Example 2: Joe DiMaggio. DiMaggio's 1941 streak of safely hitting in 56 consecutive baseball games is widely considered the most outstanding record in the history of sport (cf. [8, p. 467]) What role did luck play? Arbesman and Strogatz [2] conducted a Monte Carlo simulation on the history of baseball, using a comprehensive baseball statistics database (from 1871 to 2004). They constructed a variety of different mathematical models of alternate possible histories of baseball, taking into account for each player the number of games played, number of at-bats, times walked, being hit by a pitch, sacrifice hits, and so on. Their five models varied the minimum number of plate appearances and a few other variables, and they ran 10,000 computer simulations for each model. These simulations amounted to complete alternative histories of baseball. One of the results was that there was only between a 20 and 50 % chance that anyone would have safely hit in 56 or more consecutive games. DiMaggio, who in the actual world did have a 56 game hitting streak, was barely in the top 50 of the most probable players to hold that record. In fact, they write that, "while no single player is especially likely to hold the record, it is likely that an extreme streak would have occurred" (p. 11). The probability of someone or other having a long hitting streak is high, but the probability of DiMaggio in particular having the record is low.

So, given the Arbesman and Strogatz analysis, DiMaggio was hugely lucky to have the streak, and the probability, modal, and control accounts all line up in agreement. In fact, the longer the streak went on, the luckier he was to keep it up. Was DiMaggio unlucky on July 17, 1941, the date that his streak ended? In that game, Indians third baseman Ken Keltner made two terrific backhanded stops to prevent DiMaggio from hitting successfully in what would have been the 57th game of the streak. The day after the streak ended, DiMaggio began another hitting streak that lasted 17 games. Surely it was terrible luck that Keltner made such good plays and prevented DiMaggio from getting a hit in game 57. If he had, then instead of being 56 games long, DiMaggio's hitting streak would have been a stunning 74 games in a row (NB: safely hitting 73 of 74 games in a row is also still a record). Viewed diachronically, as an element of a series, DiMaggio had bad luck against the Indians in game 57. Had he managed to get even one hit that game, then he would have the untouchable mega streak of hitting in 74 games in a row.

Considered synchronically, however, it was not bad luck at all that DiMaggio failed to hit. As was argued earlier, on all three theories of luck it is simply luck when a baseball player gets a hit—it is always improbable, modally frail, and not really within their control. Even during his streak DiMaggio missed most of the time (batting .409). The fact that he failed to hit in game 57 was a wholly ordinary, routine part of baseball. If it were any other game, no one would think DiMaggio was unlucky; it is only because of its location between his two streaks that it seems that way.

I've presented the DiMaggio case as one in which he was diachronically unlucky (which the theories of luck under consideration cannot accommodate) but synchronically his performance was not a matter of bad luck at all (both intuitively and according to the three theories of luck). Here is one final case to make the point, this time of a streak in which a player is diachronically lucky but viewed synchronically his performance is not due to luck.

Example 3: Micheal Williams. Micheal Williams, a point guard for the NBA Timberwolves, holds the NBA record for a free throw streak: over a period of nine months in 1993 he sunk 97 free throws before missing.[1] Since his career free throw percentage was .868, on the probability account no individual free throw was lucky —Williams was very likely to sink it. On the modal account it is difficult to judge how distant the closest possible world is in which Williams misses any particular free throw. Unlike baseball, where each pitch is different and more generally the playing conditions vary, free throws have replicable conditions. Players shoot from the same spot, and no one else interferes with or controls the ball prior to their shot. So it may be that the world would have to be considerably different for Williams to miss a free throw that he made in the actual world. On the control account Williams has significant control over the basketball—he is a pro ballplayer and is shooting without interference or unusual distractions. So it seems that all three theories of luck are in agreement: Williams is not lucky when he hits any particular free throw. Nor is he unlucky; successfully sinking a free throw just isn't a matter of luck at all. His success seems properly assignable to skill, not luck.

Nevertheless, when Williams was lucky to hit the 79th shot in a row, the one that broke Calvin Murphy's old record, since that was the shot that cemented his place in the record books. As NBA hall of famer Rick Berry writes, "all great free-throw shooters must possess technique, confidence, routine and a little luck."[2] Calvin Murphy agreed, complaining at the time that, "what really bugs me is Micheal Williams breaking my consecutive streak and now he's shooting 83 %. That tells me he was lucky."[3] Williams was lucky to make the record-breaking shot, despite the fact that he was in control of the ball, very likely to make it, and apart from the streak it was more-or-less an indistinguishable shot from any other free throw. As in

[1]http://www.basketball-reference.com/players/w/willimi02.html,   http://www.nba.com/history/records/regular_freethrows.html.

[2]http://www.nba.com/2009/news/features/01/14/barry.011409/index.html.

[3]http://community.seattletimes.nwsource.com/archive/?date=19940410&slug=1904894.

the slot machine case, viewed synchronically no particular free throw was lucky, but when seen as a part of diachronic series, the third cherry or the record-breaking shot is lucky indeed.

4 Conclusion

None of the three accounts of luck—probability, modal, or control—can consistently provide the intuitively correct results when confronted with synchronic and diachronic cases. That is, the theories either rule that considered synchronically, there is no luck of any sort present, but diachronically there is luck, or conversely. There are three options now available. The first option purveyors of these theories of luck might pursue is to throw in the towel and concede that we do not yet have the right account of luck. Luck is a real phenomenon, but since the extant theories yield inconsistent results, they are mistaken. The second option is to continue to defend one of the three theories of luck, but supplement that theory with a perspectival parameter that modifies the theory's output depending on temporal point of view. To take one example, from the diachronic perspective, it was not modally robust for Williams to break the free throw record, even though from the synchronic perspective he did sink the free throw in relevantly close possible worlds. The third option is to conclude that we have been pursuing a red herring all along, and that there is no such thing as luck after all. We should expect any theory of luck to have fatal flaws just as we should expect any theory of phrenology to incompatible with the evidence. Perhaps luck attributions are no more than a narrative frame we hang around stories of success (like three reels lining up on a slot machine, or sinking a free throw at an opportune time) or failure (like missing a hit in baseball when it would have been very nice to get one). Pursuing any one of these three alternatives is a much larger project; but the problem of synchronic and diachronic luck shows the need for such an endeavor.

References

1. Ambegaokar, V. (1996). *Reasoning about luck: Probability and its uses in physics*. Cambridge: Cambridge University Press.
2. Arbesman, S., & Strogatz, S. H. (2008). A Monte Carlo approach to Joe DiMaggio and streaks in baseball (pp. 1–14). arXiv:0807.5082v2 [physics.pop-ph].
3. Arneson, R. J. (2011). Luck egalitarianism—A primer. In C. Knight & Z. Stemplowska (Eds.), *Responsibility and distributive justice*. Oxford: Oxford University Press.
4. Bewersdorff, J. (2005). *Luck, logic, and white lies: The mathematics of games*. Wellesley, Mass: A. K. Peters.
5. de Moivre, A. (1718). *The doctrine of chances, or, A method of calculating the probability of events in play*. London: W. Pearson for the author.
6. de Rond, M., & Morley, L. (Eds.). (2010). *Serendipity: Fortune and the prepared mind*. Cambridge: Cambridge University Press.

7. Eco, U. (1998). *Serendipities*. New York: Columbia University Press.
8. Gould, S. J. (1991). *Bully for brontosaurus*. New York: W.W. Norton and Co.
9. Greco, J. (2010). *Achieving knowledge: A virtue-theoretic account of epistemic normativity*. Cambridge: Cambridge University Press.
10. Levy, N. (2011). *Hard luck: How luck undermines free will and moral responsibility*. Oxford: Oxford University Press.
11. Mazur, J. (2010). *What's luck got to do with it?: The history, mathematics, and psychology of the gambler's illusion*. Princeton: Princeton University Press.
12. Mele, A. R. (2006). *Free will and luck*. Oxford: Oxford University Press.
13. Nagel, T. (1979). Moral luck. In *Mortal questions* (pp. 24-38). Cambridge: Cambridge University Press.
14. Pritchard, D. (2005). *Epistemic luck*. Oxford: Oxford University Press.
15. Pritchard, D. (2012). Anti-luck virtue epistemology. *The Journal of Philosophy, 109*(3), 47–79.
16. Pritchard, D. (2014). The modal account of luck. *Metaphilosophy*, forthcoming.
17. Rescher, N. (1995). *Luck: The brilliant randomness of everyday life*. New York: Farrar Straus Giroux.
18. Riggs, W. D. (2007). Why epistemologists are so down on their luck. *Synthese, 158*(3), 329–44.
19. Tan, K. C. (2012). *Justice, institutions, and luck*. Oxford: Oxford University Press.
20. Teigen, K. H. (2003). When a small difference makes a large difference: Counterfactual thinking and luck. In D. R. Mandel, D. J. Hilton, & P. Catellani (Eds.), *The psychology of counterfactual thinking* (pp. 129–46). London: Routledge.
21. Williams, B. (1982). *Moral luck*. Cambridge: Cambridge University Press.

Chapter 10
Presentism, Non-presentism and the Possibility of Time Travel

Juan J. Colomina and David Pérez Chico

Abstract This chapter argues for a notion of time that allows time travel. In order to time traveling to happen, in contrast to Presentism, the chapter demonstrates that we can change the past and we have some place where to travel. It shows the advantages of a non-presentist ontology that advocates for indeterminacy of future facts based not on its absence of truth-value, but on the overdetermination of future facts. The conclusion is that to break the causal chain is impossible in we are placed in the same causal line. But if we rethink the time traveling as a trans-world traveling, it is possible to open a new causal line anytime that someone travel in time, to the past as well as to the future.

Even if some time soon our physicists find that the laws of physics support time travel, it might nonetheless never occur because, say, it will never be technologically feasible. Moreover, problems connected to time travel are not just empirically motivated, but they are also metaphysically and logically driven. In such a way that even those who see how deep and serious these problems are must admit that the existence of scenarios in which time travel is a possibility, it brings light over questions such as the nature and the topology of time, the time-asymmetry of causation, of backward causation, and some others.

Time is ubiquitous: all our experience is placed in time. Time is what gives an order to our lives, it is the background of our experience, but what exactly is its nature is far from being clear. Ordinarily we assume a conception of time that seems

This work has been granted by Spanish Government, "Ministerio de Economía y Competividad", Research Projects FFI2008-01205 (*Points of View. A Philosophical Investigation*), FFI2011-24549, (*Points of View and Temporal Structures*), and FFI2014-57409-R (*Points of View, Dispositions, and Time. Perspectives in a World of Dispositions*).

J.J. Colomina (✉)
The University of Texas at Austin, Austin, USA
e-mail: Colomina-Alminana_Juan@austin.utexas.edu

D.P. Chico
Universidad de Zaragoza, Saragossa, Spain
e-mail: dcperez@unizar.es

M. Vázquez Campos and A.M. Liz Gutiérrez (eds.), *Temporal Points of View*,
Studies in Applied Philosophy, Epistemology and Rational Ethics 23,
DOI 10.1007/978-3-319-19815-6_10

to prevent time travel.[1] But at the same time, and without abandoning this ordinary level, we found that time travel is conceivable and stories about it have fueled the thought of important philosophers and scientist. Not to mention our pop-culture industry. Stories about time travelers that go back to the past (or to the future),[2] are not just fictions to feed our teenagers' imagination. These stories exist and fill our classical and contemporary sci-fi background. Be that as it may, we would like to defend here the possibility of the existence of a notion of time that makes time travel possible.[3]

Apart from the world of fiction, though, it is commonly thought that time travel is logically and metaphysically impossible. Here we will focus on two different kinds of arguments against time travel: the first kind is based on the assumption that it is not possible to change the past because the present state of the world is determined by the past; the other is known as the Nowhere Argument. The latter says that, on one hand, the past is something that has happened already, therefore the past does not exist now. On the other hand, given that the future is something that has not happened yet, then the future does not exist now. Consequently, given that the past and the future do not exist now, the possibility of time travel is nonsense because we have no-where to arrive at.

According to the first argument, there exists a (temporal) causal line (or time arrow) from the past to the future that makes it impossible to change past events because, if it were possible, we would fall in insoluble logical time paradoxes because once we have gone back in time, our actions there would have causal effects. According to our own view, arguments like these should be resisted because are based on debatable conceptions of the ontology of time and causality.

---

[1]This ordinary conception has the following features, among others: (i) time flows, and it does it in one direction: What is future will be present and will be past; what is present was future and will be past; and what is past was present and was future; (ii) time-asymmetry of causation: causes are different from effects, and past events cannot be caused by future events: (a) the causal relation is asymmetric—if A is a cause of B, then B is not a cause of A—; and (b) effects never (or almost never) occur *before* their causes [12]; (iii) the causal continuity and change through time (the time traveller that departs has to be the same person that arrives to the moment in the past).

[2]The reader can find an immensity of books and movies about time travel. Perhaps the most well known is H.G. Wells' *The Time Machine*. Lewis [6] talks about some classical works of R.A. Heinlen. The *John Carter* series written by E.R. Borroughs combines time travel and space long-distance travel. Among movies, *Back to the Future* saga is a classic, but also other films as *La Jetée, Time Bandits, The Boy, 12 Monkeys, Donnie Darko, Timecrimes,* or *Loopers*, to name just a few, include time traveling. Generally, we can distinguish two different kinds of time travel. The first one supposes just a physical lineal transportation of the time traveler from the original time to the new time (through the intentional or accidental use of some time traveling device). The second one, as Lewis [6: 148] points out, presupposes some causal loop that transports the time traveler from some concrete spot of time to a different one. We will focus on the first kind, but our conclusions here could be easily applied to the other kind.

[3]We shall accept in this paper the thesis of D. Lewis about the possibility of time travel. As him, we think that time traveling stories are, in some cases, perfectly consistent with the causal laws of our world (or some other possible world) and that "the paradoxes of time travel are just oddities, not impossibilities" [6: 145].

We will argue against presentism's ontology according to which only present events exist, the future has not existence yet and the past events existed in an earlier time but they do not longer exist. Therefore, there does not exist a past time or a future time where a time traveller could travel. We show the advantages of a non-presentist ontology (inspired by the McTaggart's B-series, without necessarily accepting the unreality of time) that defends an indeterminacy of the future facts based not on the absence of the relevant truth-makers (what it is called a 'gappy' future), but on the over-determination of future facts (a 'glutty' future).

One immediate consequence of our view is the possibility of a 'glutty' past. The idea of changing the past by subtraction is still impossible: you cannot change what has happened already. But with the new ambiguity provided by the 'gappy' past in hand, we can rethink the possibility of changing the past by addition. That is to say, a time traveller could change the past by adding a new fact that changes something that did happen (or did not happen). If this possibility is successful, the notion of causality has to be rethought in a way that time travel is a real option. To reiterate, we don't mean that it is possible to break a causal chain. What we mean instead is that whenever one travels in time, she is creating a new causal chain. So, time travel is only the new origin of a trans-world travel.

For many people then, time traveling is just an illusion, an empirical impossibility. It gets worse, though, because it is also commonly thought that time travel is logically and metaphysically impossible,[4] and for this reason, it is thought philosophically irrelevant. Those who think this way put forward at least two different kinds of arguments: one is known as the Nowhere Argument; the other is based on the impossibility of changing the past: the Causality Argument.[5] The first argument says that given that, on one hand, the past is something that has happened, it does not exist right now. On the other hand, given that the future is something that it has not happened, the future does not exist yet. Consequently, given that the past and the future do not exist, time traveling is nonsense, because we have nowhere to go.

According to the second argument, there exists a (temporal) causal line (from the past to the future, according to the standard time arrow direction) that makes it impossible to change past events because, if it were possible, we would be faced with insoluble logical time paradoxes. According to our view, these arguments are based on a wrong conception of the ontology of time and one debatable conception of causality.

---

[4]We will avoid here the question about its empirical impossibility or whether time travel is even permitted by the stipulated scientific laws of nature because these are matters to be treated by empirical sciences. According to actual relativistic physical theories, nevertheless, time travel is theoretically possible.

[5]Notice that the impossibility of changing the past is not per se a reason against time travel. It has to be added to the principle of autonomy: on arriving in the past, a time traveller can locally engage in acts other that those ones that history records without being inconsistent with the way reality is, as long as this other acts are compatible with the laws of physics. Therefore, according to the autonomy principle, if time travel is possible, time travellers have the possibility of causally affecting past events. Consequently, we will have every right to ask how things could have been.

The first one is just a defense of presentist ontology. We could counter it with an alternative ontology of time. The second argument is more difficult to counter. In this case we will make some comments on the difference between objective and subjective time, about a tenseless-reductionist-relationalist conception of the nature of time and, above all, we will follow Markosian [9] in that if we suppose that some propositions about the future can have an indeterminate true-value, then we should suppose that some propositions about the past are indeterminate too. The key point here is the following: the future allows for the existence of temporal branches that go from being possible to get actualized and others that do not get actualized (but it is not as if they disappear). Well, let's imagine that the same takes place in the past: some branches are not concrete because they never got actualized, but it could be the case that they would end up being actualized if somebody travels to the past and generates a new branch. The moment changes, and the object that persists in different moments gets actualized in the future, but not in the past. The state function changes. Consequently, time travel would be more like traveling among possible worlds than traveling in time because we are supposing that the possible worlds are located in time.

In what follows we shall put in place what we mean when we talk about time travel. We will describe the contradictions that must be faced anytime that time travel is taken into account and how they are usually interpreted. After that, we will comment on the presentist view about time and time traveling. It is pretty straightforward that if we only accept the existence of present facts (this is to say, if we support Presentism), it is impossible to find an interpretation that fits really well with the possibility of time travel. In order to prove the presentist is wrong when she denies the possibility of time travel, we will show that her conception of the metaphysics of time is also wrong.

We will support the four-dimensionalist view about time [18]. This is to say, we shall accept that events exist in different slots of time. In the same way that the same thing can exist in different points of space, also the very same event can exist in different times. Thus, every slot of time has what we can call indexical embedment: we cannot isolate a concrete slot of time from its experiencers. Nonetheless, this does not mean that time is mind-dependent. Time is objective, but the subjective time has to be taken into consideration if we want to solve the apparent time-paradoxes.

Based on this four-dimensionalist notion of time, then, we shall defend the possibility of an open future. Given that there is not a privileged slot of time (like Presentism insists), every future event is real but still not concrete. Following Lukasiewicz's thesis about the indeterminacy of future, we will argue that the future can be interpreted as a function where some possibilities are offered according to a previous causal chain of events.

It is true that we can see this thesis under a presentist view, as Prior [14] did. He concluded that (at least) some utterances about future events (the future contingents) lack truth-value because they are not still actual (this is called a 'gappy' future). But for us, since we have denied the presentist view and accepted the four-dimensionalism, the only possible interpretation is a realist account about the indeterminate future events. According to this view, every possible future event exists right now, even though its truth-value needs to be stipulated by the instantiation of the relevant truth-makers. This is to say, the future is over-determined (it is a 'glutty' future) and so, all we need to know is which are the truth-makers that enables us to instantiate future events and therefore to get the future determinated.

An immediate consequence of this is the emergency of the possibility of a 'glutty' open past. The idea of changing the past by subtraction continues to be an impossibility: there is no way of changing what has already taken place. But with the new ambiguity provided by the 'gappy' past in hand, we can rethink the possibility of changing the past by addition. It is possible to change the past later on by adding a new fact that changes something that has happened (or that did not happen). All what is needed in order to achieve this oddity is the instantiation of a new past fact that plays the role of the truth-maker that allows to determine a non-actualized past chain of events. If we can accomplish this, we will have some chance to change the past by addition. In order for this possibility to have some success, we need to rethink the notion of causality in a way that makes room for the possibility of time travel. This could be achieved if time travel is seen not as a movement in the same time line, but as a displacement among possible worlds and different causal chains.

Think about the following situation.[6] Tim considers that his grandfather ammunition factories are not a good business. A lot of people were killed by the tones of ammo produced by those factories. Tim hates his grandfather for that reason and we wish his grandfather had died before he could open his factories. So Tim toys with the idea of traveling to the past and assassinating his grandfather. Now, the fortune made by his grandfather's factories has provided Tim the money enough to complete an engineering degree in the MIT. There he devoted himself to study the physics of time and, after several years of hard work, finally he builds a time machine that works. He could just now go back to the past, to the thirties, when his grandfather opened his first ammo factory. Tim could buy the best rifle that money could pay in this age. He can devote his time to learn his grandfather's habits, his frequented places, and his schedules. He could even spend some time taking shoot-training lessons. Everything seems to indicate that Tim could kill his grandfather.

But, wait; something is wrong with this picture. Tim has the time machine. He has time in his hands now. He can travel to the past, take the best gun possible, take some shooting lessons, and make a good plan to kill his grandfather before he has the opportunity of opening his first ammo factory. Everything seems pretty straightforward. The target is right there. He just needs to shoot him. But the thing is that, if this would be possible, he could not be there to shoot his grandfather in the first place! If Tim kills his grandfather, he had never had success in ammo business, had made money, and provided to Tim a good education that had permitted him to construct a time machine. Even more, his grandfather would have

---

[6]This example is basically the same that can be found in Lewis [6: 149].

never met his grandmother. Therefore they wouldn't have been married. Tim's father wouldn't have been born. And so, Tim's own existence would be in jeopardy, since he could shoot his grandfather in the past. However, it is still the case that he is there, with the possibility to kill him.[7] What can be said about this?

The most popular response is to think that some kind of impossibility is involved here. It cannot be held that something both happened and didn't happen in the past. In other words, the mere possibility of asserting the very same fact sentences with different truth-values is contradictory. (Something similar can be said about the problem of temporary intrinsics and other time paradoxes).[8] Contradictory facts like these have led many philosophers to affirm that time travel is impossible.

The problem we face here is the impossibility of changing past facts. According to a perdurantist view (which defends that objects and events are wholes composed by temporal parts located in different places and different slots of time), change is the difference between different stages (or temporal parts) of the very same object or event connected in a causal way. Something that is not composed of different parts cannot change. The point here is that, according to Perdurantism, past facts are not composed of different parts because they are something that was concrete. Then, if they cannot change, changing the past is impossible, and time travel is, thus, nonsense.

Even under an endurantist view (which defends that objects and events have a total and complete existence in time right now, in the present), it is impossible to change the past events because it would be incompatible with the present: what have taken place in the past cannot be changed afterwards. Given the impossibility of changing the past, it seems plausible to affirm that the only events that can be said to exist are present events. Past events existed some time before, but they do not exist right now. This is the reason why they cannot be changed. If changing the past does not have sense, then neither does it time traveling.

There is still another argument by the same endurantist view. This is the so-called Nowhere Argument. According to this argument, given that the only events that exist are the actual events, and the past and future events do not exist right now, we cannot travel to some place that does not exist. Again, time travel would be impossible.[9]

As we will argue in the following sections, we think that this kind of argumentation is too fast and it is based on wrong images of the nature of time. Would the nature of time provide an adequate answer to the problem of causal chains? It would be necessary to think that going back in time is not like going back to some

---

[7]This story is popularly known as the "Grandfather Paradox".

[8]The problem of temporary intrinsics says that sometimes we can find objects with apparently contradictory or conflicting properties. This is to say, one object can seem, for instance, to be bent at one time and not bent at another. This is, obviously, a contradiction because supposedly the object needs to have the same properties to be the same object, to have some continuity. Lewis addresses this problem in Lewis [8]. See Footnote 10 for other examples of time paradoxes.

[9]Our use of the terms "Endurantism" and "Perdurantism" is very straightforward and is meant to be that way.

point in the same causal line. Rather, if we think that it is possible to open new causal lines in the non-concrete past events, then it is possible to say that in some sense we can change the past.

According to our view, then, it is also possible to provide a notion of time that allows us to avoid the difficulties that Presentism faces in connection with time travel. If we think about time in terms only of present events, then it is impossible to have some place where to go if our intention is to travel in time either to the past or to the future. But if we could advocate for a notion of time that accepts the real existence of past and future events right now, then there would be no reason whatsoever to affirm the impossibility of time travel.

We need to acknowledge that, inevitably, time travel supposes some asymmetry because the duration of the time travel and the interval of time traveled cannot coincide. This is to say, the elapsed time from the original spot of time to the arrival spot of time is less[10] than the total time between both spots. As Lewis [6: 145] puts it, "he (the time traveler) departs; he travels for an hour, let us say; then he arrives. The time he reaches is not the time one hour after his departure. It is later, if he has traveled toward the future; earlier, if he was traveled toward the past. If he traveled far toward the past, it is earlier even than his departure."

In this definition, what counts as causally relevant is not the possibility to identify an objective or personal operational time. What really counts is a functional definition of personal time that gives us the possibility of establishing a real attribution of properties to the time traveler similar and adequate to the role that time usually plays in the life of someone else. This is to say, the time traveler has the same first-person experience that everyone has. The difference between all of us and the time traveler is that for her all these personal experiences have a lineal movement (her personal spots of time follow one after another), but for an external observer her experiences jump among different spots of time (they do not follow one after another in the objective/external time).

The notion of time behind time traveling stories then is that provided by non-classical physics: time is a four-dimensional manifold of events. Under this interpretation, time is a fourth dimension orthogonal to the other three spatial dimensions. According to this view, in the same way that all the things in this world exist in some place (and they can exist in different places or points of space), things also exist in some time (and they can exist in different times or spots of time). Then, there is not a privileged time where things exist or, expressed in other words,

---

[10]We are supposing here that the time of the travel is always less than the period of time between the departure and the arrival. This means that a cryogenized traveler does not count as a time traveler. We are neither having into consideration the possibility that the duration of travel would be more than the time traveled, as in cases where the relativity of time can be involved. If curiosity is biting you, just think, for instance, in a cosmonaut that was traveling in the space with a velocity close to the light-speed for two years and when she goes back to the Earth discovers that here have passed more than fifty years and her grandsons are older than her.

non-present things exist except that they do not exist in the present moment. If this is the case, we have reasons to defend the possibility of time travel between different spots of time.

Unfortunately, as we have said before, some scholars have argued that time travel is not possible because there are no places where we can go because only the present exists. This view is called Presentism. As Markosian [10] asserts, this kind of rejection of the existence of time travel is based on two suppositions: first, we find the principle of non-contradiction. It sustains the impossibility of the very same thing having different truth-values. Second, we find the idea that says that only the present facts or objects exist. In the next section, we confront this view and its conclusions with a four-dimensionalist notion of time.

Presentism is viewed as a version of what is called A-series [11].[11] According to the A-series, the different slots of time are ordered according to their possession of different properties as, for example: 'being two days past', 'being one day past', 'being present', 'being one day future', and 'being two days future'.[12] If this is how we think, then we found ourselves in the middle of a contradiction since the very same slot of time has different A-properties along time. It can be past, present and future all together, and this is not possible. Then, the presentist says, the best way to think about time is to consider that just the present moment exists, and tenses the other moments. This is to say, according to the presentist conception of time, past existed in other periods but not anymore. In a similar way, the future will exist but it does not exist now. Anyway, everything that exists right now (this is, present objects and events) has some relation with tensed objects and events (whatever this relation could be).[13]

Presentist face one important problem: there are not truth-makers for things that happened in the past, or that will happen in the future, because Presentism hold that the past and future are inexistent right now. The problem derives from an

---

[11]Defenses of this view can be found in Markosian [10] and Zimmerman [20], and in a number of books and papers.See also [3–5]. Rasmussen [16], nevertheless, thinks that Presentism is compatible with a tenseless theory of time by contradicting one of the principles of A-theory and reducing the A-properties (or determinations) to facts about B-relations. This is something that McTaggart himself rejects in his paper. Unfortunately, since this debate is not important for our argumentation here, we will say nothing about that.

[12]In contrast to what McTaggart calls the B-series: a tenseless ordination of slots of time characterized by two-place relations as 'two days later than', 'one day later than', 'simultaneous with', 'one day earlier than', and 'two days earlier than'. The important thing in McTaggart's argument is that the B-series does not constitute a proper time series by itself. Because B-series does not involve genuine change without the A-series, and change is essential to time. Then, he concludes, the A-series is essential to time. But the A-series is, according to him, contradictory because the different A-properties that a very same slot of time can have (past, present, and future) are incompatible with one another. Then, he concludes, time is unreal. According to our own point of view, denying both the A-series and Presentism does not necessarily imply that time is unreal.

[13]For instance, Bigelow [2] talks about some tensed properties that the present objects have. Augustine of Hippo appeals in his *Confessions* to some relationship between present things and memories and predictions present in the mind of God. See also Prior [13] and [14].Examples are really numerous.

interpretation of Lukasiewicz's indeterminacy argument about the future in a gappy way (against Fatalism): utterances about future events (the future contingents) lack truth-value because they are not still actualized. Then, the future does not exist.

Other possible interpretation is a realist account of the indeterminate future events: every possible future event (even though it is indeterminate) exists right now, but we still need to estipulate its truth-value by instantiating its truth-makers. This is to say, we have a totally overdeterminate future (a 'glutty' future), and we just need to wait until we have some news from the truth-maker that allows the instantiation of future events in order to establish its truth-value.

Someone can object that if we reject Presentism, we cannot possibly explain change and time. Or even that we are advocating for a tenseless theory. Nonetheless, by rejecting Presentism we are neither forced to embrace a tenseless B-series nor to reject the reality of time. In fact, we can avoid both Presentism and the unreality of time. As Zimmerman [20] says, all that is needed is to 'take tense seriously' (and he is a presentist). We need to think that everything that happens, happens in time (something very intuitive).

Let us explain our point. We can say that any theory that accepts A-series is invalid because it infers from the fact that "x is (was or will be) F" that "x is F" *simpliciter*, and it is our contention that everything that happens is context dependent, and in this context some time is always involved. Furthermore, it is also invalid because the A-series supposes a premise that makes use of tenseless predication, and this fact is just absurd because every truth and every event happens in some time (is a "tensing the copula" like-argument). In other words, events happen in different times, and events happen always in time [17].

An immediate consequence is the possibility of a 'glutty' open past [8]. If the future is open it is because we accept some indeterminacy in the laws of nature [1]. Laws of nature are time-symmetrical: given a particular sequence of world states WS1, WS2, WSn allowed by a concrete law, then WSn', WS2', WS1' is still allowed by the same law (in virtue of the reverse-state principle). If we accept that the future is open and the facts of the future are indeterminate (as a law), necessarily we have that the past is also indeterministically open. If we add to this conclusion the truth-maker thesis, as we do, we can move forward and backward in time because of the instantiation of the fact that allows defining future and past events.

Now, the idea of changing the past by subtraction is still impossible: you cannot change what has happened. But with the new ambiguity provided by the 'glutty' open past in hand, we can rethink the possibility of changing the past by addition. You can still change the past by adding a new fact afterwards that changes something that has happened (or that adds something that has not happened). The only thing that you need as a requisite to complete this oddity is to instantiate a new past fact that plays the truth-maker role that allows to define a non-actualized past chain of events. If we can accomplish this, we have a very good chance to change the past by addition.

State functions applied at the different times. When we apply a function to a past event, we are generating a new fact that is a truth-maker. This fact starts a new time sequence in which the past is modified. (After killing his grandfather, Tim does not return to the original present. He is still there and he can continue with his life and with his causal chain of events through this moment, creating a parallel universe/reality/possible world).

Time traveling stories are oddities, not impossibilities. We can change the past by addition. Time traveling is more like a trans-world travel between possible worlds.

Our aim from the beginning has been to argue for a realistic notion of time that accepts the possibility of time travel in such a way that the grandfather paradox does not arise. We based our intuition on the (modal) possibility that it is possible to change past and future facts. We acknowledged that the presentist approach poses a big issue. Since according to Presentism only the present exists, there are no facts that instantiate historical events and properties. This is to say, there is not truth-maker for events of the past and the future. According to our view, other interpretation is possible.

Every possible past and future event (even though it is indeterminate) exists now, but we still need to set its truth-value. Every possible truth-value is available now, and after the instantiation of the adequate properties, the truth-value will be concrete. Following Lewis, we maintain that past is overdeterminate. But contrary to what Lewis thinks, we maintain that possible worlds can be similar to the actual world even when they are constituted by events with different causes (when the future is overdeterminate). Lewis cannot accept this possibility for empirical reasons: in the external time it is not permitted to change the events that have already taken place. However, by recurring to personal time, we can reconstruct those events in a different way. By personal time Lewis means a subjective perspective. We do not think that this has to be necessarily so. We can turn to personal time and at the same time be realists about time and about the causal relations between events even though these could have been different to the ones that have taken place in the actual world.

According to our view then, it is possible to open new causal lines in the non-concrete past and future events. Our time traveller disappears in her original causal line and appears with all her memories and knowledge in the very causal line that she opens upon her arrival. Both causal lines are temporally unrelated. Time travel, then, is a one way travel. Changing the past by subtraction is still impossible: what has happened cannot be changed. But it is possible to open a new causal line anytime that someone travels in time, to the future as well as to the past, which allows the occurrence of different and new facts. It also opens new causal chains. And this will necessarily require rethinking the theory of counterparts (see [7] and [19]). In other words, according to our new view, the past and the future can change by addition. If correct, then it can be concluded that past as well as future events can be changed, and time travel will become then non-contradictory.

References

1. Beall, J. (2012). Future contradictions. *Australasian Journal of Philosophy, 90*(3), 547–557.
2. Bigelow, J. (1996). Presentism and properties. In J. E. Tomberlin (Ed.), *Philosophical Perspectives 10: Metaphysics* (pp. 35–52).
3. Godfrey-Smith, W. (1980). Traveling in time. *Analysis, 40*(1), 72–73.
4. Keller, S., & Nelson, M. (2001). Presentists should believe in time-travel. *Australasian Journal of Philosophy, 79*(3), 333–345.
5. Lewis, D. (1973). Causation. *Journal of Philosophy, 70*(17), 556–567.
6. Lewis, D. (1976). The paradoxes of time travel. *American Philosophical Quarterly, 13*(2), 145–152.
7. Lewis, D. (1979). Counterfactual dependence and time's arrow. *Noûs, 13*(4), 455–476.
8. Lewis, D. (1986). Causal explanation. In D. Lewis, *Philosophical Papers, Vol II* (pp. 214–240). Oxford: Oxford University Press.
9. Markosian, N. (1995). The open past. *Philosophical Studies, 79*(1), 95–105.
10. Markosian, N. (2010). Time. *Stanford Encyclopedia of Philosophy*.
11. McTaggart, J. (1908). The unreality of time. *Mind, 17*(68), 457–474.
12. Price, H. (1992). Agency and causal asymmetry. *Mind, 101*(403), 501–520.
13. Prior, A. N. (1955). Diodoran modalities. *Philosophical Quarterly, 5*, 205–213.
14. Prior, A. N. (1957). *Time and modality*. Oxford: Oxford University Press.
15. Prior, A. N. (1967). *Past, present and future*. Oxford: Clarendon Press.
16. Rasmussen, J. (2012). Presentist may say goodbye to A-properties. *Analysis, 72*(2), 270–276.
17. Rettler, B. (2012). McTaggart and indexing the copula. *Philosophical Studies, 158*(3), 431–434.
18. Sider, T. (1997). Four dimensionalism. *Philosophical Review, 106*, 197–231.
19. Sider, T. (2002). Time travel, coincidences and counterfactuals. *Philosophical Studies, 110*, 115–138.
20. Zimmerman, D. (2005). The A-theory of time, the B-theory of time, and 'taking tense seriously'. *Dialectica, 59*(3), 401–457.